SOIL GENESIS AND CLASSIFICATION

Soil Genesis and Classification

S. W. Buol

F. D. Hole

R. J. McCracken

The Iowa State University Press, Ames

1973

STANLEY WALTER BUOL, professor of Soil Science at North Carolina State University, Raleigh, holds the B.S., M.S., and Ph.D. degrees from the University of Wisconsin. He is a member of Alpha Zeta, Sigma Xi, Soil Science Society of America, American Society of Agronomy, North Carolina Soil Science Society, and the Soil Conservation Society of America. Besides this book, Dr. Buol has authored and coauthored Agricultural Experiment Station Technical Bulletins at the University of Arizona and North Carolina State University and published articles in various journals, including *Soil Science, Soil Science Society of America Proceedings, Southeastern Geology, Journal of Soil and Water Conservation,* and *Progressive Agriculture in Arizona.*

FRANCIS D. HOLE, professor of Soil Science and Geography and chairman of the Soil Survey Division of the Wisconsin Geological and Natural History Survey at the University of Wisconsin, Madison, holds the B.A. degree from Earlham College, the M.A. degree from Haverford College, and the Ph.D. degree from the University of Wisconsin. Dr. Hole is a Fellow of the American Society of Agronomy, the Geological Society of America, the Indiana Academy of Science, the American Association for the Advancement of Science; a member of the Soil Science Society of America, the Society for Social Responsibility in Science, and the Wisconsin Academy of Sciences, Arts and Letters. Besides this book, Dr. Hole has coauthored *Soil Resources and Forest Ecology of Menominee County, Wisconsin* and many other county soil survey maps and reports, and has published articles in various journals, including the *Soil Science Society of America Proceedings, Soil Science,* and the *Transactions of the Wisconsin Academy of Sciences, Arts and Letters.*

RALPH J. MC CRACKEN is currently assistant director of research in the School of Agriculture and Life Sciences, North Carolina State University, Raleigh. During the period of 1963 to 1970, he was head of the Department of Soil Science; from 1956 to 1963, he was associate professor and professor of Soil Science at that University. He was associate agronomist, University of Tennessee, 1954 to 1956. He served as soil scientist with the United States Department of Agriculture, 1947 to 1954, with leaves of absence for graduate training. During this period, he served as soil surveyor in California, Western Pacific Islands, New York, and Iowa. He is past president of the Soil Science Society of America and the Soil Science Society of North Carolina. He is a Fellow of the American Society of Agronomy and the American Association for Advancement of Science.

Library of Congress Cataloging in Publication Data

Buol, S   W
    Soil genesis and classification.
    Includes bibliographies.
    1. Soil science. 2. Soil formation. 3. Soils—Classification. I. Hole, Francis Doan, 1913–   joint
author. II. McCracken, R. J., joint author. III. Title.
S591.B887   631.4   72–3434
ISBN 0–8138–1460–X

Composed and printed by
The Iowa State University Press

First edition, 1973
Second printing, 1973
Third printing, 1974

We dedicate this book to
JAMES THORP,
soil scientist, geographer,
teacher, scholar, man of arts
and letters, philosopher,
world citizen, and friend.

CONTENTS

vii

PREFACE

WE HAVE INTENDED in this work to summarize the body of knowledge called pedology, to direct readers to sources of additional information in the literature, and to encourage students to learn directly from the soil in its natural setting. This book is one in a succession of periodic reviews of soil morphology, genesis, and classification that may serve as stepping stones across the seeming morass of terminology and information.

These are exciting times in pedology. The "information explosion" in the several fields of soil science has enabled us to better understand soils and define them more quantitatively. The adoption and use of the Comprehensive Soil Classification System has provided new concepts and nomenclature. And the uses of soil survey, the end product of our classification, have greatly increased through their interpretation for application to land use and productivity studies, especially for nonfarm land use problems.

Yet the student, formally enrolled in college or a self-taught learner, has not had access to any summary in the form of an up-to-date reference or text. We have designed this book with the hope and desire that it will be equally useful to graduates and advanced undergraduates, professional pedologists and geographers, ecologists, and all others interested in or involved with the land. Portions of this book we hope and expect will be of use to planners, to highway engineers, and to sanitarians involved with disposal problems. An understanding of soil genesis and classification is a prerequisite to sound land use planning and land management. Soil science can help people learn to live adequately and significantly in a varied ecosystem and derive necessities from it without damaging it.

Limited use has been made of sketches to illustrate soil profiles. It is our intent that these be supplemented by color slides to illustrate classroom lectures. The Marbut memorial collection of 2-by-2-inch color slides assembled by the Soil Science Society of America is excellent for this purpose.

Our effort is in appreciation of the direction and motivation of which we have been the fortunate recipients in the past, and an attempt to share what we have learned. We are grateful to the many persons who helped us in preparing this manuscript. Ours is the responsibility for any oversights.

S. W. Buol
F. D. Hole
R. J. McCracken

❧

P R O L O G U E

In THIS BOOK we have treated soil as an object
of inquiry. At times our mood may have been in the tradition of the poet,
Tennyson, who investigated the "flower in the crannied wall" by pulling it
out, "root and all," and reflecting on it and his own growing awareness.
Our comments might now follow in a similar vein:

> *Soil on the hilltop and in the valley,*
> *I dig deep and lift you out of the cool darkness*
> *And hold you here, blocks and crumbs of soil in my handclasp,*
> *Moist earth or hard—but if I could somehow grasp*
> *Why you grew red on the crest and black in the lowland,*
> *How you distribute the rainwater, nourish crop and forest stand*
> *And the city man's* Ailanthus *tree in the alley,*
> *Then I should become more aware of the mystery of the universe.*

A sense of wholeness lies in our experience as pedologists which balances
the pulling apart, labeling, classifying, and interpreting. After all our study,
one glance at the landscape may give us the essence of all things pedologic:

> *Sunshine on plowed fields*
> *Shows hilltops bright, footslopes dark,*
> *All moist from Spring rains.*

1

Introduction

$SOIL$ GENESIS is that phase of soil science (sometimes referred to as pedology when combined with soil classification activity) that deals with factors and processes of soil formation. It includes description and interpretation of soil profiles, soil bodies, and patterns of soil on the surface of the earth. Generally, it is thought of as the study of the formation of soils on the land surface of the earth, but some soil scientists extend it to include subaqueous materials which support plant and animal life. Some early geologists included many unconsolidated deposits with soil. Hence arose such general terms, considered erroneous here, as "glacial soils, loessial soils, colluvial soils." The term "alluvial soils" still persists in pedologic literature. Soil genesis is the study of the development of soil from geologic materials such as granite, limestone, glacial drift, loess, colluvium, and alluvium. Because soils are developed from these geologic materials, they are referred to as "soils formed from glacial drift, from loess, from colluvium," and the like. In soil genesis we concern ourselves not with entire geologic deposits but with that upper portion which has been invaded by organic matter or otherwise altered by processes engendered by its position at the earth's surface. Soil genesis includes the weathering of both surficial mantle rock and alteration of organic compounds.

Soil genesis is the study of changes in soil bodies. It is the science of the evolution of soils which are conceived of as natural units (Pomerening and Knox 1962) and as somewhat arbitrary units (Taylor and Pohlen 1962) in the mosaic of the landscape.

The word "pedology"[1] (Editorial Staff 1940; Gibbs 1955; Leeper 1953, 1955; Northcote 1954) has been used both as a synonym for soil science (Sigmond 1938) and as another name for soil genesis (Vilenskii 1957). Soil genesis is the main repository, if not the refuge, of the concept of soil as "a natural entity to be studied as a thing complete in itself" (Cline 1961). This concept has survived the fragmentation of soil science into soil chemistry, soil mineralogy, soil physics, and soil fertility. The subscience, soil genesis,

---

1. Greek: *pedon,* ground + *logos,* word, discourse, science. Contrast: Paedology, pedology, *paido,* boy + *logos,* word, discourse, science—the science of child development.

may be likened to a system of bridges connecting eight islands called chemistry, physics, geology, biology, climatology, geography, anthropology, and agriculture. The interdisciplinary nature of the subject gives it added importance in the training of scientists (Abelson 1964) and in the solving of problems of human ecology.

Although soil genesis is interdisciplinary, it has been taught in this country chiefly on the campuses of agricultural colleges. Much of our understanding of soils is based on observations made by soil surveyors in the course of field mapping and classification. The usefulness of soil maps for practical purposes has attracted the attention of engineering and planning experts. They are representative of the increasing variety of soil-map users who are taking an interest in soil genesis.

HISTORICAL DEVELOPMENT
OF SOIL GENESIS THEORY. The historical approach to science is fruitful in yielding insight into the scientific method. In the process of reviewing the history of soil genesis, we gain perspective regarding modern concepts, we become aware that this science is not static, and we come to appreciate the resistance which new ideas have encountered.

Aristotle (384–322 B.C.) and his successor, Theophrastus (372–287 B.C.) considered soil in relation to plant nutrition. Roman writers who discussed soil in a similar vein included Cato the Elder (234–149 B.C.), Varro (116–27 B.C.), Virgil (70–19 B.C.), Columella (about A.D. 45), and Pliny the Elder (A.D. 23–79).

In 1563 Bernard de Palissy (1499–1589) published *On Various Salts in Agriculture* in which he wrote of soil as the source of mineral nutrients for plants. Van Helmont, in 1629, proposed that the nutrition of plants was solely by water. In the early nineteenth century, A. Thaer suggested that plants directly assimilate rotting organic matter. In 1840 Justus von Leibig (1803–1873) published *Chemistry Applied to Agriculture and Physiology* in which he reasserted that plants assimilate mineral nutrients from the soil and proposed the use of mineral fertilizers in agriculture. He regarded the soil as a passive reservoir of plant nutrients. In the middle of the nineteenth century several German scientists, including Ramann and Fallou, developed agrogeology which viewed the soil as weathered, somewhat leached surficial mantle rock. Fallou suggested that "pedology," which signified theoretical geological soil science, be distinguished from "agrology," the practical agronomical soil science.

In Russia, Lomonosov (1711–1765) wrote and taught about soil as an evolutionary rather than a static body. In 1883 V. V. Dokuchaev (1846–1903) published a report of a field study of Chernozems in which he applied principles of morphology to soils, described major soil groups, produced the first scientific classification of soils, and developed methods of soil mapping in the field and soil cartography in the laboratory. He laid the foundations for the sciences of soil geography and soil genesis. Dokuchaev, in 1886, proposed that the word "soil" be used as a scientific term to refer to "those horizons of rock which daily or nearly daily change their relationship under

FIG. 1.1. V. V. Dokuchaev

FIG. 1.2. C. F. Marbut

the joint influence of water, air, and various forms of organisms living and dead" (Vilenskii 1957). He later defined the soil as an independent natural evolutionary body formed under the influence of five factors, of which he considered vegetation the most important. K. D. Glinka (1867–1929) and S. S. Neustruyev (1874–1928) reemphasized the concept of soil as a surficial geologic entity, a weathered crust that exhibits zonal features corresponding to climatic zones. V. R. Williams (1863–1939) developed the concept of soil genesis as essentially a biologic process rather than a geologic one. He emphasized that phytocycling (withdrawal of nutrients from the soil by plants and their return to the soil surface or topsoil in leaves, litter, and dead root tissue) accounts for progressive increase in soil fertility, and that soil synthesis takes place at its best in grasslands. P. E. Müller, in 1878, wrote a monograph on soil humus, elucidating the biological character of forest soil genesis. In 1912 Gedroiz introduced the concept of cation exchange in soils.

In the United States of America, E. W. Hilgard (1833–1916), in his career as geologist and soil scientist, published works on alkali soils and on the relations between soils and climate (Hilgard 1892). "He saw the farmer's dirt as a richly embroidered mantle of earth, whose design and fabric were deserving of scientific zeal and quest" (Jenny 1961a). He did a monumental work on alkali soils in which "his blending of land forms, water tables, rock weathering alkali processes, plant responses and cultural practices—including irrigation and drainage—offered a unified approach that is now seldom matched." At age sixty, he produced a treatise on soil as related to climate (1892). C. F. Marbut (1863–1935), while director of the U.S. Soil Survey, read Glinka's work in a German translation and brought his concepts to the attention of soil scientists in the United States (Krusekopf 1942). Charles E. Kellogg (1902–   ), who succeeded Marbut as director of the U.S. Soil Survey, has with co-workers continued the development of soil classification, building on the foundation laid by Dokuchaev and Glinka (Soil Survey Staff 1951, 1960).

Hans Jenny (1941) wrote a masterful treatise on the five factors governing the development of soil profiles. He noted that quantitative elucidation of the processes of soil formation could not proceed without a large body of data which was not available at the time of his writing.

Soil survey specialists the world over, particularly in the USSR, Australia, New Zealand, Europe, and former colonies of the British Empire, have been developing regional soil science in terms of soil geography and soil genesis.

PERSPECTIVE ON THE ORIGINS OF
THE SCIENCE OF SOIL GENESIS. "It requires genius to create a subject as a distinct topic of thought. . . . It takes an unusual mind to undertake the analysis of the obvious" (Whitehead 1925). As we have said, if we can credit any one man with the creation of the subject now designated soil genesis, it is V. V. Dokuchaev (1846–1903). He conceived of the soil as a natural, organized body, evolved and evolving under the influence of factors of soil formation. The environment of soil, including that part called "modern man," shapes the soil and to a considerable extent, soil shapes many aspects of its environment. Dokuchaev was able to establish the science of soil genesis not only because of his unusual abilities and energy as a scientist but also because of the challenge of a practical assignment which his government gave him, namely, to improve soil productivity on the Chernozems and other soils of the steppes of Russia, where drought and famine struck periodically. He was able to establish soil science because he had inherited the attitudes and habits of modern science.

Previous to A.D. 1600, investigations of soils were elementary from the scientific point of view. About 4,000 years ago, the Chinese engineer Yu classified soils according to color and structure (Thorp 1936). Various Greek and Roman writers had discussed soils from both theoretical and practical standpoints. But the epoch of modern science, beginning in about 1600 and spreading throughout the population of Europe, created a passionate interest in the relation of general principles to irreducible, stubborn facts. The approach of modern science was a result of a remarkable and unique combination of (1) clear, analytical, deductive reasoning, as practiced by Aristotle, by the medieval religious scholars, and by Roman lawyers; (2) the medieval, Christian, instinctive, uncritical faith in a rational all-powerful, personal God who created and controls an orderly world; and (3) an active interest in simple, everyday occurrences. Powered by a faith in the existence of an order in natural phenomena, inductive reasoning became an important tool in discovering relationships between detailed facts of nature and general principles. Whereas the Greeks regarded fate as the force controlling tragic, solemn, inescapable events, modern man widely believes that the order of nature controls scientific events (Whitehead 1925). The now familiar process of classifying soils within small regions and also across continents is a remarkable demonstration of the interweaving of general principles and detailed observations.

The discovery made by Dokuchaev and a group of his colleagues (Sibirtzev, Ototzky, Kostytchev, and others) and their students, as well as certain predecessors, including M. V. Lomonosov, was that of the relationship between multitudinous facts about soils and general scientific principles. This was a pioneer application in the 1870s of the outlook, which by 1900 had made soil science and the ultimate production of a large number of soil scientists possible. Today science may be only one useful subculture among many subcultures; one among many methods whereby images change and develop (Boulding 1956). But, it is a very significant subculture.

The generalization that soils are natural bodies resulting from the influences of certain factors of soil formation, particularly climate, could most easily be made by scientifically trained observers working over a large territory, larger than a single nation of western Europe, for example. In Russia, a vast terrain was available, offering broad zones of contrasting soils and environments to the inquiring eyes and minds of scientists, who were interested in both detailed observations and general principles.

Soil genesis was perhaps slow in developing because the "soilsphere" or "pedosphere" was not generally recognized as a distinct sphere, as were the lithosphere, hydrosphere, and atmosphere. Indeed, the pedosphere is a rather arbitrary slice taken out of these spheres and the biosphere along major interfaces. The complexity of soils also made the pedosphere a somewhat forbidding subject for investigation.

We inherit from the faith of the Middle Ages in Europe the conviction that everything in the universe holds a secret that can be unveiled and dissected rationally. This conviction or tone of thought makes research possible. We have come to regard nature not just as meaningless matter but as the locus of organisms or units of organization in the process of development. Therefore, attention to everyday occurrences for their own sake has become significant to us. A soil scientist who is digging in the soil in an exposed pit face may be approached by a local landowner with the question, "Looking for oil?" or "Looking for gold?" However, because of the widespread faith that research can discover new truth, and that this is significant, the work of the soil scientist is usually accepted by a landowner as reasonable and valuable.

Three stages in the development of a discipline, enumerated by Whitehead (1925), may be applied to soil genesis:

Stage 1: Location in space and time. The basic operation of mapping soils to record their positions in space is a prerequisite to the development of the science of soil genesis. The description and characterization of soil profiles are processes of reporting what soil is and where it is. The location of soils with respect to environmental factors, of which time is one, is an essential feature of this stage.

Stage 2: Classification. Whitehead calls this a "halfway house between the immediate concreteness of the individual things and the complete abstraction of mathematical notions." A great variety of genetic and descriptive classifications has been constructed. Terminology ranges from symbols to synthetic terms based on classical languages, and finally to a hodge-podge of folk terms (Wilde 1953), nouns, and adjectives. Varieties (soil phases), species (soil types), and various broader groupings (great soil groups in the United States and soil types in the USSR) are abstractions of defined entities necessary in soil classification (Cline, 1949).

Stage 3: Mathematical abstractions. More highly developed abstractions are possible by mathematical means. Relationships between soils and other phenomena can be expressed statistically (Bidwell and Hole 1963, 1964; Hole and Hironaka 1960; Jenny 1941, 1961b). Considerable subjectivity and variability of human judgment can thereby be eliminated or at least evaluated so that the facts can better speak for themselves, often in unexpected ways.

It is not enough to state current definitions of soil entities. A consideration of other ways of thinking about soils is illuminating. For example, soil has been defined as a medium for plant growth. Such a definition is unsatisfactory in that it is dependent upon something besides the soil. Furthermore, lakes, oceans, living sphagnum moss, and even glaciers and human beings are media for growth of molds, algae, lichens, mosses, or higher plants. Nikiforoff (1959) defines soil as the "excited skin of the sub-aerial part of the earth's crust." Soil is an unstable entity, a very slow-motion "whirlpool" or short segment of a river of mineral and organic matter. Soil is a turnstile through which passes an endless procession of atoms. A soil is a natural body of mineral and organic matter which changes or has changed in response to climate and organisms. The change is called soil genesis.

From the genetic point of view, we need not consider whether a soil is productive of plants or moulding sand or china clay. We consider the soil as a collection of natural bodies worthy as such, and related to each other and other factors and phenomena in ways which can be evaluated or measured. Soil survey has a genetic outlook, as contrasted with the practical outlook of land capability survey.

Various soil scientists have compared a soil to an animal: additions to the soil of mineral and organic matter analogous to feeding; changes in these materials in the soil analogous to digestion; loss of materials by leaching or escaping gases analogous to elimination and exhaling. An upland soil can even be said to produce offspring if erosion has carried portions of it to a valley bottom where the resulting alluvial soil still exhibits some of the characteristics of the parent soil. In this sense, a young soil may "grow up." The designations of young, mature, and senile soils parallel

Davis's (1902) classifications of topography into stages of youth, maturity, and old age.

Geologists may regard a soil as an epidermal unit of a geologic body through which all material must pass in the erosion cycle from rock to sediments laid down in the oceans. Chemists may consider the soil, as did Liebig (1840), to be a vessel or test tube into which mineral matter has been placed by natural forces and agencies, and into which fertilizers are poured by men to provide nutrients for plant growth. Physicists see soil as a physical mass whose characteristics and behavior change with variations in temperature and moisture content. Ecologists see the soil as part of the environment which is conditioned by organisms, and which in turn influences the organisms. Agriculturalists and industrialists may describe the soil as a machine, whose principal parts are aggregates and roots and which manufactures crops and livestock. To the historian, the soil is a record of the past. The artist and philosopher see in the soil a beautiful, if not mystical, blend of the forces of furniture of life and death. They may even see simplicity in it, as if anything in this universe were simple enough to serve as an example of that quality. Soil is a coincidence of materials and arrangements related to the "factors of soil formation."

A soil body can be said to be a "synthograph," a natural device which records a synthesis of much that has happened at the site. A soil may contain billion-year-old grains of quartz, fresh crystals of calcite, one-thousand-year-old pottery fragments, some organic matter inherited from a plant five thousand years ago, and incorporated forest litter a few weeks old. In this sense, a soil body is indeed a synthograph. The challenge to the soil scientist is to learn to read this graph.

SOME FUNDAMENTAL CONCEPTS

OF SOIL GENESIS. Some fundamental concepts that have proved to be useful in the understanding of soil genesis are:[2]

Concept 1. Present-day pedogenic processes operate through space and time. This includes the principle of uniformitarianism of Hutton, Playfair, and Lyle. During those periods of earth history in which there was no life on land, there was no soil in the modern sense any more than there is a soil on the moon at the present moment. Rocks were exposed to the agents of weathering and erosion without organisms. "Protosoils" may have existed, including some behaving like the "self-plowing soils" of today. During the time since the appearance of organisms over the land surface, possibly from the beginning of the Devonian geologic period, different kinds of soils have formed under the influence of flora and fauna. Volcanic activity and periods of glaciation have affected the land surface, and hence soil parent materials, in different parts of the globe. But such phenomena have followed the same natural laws as they do today.

The principle also includes the idea that the old terms for the processes of soil formation ("podzolization," "laterization") are actually bundles of

---

2. Soil names in this and other early chapters are defined in later chapters.

similar processes acting in different proportions and intensities at different times and places (Simonson 1959).

Concept 2. Distinct regimes of processes of soil formation produce distinctive soils.

It is recognized that regimes of soil genetic processes are distinct not only because they differ in proportions and intensities of those processes which are common to them all but also because certain regimes include processes which are unique. For example, soil developing in regions of the earth where temperatures fall below freezing may undergo frost action which is absent in other regions.

Concept 3. Soil and its vegetative cover modify processes of degradation of land.

By concentrating plant nutrients and organic tissues at the surface of the soil, plants counteract leaching processes and to a considerable extent protect the soil against agents of erosion. Some soil horizons, known as "pans," act as resistant strata in withstanding erosion where the pans are underlain by highly erodible formations. Organic weathering contributes to the decomposition and eluviation of material.

Concept 4. Clay[3] is produced in soil.

Soils containing minerals which are weatherable to layer silicate clays are natural bodies in which clay forms. They are "clay factories."

The resistance of soil systems to erosion, referred to in Concept 3, allows time for some weatherable minerals in the soil to alter to clay, and for clay to be synthesized from weathering products. It is likely that the clay particles in the extensive shales of the sedimentary column are to a considerable extent products of soil formation.

Concept 5. Organomineral complexes are produced in soil.

During the pedogenic processes in soil, organic compounds enter into complex relationships with the clay. Earthworms as promoters of this process have been called "soil factories" (Jacks 1963).

Concept 6. In the course of pedogenesis, soil succession occurs.

As successive regimes of pedogenic processes act on a soil parent material, a sequence of soils having distinctive characteristics is produced. The concept of a pedologic cycle includes "young soils," "mature soils," and "senile soils." Biopedogenetic and catapedogenetic successions occur. Plant succession dominates in the first instance; weathering and soil "degradation" in the second.

Concept 7. Complexity of soil genesis is commoner than simplicity.

A simple soil is one formed under the influence of a single dominating process, as in the case of a young soil in acid sand dune material. Invasion of the mineral material by organic matter as "pioneer" plants take root is the dominating process. Simple soils are rare. A complex soil, the common case, is one formed under the strong influence of two or more processes.

Concept 8. Little of the soil continuum of the earth is older than Tertiary and most of it no older than Pleistocene. This is a measure of the limited stability of the land surface.

---

3. The concept of clay here includes both clay mineral structure and particles less than 2μ in diameter.

Concept 9. A knowledge of climatology is a prerequisite to an understanding of soils.

The Russian climatic-vegetative zonal concept of soil genesis is fundamental. Climate and organisms leave their imprint on the mantle rock in the form of soils.

Concept 10. A knowledge of the Pleistocene is a prerequisite to an understanding of soils.

Changes in climate, plant communities, action of geologic agents, sea levels, cycles of wind erosion, and deposition during the Pleistocene, which according to Flint (1947) still continue, have had profound influence on soil parent materials.

Concept 11. There are observable points of marked change in rates and degree of response of soil to environment.

The influence of external factors on soil characteristics is not exerted uniformly throughout the total environmental range, but is concentrated in one or more portions of the range of the factors.

Concept 12. A knowledge of soil genesis is basic to soil management.

Man's influences on the factors of soil formation can be better controlled and planned in the future in the light of discoveries by soil geneticists.

Concept 13. Paleopedology is a growing aspect of the science of soil genesis.

Although the science of soil genesis is primarily concerned with present-day landscapes, it attains its historical usefulness by extension into the past.

A soil body may be studied in at least three ways: as an anatomical specimen, as an energy transformer, and as an open system.

A SOIL AS AN ANATOMICAL SPECIMEN. Just as Louis Agassiz (1807–1873) taught his students to learn about fish by making accurate drawings of specimens of fish, so Dokuchaev, Hilgard, Marbut, Kellogg, and others have taught us to learn about soils by describing them carefully (Marbut 1935). The *Soil Survey Manual* (Soil Survey Staff 1951), which Cline (1961) has considered to be one of the major pedologic works of our time, is a detailed guide to the scientific description of soil profiles. Soil morphology treats the structure of soil. The morphology of soil bodies and larger soil regions is less systematically studied by pedologists at present than the soil profile. Description of soil anatomy is fundamental. It provides us not only with exact information about soils but also supplies us with corresponding questions as to genesis. Observations usually represent the state of the soil body and surrounding landscape at a particular hour on a particular day. Brewer (1964) used the term *pedography,* defined as "the systematic description of soils based on observations in the field, on-hand specimens and thin sections, and on data from other techniques on the size, shape, arrangement, and identification of the constituents."

A SOIL AS AN ENERGY TRANSFORMER. The soil and landscape are continually changing physically, chemically, and biologically. Soil physicists have studied soils and their vegetative covers as "energy transformers," receptors, and transmitters of solar radiant energy and energy emanating from inside the earth. The energy transformations in the soil are accomplished through wetting and drying, heating and cooling, evapotranspiration, weathering, erosion (including leaching) of material, and deposition of material (Fig. 1.3). In the soil system, photosynthesis is important. Heat and light are converted through evapotranspiration, photosynthesis, and decomposition. Mobile constituents involved in the processes named are gases, leachates (in solution and suspension), and biological fluids. Exothermic

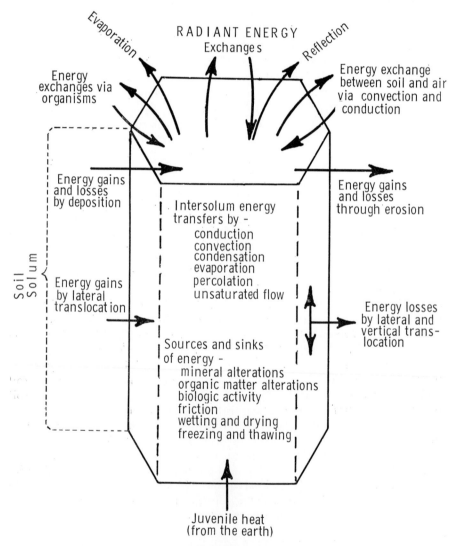

FIG. 1.3. Schematic representation of energy sources and transformations in the solum of a soil pedon.

reactions are dominant in weathering; endothermic reactions are dominant in the growth of organisms. In the soil, organisms and minerals compete for mobile materials. Exchanges or transactions (Kuhn 1963) take place between plants and soils.

A SOIL AS AN OPEN SYSTEM. A soil is an evolving entity maintained in the midst of a stream of geologic, biologic, hydrologic, and meteorologic material (Fig. 1.4). Individual soil bodies and individual horizons within them play different roles because of the unequal distribution of materials in and between them. Some soils and horizons become enriched in certain substances, others impoverished. The surface mineral horizon of a deciduous forest soil becomes enriched in bases while the subsoil (B horizon) becomes relatively impoverished. There is also exchange of material between soil bodies not only by wind but also through action of water and organisms. Some soils occupy a depressional niche in the landscape and fulfill a role as receptor of more water and leachates and even of eroded soil material than many other soil bodies in the same landscape. A soil body or even a community maintains a "role" as an entity. A soil survives because of (1) a protective covering in the form of vegetation or of a cemented surficial layer; or (2) because in its "steady state" it forms as fast as it is dissipated.

A soil has a budget of inputs and outputs. Once we have defined a soil body by means of a map, a formula, a morphological and organizational chart, and description, we can consider the soil as an entity which endures in the midst of complex and perpetually dynamic processes, which include:

1. Exchange between the soil and the environment of materials, such as oxygen, water, and carbon dioxide; and exchange of materials within the soil, as in base exchange.

2. Automatic control responses, as exemplified by the swelling and contracting of clay masses of a "self-plowing" soil and the freezing and thawing of Tundra soil.

3. Production and consumption of new organic and mineral materials. The soil system produces and cycles new forms of organic and mineral materials all the while that matter is being lost from the system to its environment and substratum.

The soil is part of a symbiotic community in which human beings, plants, and animals supply each other's needs (Bidwell and Hole 1965). The social ecosystem in southern Wisconsin, for example, determines that milk cows, not buffalo, shall be numerous; that grain, not prairie vegetation, shall be extensive; that hay fields, not forest, shall flourish over large acreages.

Soils are complexes which are as yet not subject to controlled experiments. The climatron in the Shaw Gardens in St. Louis is successful in maintaining a variety of tropical and subtropical climatic conditions within a single large space, but controlled experiments in soil development would be difficult there. The soil is not left undisturbed; the contrast in factors of soil formation is insufficient from one place to another; and the span of time since the building of the climatron is too short to have produced recognizable soil profiles. It is conceivable that laboratory experiments with

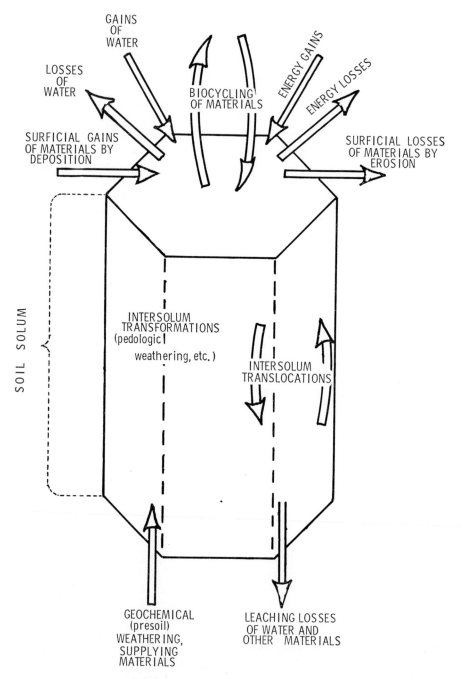

FIG. 1.4. Schematic representation of the solum of a soil pedon as an open system.

columns of soils could be extended to miniature landscapes under artificially controlled conditions, but no examples of such experiments are known to the writers.

METHODS OF SOIL GENESIS STUDY. Although approaches to soil genesis study are varied, depending on various circumstances, four general methods of analysis may be distinguished.

INDEPENDENT VARIABLE METHOD. This is a simple method insofar as it handles one item at a time (Jenny 1958). The intellectual assumption is made that all conditions are constant except one variable. An example from French Equatorial Africa is in a portion of the tropical rain forest that was cut and replaced by agricultural experimental fields. This allowed the soil, which had remained moist under forest conditions, to desiccate during dry seasons, with the result that the sesquioxide-rich surface soil hardened irreversibly. The factor which had been changed was the micro-climate above the soil and the climate in the soil itself. The danger in the independent variable method is that the observer may be led to suppose that a relationship which is obtained in one place will hold elsewhere. It is obvious that the cutting down of a forest and consequent exposure of the soil to extreme conditions of temperature and humidity do not cause the soil to irreversibly harden throughout the world, else we would have little agriculture as we know it today.

DEPENDENT VARIABLE METHOD. The soil complex is considered as a function of n variables, each of which can be written as a function of all the others, yielding n equations. This has the advantage of generalizing and of dealing with a whole system. Results may be quite unreal. For example, if the result of an analysis of a large soil region states that soils are extremely variable because conditions affecting the soils are extremely variable, then no specific image of the nature of the soil complex is forthcoming.

MACROANALYSIS METHOD. By this method the whole soil complex is divided into macrogroupings. This is actually a compromise between the oversimplicity of the independent variable method and the overgeneralization of the dependent variable method. Examples of macrogroupings of soils are great soil groups, great soil group complexes, and soil associations and catenas such as are found in the associations of soils in the north central region of the United States (Simonson, Riecken, and Smith 1952). The danger in this method is that we tend to forget that our units are really complexes and we may consider them as definite and stable entities.

NUMERICAL ANALYSIS METHOD. A useful method of considering data about soils is "ordination," which is "an arrangement of units in a uni- or multidimensional order" (Curtis 1959; Goodall 1954). In one such method, the investigator assembles data for the soils in which he is interested; on the basis of range of values for each kind of data, he scales the values onto a scale of 1 to 100; he sums the scaled values for each soil *(A)*, and then compares them in turn with the sum for every other soil *(B)*, by means of the formula

$$\frac{2w}{A + B} \times (100) = I \text{ (index of similarity)}$$

where w is the sum of the lesser figures in the pairs of scaled values. From a matrix table of indices of similarity and dissimilarity ($100 - I =$ index of dissimilarity) one can construct two-dimensional "dendograms," or three-dimensional constellation models. In the latter, those soils which cluster at the center may be called "centric," and those which lie toward the periphery may be called "eccentric." There is a growing literature on this subject (Bidwell and Hole 1964). In addition to providing classification guidelines, this method may evolve into computer simulation of soil genesis.

SOIL PROFILE. As used in this discussion, a soil profile is taken to mean vertical exposure of the horizons of a soil individual. A soil individual is a soil body that may be defined in terms of profile features whose arrangements and combinations over a geographic area are unique.

DEFINITION AND CONCEPT. A complete soil profile is a vertical exposure of a surficial portion of the earth's crust that includes all the layers that have been pedogenically altered during the period of soil formation and also deeper layers that influenced pedogenesis (Fig. 1.5). A soil profile may be observed in a freshly dug pit, along a road bank, or in many other places. In practice, many incomplete profiles are examined in shallow pits that reveal only a few surface layers of the soil. Interpretations based on incomplete soil profiles are not valid unless complete profiles of the same soil have been studied.

The concept of the soil profile, however, is separate from that of a soil sample in that it includes more than one layer of soil while the soil sample may be taken from only the surface layer or some subsoil layer.

SOIL SOLUM. The soil solum is an incomplete soil profile that may be simply defined as "the genetic soil developed by soil-building forces" (Soil Survey Staff 1962). Although this definition seems simple enough, much confusion can arise when application is made in the field.

Determination of the lower boundary of the solum is difficult (Chizhikov 1968). Several somewhat arbitrary definitions of this boundary have been made (Soil Survey Staff 1962). It is important from an operational viewpoint that the lower limit of the solum be near enough to the surface for the soil surveyor to observe it in the course of traverses. For the purpose of making a utilitarian soil map, observations of the solum may, in some cases, be limited to the soil within a few feet of the surface. However, the capacity of the soil surveyor to interpret observations of incomplete sola depends on his experience as a soil geneticist who has examined many soil

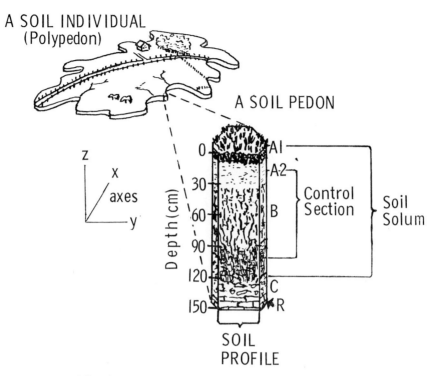

FIG. 1.5. A soil individual is a natural unit in the landscape, characterized by position, size, slope, profile, and other features.

profiles to the lower limit of the perennial plant roots. This is indeed the lower limit of the solum. The primary difference between soil and geologic material is the presence in the soil of living plant roots and deposits of organic and mineral materials originating in the rooting zone. The plant acts as a pump by which ions are taken up by the root, transported to the aerial portions of the plant, and subsequently deposited on the soil surfaces. Therefore, it can be reasoned that any portion of the earth's crust that is reached by the plant roots has been changed from geologic material into soil.

The solum cannot be properly understood without observing at least one layer below it, in order to discover, if possible, why the rooting zone stops where it does and, in some cases, why materials that have been leached out of the rooting zone have been deposited just below it. For example, if the roots are stopped at a particular depth by a permafrost layer, it is important to the understanding of the solum to describe the permafrost layer.

In conclusion, it appears most reasonable for a complete understanding of the soil solum to consider it to be that part of the soil profile which is influenced by plant roots. A complete soil description, however, does not stop with the soil solum but includes one or more underlying horizons.

SOIL SEQUUM. Soils are bodies that exhibit three-dimensional sequences of characteristics. First, there is the familiar vertical sequum of properties from the soil surface, or contact with the free air, to a depth where the material can no longer be considered as influenced by pedogenic processes and is therefore geologic material. The lateral sequum is a succession of contiguous soil bodies in horizontal direction from the soil body in question (Fig. 1.5).

The vertical sequum in most soils exhibits a decrease in content of organic matter with depth. Clay content usually increases to a maximum within a few feet of the surface and then decreases. These and other changes in soil properties with depth are usually gradual in nature. Abrupt boundaries to major soil horizons do occur in stratified geologic materials, in soils disturbed by man, and in some soils which have low biologic activity.

The sequum formed laterally (along the x and y axes, Fig. 1.5) is likewise generally gradual except where interrupted by man or geologic influence.

CONTROL SECTION. The control section is a portion of the soil profile delimited in terms of an arbitrary depth or depth range in centimeters or inches.

Some soil sola are so deep or have such subtle lower boundaries that in practice the soil is classified and mapped without knowledge of or reference to the complete solum. In soil survey work in the United States, the layer of soil and/or parent material at a depth of 25 to 100 cm (10 to 40 in.) is considered as the control section in soils with no textural B horizon (argillic horizons) or the top 50 cm (20 in.) of that argillic horizon if present. In regions where irrigation is practiced or where drainage tile needs to be employed, 1 m (40 in.) is not deep enough to describe the soil adequately, and hence a special, deeper control section must be used. However, where these practices are not common, it is not considered time well spent to describe a soil to any depth over 1 m (40 in.) when only practical interpretations are to be made of the survey. This arbitrary working rule has no place in a study of soil genesis, even though reference is made to control sections in the classification of certain soils.

SOIL INDIVIDUAL: SOIL AS A THREE-DIMENSIONAL
COMPONENT OF THE LANDSCAPE. *Soil* is located in the space along
a vertical axis between the free air above and the geologic substratum below and in a horizontal plane to material that is not considered soil; that is, deep water, ice, rock outcrop, etc. (Fig. 1.5).

An *individual soil body* is bounded laterally by other soil bodies or by nonsoil material. Adjacent soil bodies may be differentiated on the basis of depth of solum. For example, soil individual *A* may be recognized as having a depth range of 61 to 100 cm (2 to 3 ft) along the z axis. By making this definition, two other kinds of soil individuals are associated with the first, one with a solum more than 1 m (3 ft) thick and another with a solum less than 61 cm (2 ft) thick.

Thus, each actual soil individual as observed, described, and mapped from other soil bodies in the field is considered within the framework of the abstract concepts of soil, soil profile, and soil body already discussed. We must apply these concepts and obtain a minimum volume of observation in establishing soil individuals. What is the minimum size of an individual soil body?

THE PEDON. A pedon is the smallest volume that can be recognized as a soil individual.

The pedon has been described as follows (Soil Survey Staff 1960):

A pedon is the smallest volume that can be called "a soil." . . . A pedon has three dimensions. Its lower limit is the vague and somewhat arbitrary limit between soil and "not soil." The lateral dimensions are large enough to permit study of the nature of any horizons present, for a horizon may be variable in thickness or even discontinuous. Its area ranges from 1 to 10 square meters, depending on the variability in the horizons. Where horizons are intermittent or cyclic and recur at linear intervals of 2 to 7 meters (roughly 7 to 25 feet), the pedon includes one-half of the cycle. Thus each pedon includes the range of horizon variability that occurs within these small areas. Where the cycle is less than 2 meters or where all horizons are continuous and of uniform thickness, the pedon has an area of 1 square meter. Again, under these limits, each pedon includes the range of horizon variability associated with that small area. The shape of the pedon is roughly hexagonal. One lateral dimension should not differ appreciably from any other.

Thus, the concept of the pedon enlarges upon that of the soil profile to include a lateral as well as a vertical extent of a soil and puts limits on the volume to be considered. It is seldom, however, that a soil individual can be adequately described and defined from one pedon.

THE POLYPEDON AND THE SOIL INDIVIDUAL. A soil body consisting of more than one pedon is termed a polypedon. The concept of the polypedon provides the essential link between basic soil entities (the pedons) and the soil individuals that form units in the taxonomic system. The polypedon is defined (Johnson 1963) as "one or more contiguous pedons, all falling within the defined range of a single soil series. It is a real, physical soil body, limited by 'not-soil' or by pedons of unlike character in respect to criteria used to define series. Its minimum size is the same as the minimum size of one pedon, 1 square meter; it has no prescribed maximum area. Its boundaries with other polypedons are determined more or less exactly by definition." Two or more pedons with a 2-square-meter smallest limit seems a more apt definition.

We thus have created concepts by which we can divide the continuum of soil into individual soils or individual units that can be variously arranged in taxonomic systems.

LIMITATIONS OF SOIL TAXONOMY. As pedologists continue to learn more about the morphology, chemistry, mineralogy, physics, and microbiology of the soil, they will be able to place better and more meaningful limits on the soil taxonomic units. A classification system must follow the discoveries of science and the limits we set must be such that they are consistently recognizable. Probably no greater problem faces soil taxonomy than this one of establishing consistently recognizable limits for soil individuals. Limits that are vaguely defined or that can be recognized only under very special conditions add confusion and distrust of the system. Even with precise definition of soil taxa, consistent mapping of corresponding soil bodies is difficult.

Consistency in soil mapping comes from the training that the inexperienced workers receive from the experienced ones. But the best training cannot make up for deficiencies in definition of taxonomic units.

One should look into the properties of the soil to find the criteria for establishing taxa. Although factors such as vegetation, topography, and parent material can indicate where soil differences are likely to occur, factors should not be used as criteria in defining soil individuals.

Criteria that are dependent on management practices are usually too time dependent to be useful. An improvement in soil management can render meaningless soil classes based on conditions that were obtained under the old practices.

One should not be dogmatic about criteria of soil classification. The best test of the criteria is their applicability in soil survey work by more than one individual or groups of individuals.

The principles and historical development of soil classification are considered in Chapter 13.

LITERATURE CITED

Abelson, P. H. 1964. Trends in scientific research. Science 143:218–23.

Bidwell, O. W., and F. D. Hole. 1963. Numerical taxonomy and soil classification. Soil Sci. 97:58–62.

———. 1964. An experiment in the numerical classification of some Kansas soils. Soil Sci. Soc. Am. Proc. 28:263–68.

———. 1965. Man as a factor of soil formation. Soil Sci. 99:65–72.

Boulding, K. E. 1956. The image: Knowledge in life and society. Univ. Mich. Press, Ann Arbor.

Brewer, R. 1964. Fabric and mineral analysis of soils. John Wiley & Sons, New York.

Chizhikov, P. N. 1968. The lower soil boundary. Soviet Soil Sci., 1968. No. 11, pp. 1489–93. (Transl. from Pochvovedeniye 1968, 11:16–21.)

Cline, M. G. 1949. Basic principles of soil classification. Soil Sci. 67:81–91.

———. 1961. The changing model of soil. Soil Sci. Soc. Am. Proc. 25:442–46.

Curtis, J. T. 1959. The vegetation of Wisconsin: An ordination of plant communities. Univ. Wis. Press, Madison.

Davis, W. M. 1902. Baselevel, grade and peneplain. J. Geol. 10:77–111.

Dokuchaev, V. V. 1883. Russian Chernozem (Russkii Chernozem). (Transl. from Russian by N. Kaner.) Israel Prog. for Sci. Trans., Jerusalem, 1967. Available from U.S. Dept. Commerce, Springfield, Va.

Editorial Staff. 1940. Obituary notice. Soils Fertilizers 3:147–48.

Flint, R. F. 1947. Glacial geology and the Pleistocene epoch. John Wiley & Sons, New York.

Gibbs, H. S. 1955. What use is pedology? Soils Fertilizers 18:188–90.

Goodall, D. W. 1954. Vegetational classification and vegetational continua. Angew Pflanzln Sociologie, Wien, Festschrift Aichinger (Vienna) 1:168–82.

Hilgard, E. W. 1892. A report on the relations of soil to climate. U.S. Dept. Agr. Weather Bull. 3:1–59.

Hole, F. D., and M. Hironaka. 1960. An experiment in ordination of some soil profiles. Soil Sci. Soc. Am. Proc. 24:309–12.

Jacks, J. V. 1963. The biological nature of soil productivity. Soils Fertilizers 26: 147–50.

Jenny, H. 1941. Factors of soil formation. McGraw-Hill, New York.

———. 1958. Role of the plant factor in the pedogenic functions. Ecology 39:5–16.

———. 1961a. E. W. Hilgard and the birth of modern soil science, Agrochimica, Pisa, Italy, and Farallon Publ., P. O. Box 564, Berkeley, Calif.

———. 1961b. Derivation of state factor equations of soils and ecosystems. Soil Sci. Soc. Am. Proc. 25:385–88.

Johnson, W. M. 1963. The pedon and the polypedon. Soil Sci. Soc. Am. Proc. 27:212–15.

Krusekopf, H. H. 1942. Life and work of Curtis F. Marbut, 1863–1935. Soil Sci. Soc. Am., Columbia, Mo.

Kuhn, A. 1963. The study of society: A unified approach. R. D. Irwin (The Irwin-Dorsey Series), Homewood, Ill.

Leeper, G. W. 1953. What use is pedology? Soils Fertilizers 16:241–42.

———. 1955. What use is pedology? Soils Fertilizers 18:375–76.

Liebig, J. 1840. Chemistry in its application to agriculture and physiology. Ed. by Playfair, Taylor, and Walton, London.

Marbut, C. F. 1935. The soils of the United States. In USDA atlas of American agriculture, part 3. Advance sheets, no. 8.

Nikiforoff, C. C. 1959. Reappraisal of the soil. Science 129:186–96.

Northcote, K. H. 1954. The place and function of pedology in soil science. Soils Fertilizers 17:305–10.

Pomerening, J. A., and E. G. Knox. 1962. A test for natural soil groups within the Willamette catena population. Soil Sci. Soc. Am. Proc. 26:282–87.

Sigmond, A. A. J. de. 1938. The principles of soil science. (Transl. from Hungarian by A. B. Yolland; edited by G. V. Jacks.) T. Murby & Co., London.

Simonson, R. W. 1959. Outline of a generalized theory of soil genesis. Soil Sci. Soc. Am. Proc. 23:152–56.

Simonson, R. W., F. F. Riecken, and G. D. Smith. 1952. Understanding Iowa soils. W. C. Brown, Dubuque, Iowa.

Soil Survey Staff. 1951. Soil survey manual. U.S. Dept. Agr. Handbook 18. U.S. Govt. Printing Office, Washington.

———. 1960. Soil classification, a comprehensive system—7th approximation. U.S. Dept. Agr. U.S. Govt. Printing Office, Washington.

———. 1962. Supplement to Agr. Handbook 18, Soil survey manual (replacing pages 173–88). U.S. Dept. Agr. U.S. Govt. Printing Office, Washington.

Taylor, N. H., and I. J. Pohlen. 1962. Soil survey methods. New Zealand Soil Bur. Bull. 25.

Thorp, James. 1936. Geography of the soils of China. Nat. Geol. Survey of China, Nanking, China.

Vilenskii, D. G. 1957. Soil science. (Transl. by A. Birron and Z. S. Cole.) Israel Prog. for Sci. Trans., Jerusalem, 1960. Available U.S. Dept. Commerce, Washington.

Whitehead, A. N. 1925. Science of the modern world. Macmillan, New York.

Wilde, S. A. 1953. Soil science and semantics. J. Soil Sci. 4:1–4.

❧

Morphology of Soils ✤

SOIL MORPHOLOGY has been studied largely under field conditions. The morphology of a soil is best evaluated from the in situ examination of the soil profile. A recently dug pit large enough for observation of a pedon is desirable. Old exposures such as road banks and ditches are acceptable only for preliminary examination because morphological features often become altered after prolonged exposure. The profile examination begins with a first approximation and marking of soil horizon boundaries on the profile. Each horizon is then carefully observed and described. Horizon boundaries are relocated as required by the detailed study. For ease in comparing soil profile descriptions, conventional methods and terms need to be followed as faithfully as possible. However, when conventional terms are not adequate to express fully what is observed, the soil scientist should append whatever additional notes he considers necessary.

SOIL HORIZONS AND THEIR
RECOGNITION IN THE FIELD.
A soil horizon is "a layer of soil, approximately parallel to the soil surface, with characteristics produced by soil-forming processes" (Soil Survey Staff 1962). Soil horizons are real bodies, subparts of the soil individual, that extend laterally in the x and y directions and vertically along the z axis (Fig. 2.1).

In the field, soil horizons are described according to the following properties: color, texture, consistence, structure, cutans, nodules or concretions, voids, pH (by field method), boundary characteristics, and horizon continuity. Thickness and depth of each horizon below the soil surface are measured and recorded. The skill with which the soil properties are recognized and recorded determines the quality of the resulting soil description. Soil descriptions are a major part of the basis of the classification of soils into defined categories. The following definitions of soil properties are based on the current practices in the United States. It should be recognized that practices vary somewhat and what is herein presented represents the format prescribed by the SCS-USDA (Soil Survey Staff 1962).

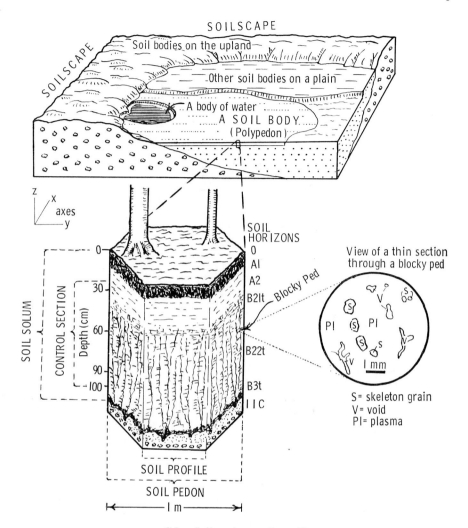

FIG. 2.1. Soil pedon and profile.

COLOR—BASE COLOR AND MOTTLING. Soil color is probably the most obvious feature of the soil and is easily seen by the layman. An observer experienced in a given area can often relate soil color to specific chemical, physical, and biological properties of the soils in that area. Some broad generalizations are also possible. Black soil color usually indicates the presence of organic matter. Red colors indicate the presence of free iron oxides common in well-oxidized soil. Upon the removal of free iron under reducing conditions, the soil mineral grains usually appear gray or bluish gray in color. In some instances relict colors, that is, those inherited from the initial materials, persist in the soil.

Color is composed of three measurable variables: hue, value, and chroma (Soil Survey Staff 1951). *Hue* is the dominant spectral color and is related to wavelength of light. *Value* is a measure of degree of darkness or lightness of the color and is related to the total amount of light reflected. *Chroma* is a measure of the purity or strength of spectral color. These three variables have been combined into books of standard colors that cover the ranges found in soils. In the soil color book, the various hues are arranged by pages, one hue to a page. The units of value are arranged vertically and the units of chroma are arranged horizontally on a page. Opposite each page of color chips is a page of color symbols and corresponding English names. An example of a color notation made for a soil horizon color is 10YR 6/3. The interpretation of the notation is 10YR (10 yellow-red) hue, a value of 6, and a chroma of 3. The proper name for this color is pale brown.

Because soil color is moisture dependent, especially with respect to color value, it is necessary to record the moisture status at the time the soil color is described. Many times it is practical to describe only the moist color. However, when the soil under observation is dry, both dry and moist colors are recorded.

Mottling is described in terms of three characteristics: contrast, abundance, and size of area of each color. The following conventions based on a paper by Simonson (1951) are quoted from the *Soil Survey Manual* (Soil Survey Staff 1951):

ABUNDANCE	Abundance of mottles can be indicated in three general classes as *few, common,* and *many* based upon the relative amount of mottled surface in the unit area of the exposed soil horizon, as follows:
Few:	Mottles occupy less than about 2 percent of the exposed surface.
Common:	Mottles occupy about 2 to 20 percent of the exposed surface.
Many:	Mottles occupy more than 20 percent of the exposed surface. The last class can be further subdivided according to whether (a) the mottles are embedded in a definite matrix or (b) there is no clear matrix color.
SIZE	Size refers to the approximate diameters of individual mottles. Three relative size classes can be used as follows:
Fine:	Mottles less than 5 mm. in diameter along the greatest dimension.
Medium:	Mottles range between 5 and 15 mm. in diameter along the greatest dimension.
Coarse:	Mottles are greater than 15 mm. in diameter along the greatest dimension.
CONTRAST	Contrast may be described as *faint, distinct,* or *prominent* as follows:
Faint:	Indistinct mottles are evident and recognizable only upon close examination. Soil colors in both the matrix and mottles have closely related hues and chromas.

Distinct: Although not striking, the mottles are readily seen. The hue, value, and chroma of the matrix are easily distinguished from those of the mottles. They may vary as much as one or two hues or several units in chroma or value. The pattern may be one of a continuous matrix with mottles or one of mixtures of two or more colors.

Prominent: The conspicuous mottles are obvious and mottling is one of the outstanding features of the horizon. Hue, chroma, and value may be several units apart. The pattern may be one of a continuous matrix with contrasting mottles or one of mixtures of two or more colors.

TEXTURAL CLASSES. Soil texture is defined as "the relative proportions of the various soil separates in a soil material" (Committee on Terminology 1956). The continuum of soil texture has been divided into various textural groups for the purpose of describing the soil horizons. Figure 2.2 is a graphic model showing the various groupings now recognized. The lower part of Figure 2.2 shows how the separates or particle sizes are grouped in three systems. For most soil purposes the USDA scale is used. In addition to the textural group, determined from the proportion of particles less than 2 mm in diameter, the textural name is prefaced by gravelly when 20 to 50% of the material is of gravel or very gravelly when more than 50% of the soil is of gravel size (2 to 76 mm). The same proportions apply to coarser material such as cobbles [7.6 to 25 cm (3–10 in.) diameter] and stones [>25 cm (10 in.) in diameter] (Soil Survey Staff 1951).

**CONSISTENCE AT VARIOUS
MOISTURE CONTENTS.** The consistence of the soil material in each horizon has a decided bearing on several characteristics of the soil. Unfortunately, field measures of this property are crude. Consistence is described at as many moisture contents as possible under field conditions. Three wetness conditions are usually attempted. The following terms are useful for describing consistence in the field. Brief descriptions of how the property demonstrates itself in a field test are made after each term.

I. Wet consistence—moisture content at or slightly more than "field moisture capacity"
   A. Stickiness—quality of adhesion to other objects
      0. Nonsticky   Almost no natural adhesion of soil material to fingers.
      1. Slightly sticky Soil material adheres to one finger but other finger is clean.
      2. Sticky   Soil material adheres to both fingers and thumb; stretches somewhat.
      3. Very sticky   Soil material strongly adheres to both thumb and finger.

   B. Plasticity—capability of being molded by the hands
      0. Nonplastic   No "wire" is formable by rolling material between the hands.

1. Slightly plastic	Only short (<1 cm) "wires" are formed by rolling material between the hands.
2. Plastic	Long wires (>1 cm) can be formed and moderate pressure is needed to deform a block of the molded material.
3. Very plastic	Much pressure is needed to deform a block of the molded material.

II. Moist consistence—soil moisture content between dryness and "field moisture capacity"

0. Loose	Soil material is noncoherent.
1. Very friable	Aggregates crush easily between thumb and finger.
2. Friable	Gentle thumb and finger pressure is required to crush aggregates.
3. Firm	Moderate thumb and finger pressure is required to crush aggregates.
4. Very firm	Strong thumb and finger pressure is required to crush aggregates.
5. Extremely firm	Aggregates cannot be broken by thumb and finger pressure.

III. Dry consistence—air-dry soil condition

0. Loose	Soil material is noncoherent.
1. Soft	Aggregate easily breaks to single grain in the hand.
2. Slightly hard	Gentle thumb and finger pressure is required to crush the material.
3. Hard	Aggregates are barely breakable by thumb and finger.
4. Very hard	Aggregates are barely breakable in both hands.
5. Extremely hard	Aggregates cannot be broken with both hands.

Cementation:

Weak	Material can be broken by hand.
Strong	Material is brittle but easily broken by hammer.
Indurated	Material is brittle, requiring a strong hammer blow to break.

STRUCTURE. Structure refers to the aggregation of individual soil particles into larger units with planes of weakness between them. Individual aggregates are known as peds. Soils that do not have aggregates with naturally preserved boundaries (peds) are considered to be structureless. Two forms of a structureless condition are recognized, that is, single grain (particles are easily distinguishable) or massive (individual particles adhere closely to each other but the mass lacks planes of weakness).

Three features of structure are usually described in each horizon: *grade, class,* and *type.*

The *grade* or strength of structure is moisture dependent but is evaluated by the soil scientist in the following fashion: the grade of structure is

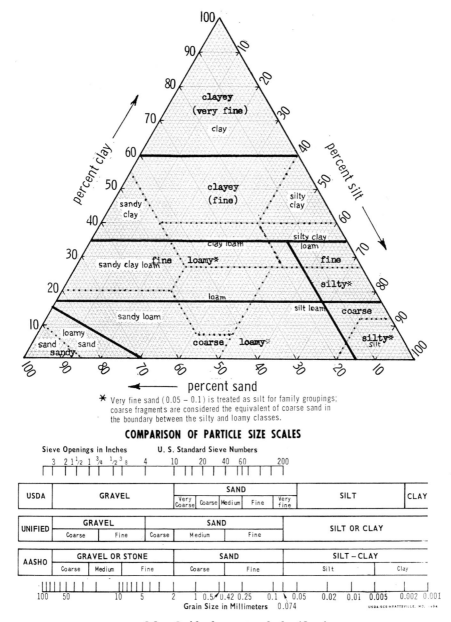

* Very fine sand (0.05 – 0.1) is treated as silt for family groupings; coarse fragments are considered the equivalent of coarse sand in the boundary between the silty and loamy classes.

FIG. 2.2. Guide for textural classification.

considered weak if the peds are observable in place but cannot be removed without being destroyed; the grade of structure is considered moderate if the peds can be removed from the profile for examination in the hand; the grade of structure is considered strong if the peds, when removed from the profile, are rigid and durable in the hand.

Class of soil structure refers to the size of the peds (Table 2.1).

Type of soil structure refers to the shape of peds and together with size of peds is used to define soil structure (see Table 2.1). These three features of structure are conventionally written in the order of grade, class, and type; for example, weak, medium, subangular blocky structure.

CUTANS—AS DESCRIBED IN THE FIELD. A cutan is a "modification of the texture, structure or fabric at natural surfaces in soil materials due to concentration of particular soil constituents or in situ modification of the plasma" (Brewer 1964). In general, this modification is considered a microstructural feature and as such will be discussed in greater detail in Chapter 3. Because of the difficulty in examining cutans, any detailed description of them is better left to laboratory studies. It is important, however, to describe as much as possible in the field. A $10\times$ hand lens is a desirable field aid for cutan examination. In general, four kinds of cutans can be recognized: *clay, stress, oxide,* and *organic matter*.

The *clay cutan* (argillan) is largely composed of clay that has been transported by water through the larger voids in the soil and deposited on their walls at a time when the water left the voids. It is, therefore, a coating of clay. It is most easily identified where it is redder or darker colored than the interior of the ped. The only certain field criteria for the identification of the clay cutan (also referred to as clay skin, clay film, or Tonhautchen) are (1) observed real thickness, (2) an abrupt boundary between the coating material and the interior of the ped, as seen in cross section on a broken surface. A hand lens of about $10\times$ power is helpful for this purpose.

The *stress cutan* is often confused with the illuvial cutan in field examination. Stress cutans are formed when peds press against each other during soil wetting. If the ped surface appears smooth and the coating has no observable thickness when viewed in cross section (indicating no illuvial cutan), it is probable that a stress cutan (pressure face) is present. A special type of stress feature called slickenside is usually present in soils that contain appreciable quantities of expanding layer lattice silicates and are subjected to a monsoon type climate. A slickenside is a smoothed surface with parallel striae and grooves. Slickensides are among the criteria for recognizing Vertisols and various Vertic intergrades.

The *oxide cutan* is a thin layer of a metal oxide, usually of iron (sesquan) or manganese (mangan). The color, usually different from that of the ped interior, is red where iron oxide is prominent and black where manganese is an important constituent.

Organic matter cutans (organs) are also dark in appearance. These cutans usually do not have the hard, smooth appearance of the argillan (clay skin) or the stress cutan (pressure face).

TABLE 2.1 ❧ Types and classes of soil structure

Class	Platy	Prismatic	Columnar	(Angular) Blocky*	(Subangular) Blocky†	Granular	Crumb
	Platelike, with one dimension (the vertical) limited and greatly less than the other two; arranged around a horizontal plane; faces mostly horizontal	Prismlike, with two dimensions (the horizontal) limited and considerably less than the vertical; arranged around a vertical line; vertical faces well defined; vertices angular		Blocklike; polyhedronlike, or spheroids, or with three dimensions of the same order of magnitude, arranged around a point		Spheroids or polyhedrons having plane or curved surfaces which have slight or no accommodation to the faces of surrounding peds	
		Without rounded caps	With rounded caps	Blocklike; blocks or polyhedrons having plane or curved surfaces that are casts of the molds formed by the faces of the surrounding peds			
				Faces flattened, most vertices sharply angular	Mixed rounded and flattened faces with many rounded vertices	Nonporous peds	Porous peds
Very fine or very thin	Very thin platy; <1 mm	Very fine prismatic; <10 mm	Very fine columnar; <10 mm	Very fine angular blocky; <5 mm	Very fine subangular blocky; <5 mm	Very fine granular; <1 mm	Very fine crumb; <1 mm
Fine or thin	Thin platy; 1 to 2 mm	Fine prismatic; 10 to 20 mm	Fine columnar; 10 to 20 mm	Fine angular blocky; 5 to 10 mm	Fine subangular blocky; 5 to 10 mm	Fine granular; 1 to 2 mm	Fine crumb; 1 to 2 mm
Medium	Medium platy; 2 to 5 mm	Medium prismatic; 20 to 50 mm	Medium columnar; 20 to 50 mm	Medium angular blocky; 10 to 20 mm	Medium subangular blocky; 10 to 20 mm	Medium granular; 2 to 5 mm	Medium crumb; 2 to 5 mm
Coarse or thick	Thick platy; 5 to 10 mm	Coarse prismatic, 50 to 100 mm	Coarse columnar; 50 to 100 mm	Coarse angular blocky; 20 to 50 mm	Coarse subangular; 20 to 50 mm	Coarse granular; 5 to 10 mm	
Very coarse or very thick	Very thick platy; >10 mm	Very coarse prismatic; >100 mm	Very coarse columnar; >100 mm	Very coarse angular blocky; >50 mm	Very coarse subangular blocky; >50 mm	Very coarse granular; >10 mm	

SOURCE: Soil Survey Staff 1960.
* (a) Sometimes called nut. (b) The word "angular" in the name can ordinarily be omitted.
† Sometimes called nuciform, nut, or subangular nut. Since the size connotation of these terms is a source of great confusion to many, they are not recommended.

Intergrades exist between these four kinds of cutans.

When describing cutans in the field, it is important to describe only what can actually be seen. The color of the cutan should be recorded as well as the thickness of the coating, if this is measurable.

NODULES AND CONCRETIONS—
RECOGNITION AND DESCRIPTION. Local concentrations of soil material constitute concretions and nodules in soil. The color, hardness, size, and relative abundance of nodules and concretions are usually reported in the field description. If possible, these features are crushed and description made of the interiors. Quick tests can be made to determine the cementing agent. Effervescence in HCl indicates carbonate cementation. Effervescence in H_2O_2 indicates manganese oxides. Silica cementation is indicated in concretions that may or may not effervesce in HCl but disintegrate after being placed in concentrated NaOH for several hours.

VOIDS (PORES).
Approximately half the volume of a soil is pore space. The size, shape, and continuity of the voids determine to a large extent the movement of air and water in the soil. The characteristics of the pores are in some measure determined by the structure. Many pores are too small for observation under field conditions. However, certain characteristics of the larger pores, those visible in the field, are included in the description of each horizon. Of particular importance are the shape of the larger pores and the degree of interconnection as these features influence infiltration and permeability characteristics.

FIELD MEASUREMENT AND
INTERPRETATION OF pH VALUES. The determination of pH values is probably the most important chemical measurement that can be made in soils. Although not a morphological measurement, it is common to make a field pH determination at the time of writing a profile description. Information about soil reaction is necessary to determine liming needs and fertilizer responses. In addition, pH values make it possible to infer many of the chemical processes that have taken place in the genesis of the soil.

In soil science, the pH value is defined as the negative logarithm of the hydrogen ion activity. By specifying activity rather than concentration, we recognize that there are other hydrogen ions in the soil system, such as in the organic matter and in the mineral structure, but that we are measuring only that H^+ active in the soil solution.

Field methods of pH determination are either colorimetric or electrometric. The electrometric methods require the use of portable battery-operated pH meters. These instruments are available from several suppliers but are not widely used because of high cost and susceptibility to damage.

Colorimetric methods can be devised for almost any soil situation by the careful selection of the proper indicators that have color changes in the pH range found in that soil. The Hellige-Truog soil pH kit, using three organic indicators, has been successfully used in many soils. It performs very well in soils where the pH value is between 4 and 7.5. For soils with pH values above 7.5, it is desirable to use an indicator such as Thymol Blue or Cresol Red. For extremely acid soils, such indicators as Methyl Yellow, Bromphenol Blue, or Thymol Blue are useful. One problem inherent to all colorimetric procedures for determining soil pH values is that of providing a uniform background color free from discoloration by suspended organic or mineral material. This problem is solved in the Hellige-Truog pH tester by dusting inert barium sulfate over the sample, after the indicator has been mixed with the soil, and allowing the colored solution to soak up into the white barium sulfate powder. This same practice can be used with other indicators.

In some soils organic indicators are absorbed by the soil material, leaving the solution colorless. This problem can be at least partly solved by placing some clean filter paper in contact with the soil solution and allowing the paper to wet by capillarity. Indicator can then be added to the paper with good results. This technique can be used either with H_2O or KCl solution. (Further discussion of pH is in Chapter 4.)

HORIZON BOUNDARIES AND HORIZON CONTINUITY WITHIN THE PEDON. Conventions have been developed to describe soil horizon boundaries. Two measurements are considered necessary, namely, the *width* or *thickness* of the boundary along the z (vertical) axis and the *topography* of the surface of the boundary in the x-y plane. The following terms have been used to describe the width of the boundary between horizons: (1) *abrupt*—the boundary is of such contrast that it is less than 2.5 cm (1 in.) wide; (2) *clear*—the boundary layer is 2.5 to 6.4 cm (1 to 2½ in.) wide; (3) *gradual*—the boundary layer is 6.4 to 12.7 cm (2½ to 5 in.) wide; (4) *diffuse*—the boundary layer is more than 12.7 (5 in.) wide.

The surface topography of a soil horizon can be described by the following terms: (1) *smooth*—the boundary is nearly a plane; (2) *wavy*—the boundary is undulating, with the pockets having a greater length in either the x or y direction than in the z direction; (3) *irregular*—the boundary topography is such that pockets vary more in the z axis direction than in either the x or y direction; (4) *broken*—the horizon is discontinuous within the pedon.

SOIL HORIZON DESIGNATIONS. After the morphological characteristics of the horizons have been described, it becomes a convenience for purposes of communication to classify and name each horizon. Horizon names are not mutually exclusive, and a great deal of interpretation on the part of the soil scientist is involved in naming a horizon. Horizon designations differ somewhat from country to country. In the United States, soil horizons are designated by a code of letter and numbers. The following system has been developed by soil scientists of the cooperative Soil Survey in the United States (Soil Survey Staff 1962).

MORPHOGENIC MASTER HORIZON
DESIGNATIONS AND DEFINITIONS. Master or major horizons are designated by capital letters. Subdivisions of the master horizons are designated by arabic numerals.[1] (Definitions quoted in part from Soil Survey Staff 1962.)

Organic Horizons on Mineral Soil Surfaces

O—Organic horizons of mineral soils include horizons (1) formed or forming above the mineral part of mineral soil profiles; (2) dominated by fresh or partly decomposed organic material; and (3) containing more than 30% organic matter if the mineral fraction is more than 50% clay, or more than 20% organic matter if the mineral fraction has no clay. Intermediate clay content requires proportional organic matter content equal to $20 + (0.2 \times \%$ clay).

The O horizons have been formed from organic litter derived from plants and animals and deposited on the mineral surface.

These horizons are measured upward from the top of the underlying mineral material, that is, 15 to 0 cm (6 to 0 in.).

O1—Organic horizons in which essentially the original form of most vegetative matter is visible to the naked eye.

Organic matter in an O1 horizon is essentially unaltered except from leaching of soluble constituents and discoloration. The source of the material should be identifiable from examination. The O1 horizon corresponds to the L layers and some F layers mentioned in literature on forest soils (Wilde 1958). These horizons were formerly called Aoo horizons.

O2—Organic horizons in which the original form of most plant or animal matter cannot be recognized with the naked eye.

Remains of parts of plants and animals often can be identified but most of the material is decomposed to a point such that the source cannot be identified. The O2 corresponds to the H layer and some F layers described in literature on forest soils (Wilde 1958). These horizons were formerly called Ao horizons.

Mineral Horizons and Layers

Mineral horizons contain less than 20% organic matter if the mineral fraction contains no clay; or less than 30% organic matter if the mineral fraction contains 50% or more clay. For soils containing 0 to 50% clay, intermediate maximum limits of organic matter contents between 20 and 30%, in proportion to the clay content, are prerequisite.

---

1. Nomenclature for organic soil (Histosols) horizons is not included. See Chapter 24.

A—Mineral horizons consisting of (1) horizons of organic matter accumulation formed or forming at or adjacent to the surface; (2) horizons that have lost clay, iron, or aluminum, with resultant concentrations of quartz or other resistant minerals of sand or silt size; or (3) horizons dominated by 1 or 2 above but transitional to an underlying B or C.

The following major subdivisions of the A horizon are recognized.

A1—Mineral horizons, formed or forming at or adjacent to the surface, in which the feature emphasized is an accumulation of humified organic matter intimately associated with the mineral fraction.

The soil is as dark or darker than underlying horizons because of the presence of organic matter. The organic material is assumed to be derived from plant and animal remains deposited on the surface of the soil or deposited within the horizon without appreciable translocation.

A2—Mineral horizons in which the feature emphasized is loss of clay, iron, or aluminum, with resultant concentration of quartz or other resistant minerals in sand and silt sizes.

An A2 horizon is usually differentiated from an overlying A1 horizon by lighter color and lower organic matter content. An A2 horizon is commonly differentiated from an underlying B in the same profile by its lighter color and/or coarser texture. Position in the profile is not diagnostic, but in cases where the surface horizon would qualify equally well as either A1 or A2, the designation A1 is given preference over A2.

A3—A transitional horizon between A and B, and dominated by properties characteristic of an overlying A1 or A2 but having some subordinate properties of an underlying B.

Several kinds of material form the transitional zone from A1 or A2 to different kinds of B horizons. They may be quite unlike one another. Inferences can be made after noticing the symbols assigned to the overlying and underlying horizons. The symbol A3 is also used to designate the transition of an A1 horizon to other subsoil horizons where no B horizon is present.

AB—A horizon transitional between A and B, having an upper part dominated by properties of A and a lower part dominated by properties of B, and the two parts cannot conveniently be separated into A3 and B1. This nomenclature is normally used only for thin horizons.

A & B—Horizons that would qualify for A2 except for included parts constituting less than 50% of the volume that would qualify as B.

Most frequently this nomenclature is used where A2 material partially surrounds thin columnarlike upward extensions of the B horizon or where bodies of B horizon material are surrounded by A2 horizon material.

AC—A horizon transitional between A and C, having subordinate properties of both A and C, but not dominated by properties characteristic of either A or C.

This nomenclature is used in a similar fashion as AB nomenclature when the A horizon rests on a C horizon.

B—Horizons in which the dominant feature or features are one or more of the following: (1) an illuvial concentration of silicate clay, iron, aluminum, or humus, alone or in combination; (2) a residual concentration of sesquioxides or silicate clays, alone or mixed, that has formed by means other than solution and removal of carbonates or more soluble salts; (3) coatings of sesquioxides adequate to give conspicuously darker, stronger, or redder colors than overlying and underlying horizons in the same sequum, but without apparent illuviation of iron and not genetically related to B horizons that meet requirements of 1 or 2 in the same sequum; or (4) an alteration of material from its original condition in sequum lacking conditions defined in 1, 2, and 3 that obliterates original rock structure, that forms silicate clay, liberates oxides, or both, and that forms a granular, blocky, or prismatic structure if textures are such that changes accompany changes in moisture.

No universal diagnostic property or location in the profile is satisfactory to identify all B horizons. Usually a B horizon is under an A horizon. However, the A horizon may have been truncated. Usually the B horizon carries some subhorizon nomenclature that, in the judgment of the person describing the soil, more fully characterizes the condition.

B1—A transitional horizon between B and A1 or between B and A2 in which the horizon is dominated by properties of an underlying B2 but has some subordinate properties of an overlying A1 or A2.

It is necessary to have an overlying A1 or A2 horizon and an underlying B2 horizon to characterize a B1 horizon unless accelerated erosion or cultivation is apparent.

B & A—Any horizon qualifying as B in more than 50% of its volume, including parts that qualify as A2.

Use of this nomenclature is similar to the use of the A & B horizon nomenclature except that the B horizon bodies make up over 50% of the horizon.

B2—That part of the B horizon where the properties on which the B is based are without clearly expressed subordinate characteristics indicating that the horizon is transitional to an adjacent overlying A or an adjacent underlying C or R.

This nomenclature is used to name the horizon that has the characteristic upon which the B horizon is most clearly based. The characteristics vary from soil to soil. In some profiles, the most strongly expressed part of the B horizon, which is what would be called a B2, has less well-expressed features than B3 or B1 horizons of other profiles. The designation of a B2 horizon is used strictly with reference to a single profile.

B3—A transitional horizon between B and C or R in which the properties diagnostic of an overlying B2 are clearly expressed but are associated with clearly expressed properties characteristic of C or R. A B3 horizon is designated only if there is an overlying B2 horizon.

C—A mineral horizon or layer, excluding bedrock, that is either like or unlike the material from which the solum is presumed to have formed, relatively little affected by pedogenic processes, and lacking properties diagnostic of A or B but including materials modified by (1) weathering outside the zone of major biological activity; (2) reversible cementation, development of brittleness, development of high bulk density, and other properties characteristic of fragipans; (3) gleying; (4) accumulation of calcium or magnesium carbonate or more soluble salts; (5) cementation by such accumulation as calcium or magnesium carbonate or more soluble salts; or (6) cementation by alkali-soluble siliceous material or by iron and silica.

Many types of alterations are permitted in material designated as C horizon. Alterations by chemical weathering deep in the soil are commonly accepted in C horizons; however, biologically influenced alteration is usually excluded. Accumulations of carbonates, gypsum, or more soluble salts are permitted in C horizons where other processes are considered inconsequential. Cementation by such materials is permitted and its presence is indicated by using the suffix m. Present usage of C may include older D and G designations (Soil Survey Staff 1951). The C horizon has often been called parent material for the solum. This is incorrect and it is more correct to say it is like the material from which the A & B horizon is presumed to have formed. The present usage of C does not make even this interpretation completely correct as will be seen in the discussion of lithologic discontinuities. The C1 and C2 designations, as defined in the *Soil Survey Manual* (Soil Survey Staff 1962), have been discontinued.

R—Underlying consolidated bedrock, such as granite, sandstone, or limestone. If presumed to be like the parent rock from which the adjacent overlying layer or horizon was formed, the symbol R is used alone. If presumed to be unlike the overlying material, the R is preceded by a Roman numeral denoting lithologic discontinuity.

SUBHORIZON SYMBOLS. To facilitate more complete expression of the horizon characteristics than is possible with master horizon designations, the following symbols are used. (Definitions are quoted in part from Soil Survey Staff 1962.)

b—Buried soil horizon.

This symbol is added to the master horizon designation of a buried genetic horizon or horizons. Horizons of another solum may or may not have formed in the overlying material, which may be similar to, or different from, the assumed parent material of the buried soil.

ca—An accumulation of carbonates of alkaline earth, commonly calcium.

This symbol is applied to A, B, or C horizons. The presence of secondary carbonates alone is not adequate to justify the use of the ca symbol. The horizon must have more carbonates than the parent material is presumed to have had.

cs—An accumulation of calcium sulfate.

This symbol is used in a manner comparable to that of ca.

cn—Accumulations of concretions or hard nonconcretionary nodules enriched in sesquioxides with or without phosphorus.

The nodules indicated by the symbol cn must be hard when dry but need not be indurated. The horizon description should characterize the nodules. Nodules, concretions, or crystals do not qualify as cn if they are of dolomite or more soluble salts, but they do qualify if they are of iron, aluminum, manganese, or titanium.

f—Frozen soil.

The suffix f is used for soil that is thought to be permanently frozen.

g—Strong gleying.

The suffix g is used with a horizon designation to indicate intense reduction of iron during soil development, or reducing conditions due to stagnant water, as evidenced by base colors that approach neutral (usually chroma ≤ 2), with or without mottles. Hues bluer than 10Y are also indicative of strong gleying in some soils. Horizons of low chroma in which the color is due to uncoated sand or silt particles are not considered strongly gleyed. Although gleying is commonly associated with wetness, especially in the presence of organic matter, wetness by itself is not a criterion of gleying. The symbol g may be applied to any of the major symbols for mineral horizons.

h—Illuvial humus.

Accumulations of decomposed illuvial organic matter, appearing as dark coatings on sand or silt particles or as discrete dark pellets of silt size, are indicated by h. This condition is expressed only as a subdivision of the B horizon.

ir—Illuvial iron.

Accumulations of illuvial iron, appearing as coatings on sand or silt particles or as pellets of silt size, are indicated by ir. In some horizons the coatings have coalesced, filled pores, and cemented the horizon. The ir designation is often used with the h as in a Bhir or B2hir to indicate both iron and humus accumulation.

m—Strong cementation, induration.

The symbol m is applied as a suffix to horizon designations to indicate irreversible cementation. The symbol is not applied to indurated bedrock. Contrary to previous usage, m is not used to indicate firmness, as in fragipans, but is confined to indurated horizons which are essentially (more than 90%) continuous, though they may be fractured.

p—Plowing or other disturbance.

The symbol p is used as a suffix with A to indicate disturbance by cultivation or pasturing. Even though a soil has been truncated and the plow layer is clearly in what was once B horizon, the designation Ap is used.

sa—An accumulation of salts more soluble than calcium sulfate.

This symbol may be applied to the designation of any horizon and in its manner of use is comparable to that described for ca or cs.

si—Cementation by siliceous material, soluble in alkali.

This symbol is applied only to the C horizon. The cementation may be nodular or continuous. If the cementation is continuous, the symbol sim is used.

t—Illuvial clay.

The presence of argillans in B horizons is indicated by the suffix t.

x—Fragipan character.

The symbol x is used as a suffix with horizon designations to indicate genetically developed properties of firmness, brittleness, high density, and characteristic distribution of clay that are diagnostic of fragipans. Fragipans, or parts of fragipans, may qualify as A2, B, or C. Such horizons are classified as A2, B, or C, and the symbol x is used as a suffix to indicate fragipan character.

OTHER HORIZON SUBDIVISIONS. It is often necessary to subdivide horizons for purposes not provided for with the foregoing designations. For example, it may be necessary to subdivide Ap, A1, A2, A3, B1, B2, B3, or C horizons for sampling, or to point out minor differences. Such subdivision may even be arbitrary with respect to differences observable in the field. Using Arabic numbers, such subdivisions are numbered consecutively from the top of the horizon downward, that is, B21, B22, B23. By convention, if suffixes consisting of lower-case letters are being used, the Arabic numbers precede all lower-case suffixes except p, that is, B21t, C1g, C2g, but Ap1, Ap2.

LITHOLOGIC DISCONTINUITIES. Where different layers are encountered in the profile and the interpretation of the soil scientist is that these are the result of geologic rather than pedologic processes, Roman numeral prefixes are used consecutively from the surface downward. A soil that consists only of one kind of material could thus carry the Roman numeral I as a prefix for all of its horizons. By convention, this numeral is therefore omitted from the symbol, as it is understood that all the material is I. Further, the top mineral horizon in a profile having two or more contrasting materials could rightly be considered to have I prior to its symbol, so this is omitted by convention, and numbering starts with the second layer of contrasting material, which is designated II. This procedure negates the use of C1, C2, C3, etc., in the sense defined in the *Soil Survey Manual* (Soil Survey Staff 1951) since in the C horizon any change can be considered geologic. The C1, C2, C3, etc., would now be C, IIC, IIIC, etc., in the case where the C horizon was thought to be of the same geologic origin as the A and B horizons. Use of the Roman numeral is not confined to the C horizon and where lithologic changes are found in the solum, the following sequence may be found: Ap, A2, B1, B2, IIB2, IIB3, IIC, IIIC where, for example, the parent material for the Ap through B2 horizons was thought to be loess, the parent material of the IIB2 through IIC horizons thought to be outwash, and the IIIC considered lacustrine sediments.

DIAGNOSTIC HORIZONS FOR CLASSIFICATION. At the present time, descriptive nomenclature of a more quantitative nature is being used in describing and naming soils. This nomenclature is an intricate part of the comprehensive classification system (Soil Survey Staff 1960, 1967). The following are brief definitions of some of the terms now being used. The reader is cautioned that these are not complete definitions but greatly abbreviated versions. In many cases definitions are still being discussed and revised, but it is doubtful that the basic meaning attached to each term contained in this material will be changed. The reader is also cautioned that before he uses a term in any written material, thesis, etc., he should refer to the original definition in the 7th Approximation and subsequent USDA material.

EPIPEDONS. Epipedons are simply the uppermost soil horizons. Epipedon is not synonymous with A horizon and may be thinner than the A horizon or include some of the B horizon.

Mollic epipedon: A surface horizon that, when mixed to a depth of 17.5 cm (7 in.), contains over 1% organic matter, with color values darker than 5.5 dry and 3.5 moist. Structure cannot be massive and hard. Base saturation is over 50%.

Anthropic epipedon: A surface horizon like the mollic epipedon but contains over 250 ppm of citric acid soluble P_2O_5.

Umbric epipedon: A surface horizon like the mollic epipedon but is less than 50% base saturated.

Histic epipedon: A surface horizon that contains more than 20 to 30% organic matter, depending on clay content, and is water saturated for 30 days at some season of the year unless artificially drained. It is thinner than 30 cm (12 in.) if drained or 45 cm (18 in.) thick if not artificially drained.

Ochric epipedon: A surface horizon that is light in color, with color values >5.5 dry and >3.5 moist and contains less than 1% organic matter.

Plaggen epipedon: A man-made surface horizon that is >50 cm (20 in.) thick, created by years of manure additions.

SUBSURFACE DIAGNOSTIC HORIZONS.

Argillic horizon: In general, this is a B horizon that has at least 1.2 times as much clay as some horizon above or 3% more clay content if the eluvial layer has < 15% clay or 8% more clay if eluvial layer has >40% clay. It is formed by illuviation of clay, and illuviation argillans are usually observable unless there is evidence of stress cutans. It should be 1/10 as thick as all overlying horizons or more than 15 cm (6 in.), whichever is thinner.

Agric horizon: This horizon is formed directly under the plow layer and has clay and humus accumulated as thick, dark lamellae to the extent that they occupy at least 15% of the soil volume.

Natric horizon: This horizon meets the requirements of an argillic horizon but also has prismatic or columnar structure and over 15% of the CEC is saturated with Na^+.

Spodic horizon: This horizon has an illuvial accumulation of free sesquioxides and organic matter. There are many specific limitations dealing with Al, Fe, O.M., and clay ratios, depending on whether the overlying horizon is virgin or cultivated.

Cambic horizon: This is a subsoil horizon with some weak indication of either an argillic or spodic horizon, but not enough to qualify as either; for example, less than 1.2 times as much clay as an overlying horizon.

Oxic horizon: This horizon is at least 30 cm (12 in.) thick. It is highly weathered, with a high content of low charge

1:1 clays and sesquioxides retaining <10 meq of NH_4^+ from $1N$ NH_4Cl or <10 meq KCl-extractable bases per 100 g of clay. Only a trace of the clay is water dispersible.

MASTER DIAGNOSTIC HORIZONS OF ORGANIC SOILS.

Fibric horizon:
: (Formerly peat.) In an unrubbed condition, fibers compose over 2/3 of the mass, and the material yields almost clear solutions when extracted with sodium pyrophosphate.

Hemic horizon:
: (Formerly mucky peat or peaty muck.) In an unrubbed condition, 1/3 to 2/3 of the total mass is composed of fibers (intermediate in decomposition between fibric and sapric horizons).

Sapric horizon:
: (Formerly muck.) In an unrubbed condition, less than 1/3 of the mass is composed of identifiable fibers and it produces sodium pyrophosphate extracts with colors lower in value and higher in chroma than 10YR 7/3.

OTHER MACROPEDOLOGIC FEATURES. In addition to the master diagnostic horizons, several other layers or horizons and macrofeatures are recognized and named in the new classification system. These features are usually used as diagnostic criteria at a somewhat lower level in the classification system than the subsurface diagnostic horizons and epipedons.

Fragipan:
: These subsoil layers are of high bulk density, brittle when moist, and very hard when dry. They do not soften on wetting, but can be broken in the hands. Air-dry peds slake in water.

Calcic horizon:
: This layer has a secondary accumulation of carbonates, usually of Ca or Mg, in excess of 15% calcium carbonate equivalent and contains at least 5% more carbonate than an underlying layer.

Petrocalcic horizon:
: This is an indurated calcic horizon. It has hardness of 3 or more (Mohs scale) and at least 1/2 of it breaks down in acid, but it does not break down in water.

Gypsic horizon:
: This is a horizon of calcium sulfate enrichment. It contains at least 5% more calcium sulfate than underlying material.

Plinthite:
: This is a humus-poor, sesquioxide-rich horizon, which hardens irreversibly to ironstone hardpans or aggregates with repeated wetting and drying. The red, indurating portions of the layer are usually mottled with yellowish, grayish, or white bodies.

Salic horizon: This is a horizon of secondary soluble salt enrichment (over 2 to 3%, depending on thickness).

Albic horizon: Typically this is an A2 horizon, with certain color limitations.

Lithic contact: This is a boundary between soil and continuous coherent, underlying material that has a hardness >3 Moh.

Paralithic contact: This is a boundary between soil and underlying coherent material, with a hardness <3 on the Moh scale.

Duripan: This is a subsurface horizon at least 1/2 cemented by SiO_2. Air-dry peds do not slake in water.

Permafrost: This is a continuously frozen layer.

Limnic materials: These are organic or inorganic materials deposited in water by the action of aquatic organisms or derived from underwater and floating organisms. Marl, diatomaceous earth, and sedimentary peat (coporgenous earth) are considered limnic materials.

In addition, the following designations are used for thick, sandy surface horizons.

Arenic horizon: This is a loamy fine sand or coarser horizon more than 50 cm (20 in.) thick over an argillic horizon.

Grossarenic horizon: This is a loamy fine sand or coarser horizon thicker than 100 cm (40 in.) over an argillic horizon.

A PERSPECTIVE OF
MORPHOLOGIC TERMINOLOGY. The terminology presented in this chapter is that used by the National Cooperative Survey of the United States. As such, it is undoubtedly the most widely used terminology in the United States. History has taught us that we can expect changes in nomenclature as the state of our knowledge changes. Also, other countries have developed nomenclature and conventions that serve equally well in naming soil horizons.

Within the United States, new proposals for nomenclature change are constantly being made. These usually take the form of research papers in the scientific journals. For example, the K horizon (Gile, Peterson, and Grossman 1965), which is a special form of what has previously been referred to as a cam (both ca and m features) subhorizon, has been used for petrocalcic layers. Whiteside (1959) proposed a rather complete system of horizon nomenclature designed to aid in the connotation of soil genesis. Upon publication, these proposals are critically examined and tested by other soil scientists and with the passing of time either adapted or ignored.

A complete history of the proposals that have been made would not only be lengthy but unduly burdensome. Throughout the development of nomenclature, however, an increasing emphasis on quantification is evident.

The development of nomenclature in many different countries has often led to misunderstandings (Dudal 1968). Undoubtedly soil science, as a scientific discipline, would be aided if universal acceptance of a common nomenclature system could be achieved. An international committee has been formed and reported horizon nomenclature that, at the time of writing, has not been adopted in the United States (Bennema et al. 1968).

Although this chapter has devoted coverage to only the present USDA system, the authors do not wish to imply that other systems should be ignored (Rozov and Ivanova 1967). At the present time, however, it appears that it is the most widely understood system. It does, however, need to be improved and further quantified as the amount of knowledge concerning soil science increases.

LITERATURE CITED

Bennema, J., J. Boulaine, R. Dudal, I. P. Gerasimov, E. Mückenhausen, and Roy W. Simonson. 1968. Soil horizon designations and definitions. Soil Sci. Soc. Am. Proc. 32:153–54.

Brewer, R. 1964. Fabric and mineral analysis of soils. John Wiley & Sons, New York.

Committee on Terminology. 1956. Report of definitions approved by the Committee on Terminology, Soil Science Society of America. Soil Sci. Soc. Am. Proc. 20:430–40.

Dudal, R. 1968. Problems of international soil correlation, in approaches to soil classification. Joint FAO/UNESCO, Rome, pp. 137–43.

Gile, L. H., F. F. Peterson, and R. B. Grossman. 1965. The K horizon: A master soil horizon of carbonate accumulation. Soil Sci. 99:74–82.

Rozov, N. N., and E. N. Ivanova. 1967. Classification of soils of the USSR (principles and a systematic list of soil types). Pochvovedeniye, No. 2, pp. 3–11 (Russian).

Simonson, R. W. 1951. Description of mottling in soils. Soil Sci. 7:182–92.

Soil Survey Staff. 1951. Soil survey manual. U.S. Dept. Agr. Handbook 18. U.S. Govt. Printing Office, Washington.

———. 1960. Soil classification, a comprehensive system—7th approximation. U.S. Dept. Agr. U.S. Govt. Printing Office, Washington.

———. 1962. Supplement to USDA Handbook 18., Soil survey manual (replacing pages 173–88). U.S. Dept. Agr. U.S. Govt. Printing Office, Washington.

———. 1967. Supplement to Soil classification, a comprehensive system—7th approximation. U.S. Dept. Agr. U.S. Govt. Printing Office, Washington.

Whiteside, E. P. 1959. A proposed system of genetic soil horizon designations. Soils Fertilizers 22:1–8.

Wilde, S. A. 1958. Forest soils: Their properties and relation to silviculture. Ronald Press, New York.

Soil Micromorphology ❧

SOIL MICROMORPHOLOGY can be viewed simply as the study of soil morphology in the size range where optical aid is needed for the naked eye. The lower limit of resolution of the unaided eye is in the range of objects 100 to 200μ in diameter.

The tools used in the study of soil micromorphology range from the simple hand lens, which will resolve to about 20μ in diameter, to the light microscope, which will resolve to about 0.52μ without oil immersion or to 0.3μ with oil immersion, to the electron microscope with resolving ability to the range of 10 Angstrom units (0.001μ).

TERMINOLOGY. Terminology for describing soil micromorphology has been presented in detail by Brewer (1964). Terminology for soil micromorphology other than that prepared by Brewer has been used and defined. However, at the present time the student of soil science should become familiar with and be able to converse in the language Brewer has developed. Basic to the terminology is the concept of s-matrix which Brewer defines as follows: "S-matrix of a soil material is the material within the simplest (primary) peds, or composing apedal soil materials, in which the pedological features occur; it consists of the plasma, skeleton grains, and voids that do not occur in pedological features other than plasma separations."

This definition outlines the subject matter of micropedology, namely the materials and fabric of the individual ped or an equivalent mass of apedal material (Fig. 3.1). Four items mentioned in the definition require introductory explanation. (1) Soil plasma is the "soil material" capable of being or having been moved, reorganized, and/or concentrated by the processes of soil formation. (2) Soil skeleton grains are soil material "not readily translocated, concentrated, or reorganized by soil-forming processes." (3) Soil voids are spaces between solid soil materials. (4) Pedological features are units distinguishable from the enclosing or enclosed materials for any reason such as origin, difference in concentration, or arrangement. Plasma separations are the only pedological features that occur in the plasma of the s-matrix itself.

43

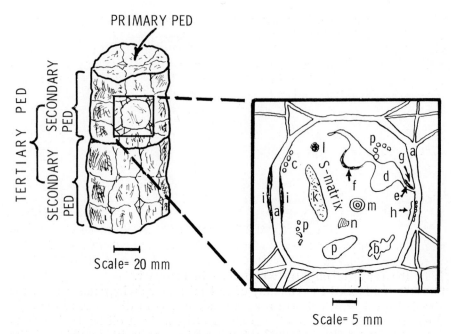

FIG. 3.1. Sketches of some micropedological features. Voids:
a packing voids; *b* vugh; *c* vesicles; *d* chamber; *e* channel.
Cutans: *f* chamber cutan; *g* channel cutan; *h* skeletans;
i argillan or sequan; *j* stress cutan. Other features: *k*
pedotubule; *l* nodule; *m* concretion; *n* papule. Note that the
S-matrix is the mass of plasma, skeleton grains *(p)*, and
voids.

VOIDS. Brewer has grouped voids into eight categories for use in describing soil micromorphology.

PACKING VOIDS. Packing voids can be further subdivided into simple packing voids, that is, those due to random packing of single grains, and compound packing voids, that is, those resulting from the packing of compound individuals (peds) that do not accommodate each other. Simple packing voids are formed by the simple mixing together of primary particles, such as sand, and thus depend on the size, shape, and packing arrangement of the primary particles. The unaccommodated aggregates or peds usually result when aggregates are formed and then deposited in the soil as with fecal pellets. They are also frequently observed in Oxisols where granular peds pack much like individual sand particles (Fig. 3.2).

VUGHS. These are unconnected voids with irregular shape and irregular walls. Several hypotheses can be advanced for their formation. In general, they are found in soil material that has a high proportion of fine-textured material capable of strong cohesion and adhesion. The vughs may be formed as the fine material cements or aggregates the primary particles. In some cases they may be solution cavities.

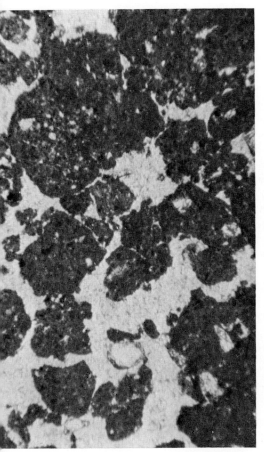

FIG. 3.2. Compound packing voids in the oxic horizon of an Eutrustox from Minas Gerias, Brazil. Thin section in plain light. ×71.

FIG. 3.3. Vesicles near the surface of a Torrifluvent from Arizona. Thin section under crossed polarizers. ×71.

VESICLES. Vesicles, like vughs, are unconnected pores. Unlike vughs they have smoothed walls, often round or elliptical in shape (Fig. 3.3). They are common near the surface in barren arid regions subject to sporadic rainfall and are often found in Ap horizons where flood irrigation is used (Evans and Buol 1968). Springer (1958) found that they could be formed in soils of certain texture by action of entrapped air bubbles during cycles of wetting and drying of soil. The vesicles appear to be rather fragile, temporary features in many soils.

CHAMBERS AND CHANNELS. Chambers can be described as vesicles connected through channels. All channels, however, do not connect chambers; that is, channels can occur without chambers.

Chambers can be formed in the same fashion as vesicles or vughs and the interconnecting channels by roots, worms, insects, etc., or possibly by gases escaping under pressure. Channels are difficult to identify in thin section because their axes may be intercepted from various angles in preparing the section. A section normal to the channel axis may cause the channel to be observed as a chamber or even a vesicle. For these reasons, it is imperative that hand lens or binocular observations be made in addition to thin section study.

PLANES. Voids elongated in one plane are known as planes. These develop as soil material contracts upon desiccation under the influence of transpiration or evaporation. Planes can be further subdivided and classified by their shape and pattern. *Joint* planes are those that are more or less parallel in orientation (Fig. 3.4). They are usually seen when a thin section is made on a plane normal to platy fabric and are often associated with microlamination of the soil material. *Skew* planes have a very irregular pattern. They are characteristically narrow, and both walls, when viewed in thin section, are parallel, showing good fit of one ped against the other (Fig. 3.5). *Craze* planes are intermediate between nonplanar voids and skew planes. They are usually wider than skew planes and less congruent. This indicates that the adjacent peds do not interlock in good fit (Fig. 3.6). They are commonest in soils of rather heterogeneous textural composition.

CUTANS. Brewer proposed the word *cutan* for use in the description of certain soil microstructures. He defines *cutan* as "a modification of the texture, structure, or fabric of natural surfaces in soil materials due to concentration of particular soil constituents or in situ modification of the plasma; cutans can be composed of any of the component substances of the soil material."

Cutan is thus a general term, and several specific varieties can be found and identified. Before Brewer's work was known, a wide variety of terms were in use to denote concentrations or modifications of materials on natural surfaces in soils. Such terms as clay skin, clay film, Tonhautchen, clay coatings, iron stain, iron coating, manganese stain, manganese coating, silica flour, slickensides, pressure faces, lime-mycelia, and possibly several other names have been used to describe soil features that can be covered in the cutan definition.

Brewer classifies cutans on the basis of three characteristics: (1) their position in relation to voids and skeletal material; (2) mineralogical and/or chemical composition; (3) arrangement of particles within the cutan.

CLASSES OF CUTANS BASED ON POSITION. Cutans are named after the voids and units with which they are associated. *Craze plane cutans, chamber cutans,* and *channel cutans* are examples of the first group. The second group includes grain and ped cutans, namely those cutans that completely coat individual sand and gravel grains and individual peds. *Free grain cutans* (simple packing void cutans) are those cutans on grains that are not separated by any other plasma. *Embedded grain cutans* are those

 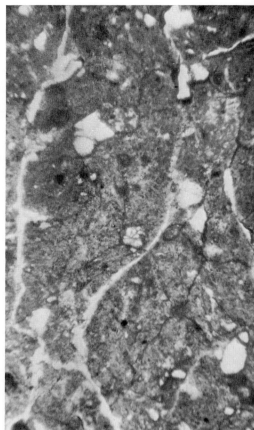

FIG. 3.4. Joint planes from near the surface of a Trorrifluvent from Arizona. Thin section under crossed polarizers. ×71.

FIG. 3.5. Skew planes in the argillic horizon of a Haplargid from Arizona. Thin section in plain light. ×71.

on coarse grains that, along with the cutans, are embedded in the plasmic matrix. Ped cutans may be subdivided according to the order of the ped (primary, secondary, or tertiary) or according to the orientation of the ped surface (horizontal, vertical).

CLASSES OF CUTANS BASED ON COMPOSITION. Terms for these classes are formed by adding the endings *ans* to an abbreviation of the dominant mineralogical or chemical component.

Dark brown or black cutans that are opaque in thin sections and effervesce upon application of 30% H_2O_2, indicating the presence of manganese oxides, are classified as *mangans* (mang = *mang*anese; ans = cut*ans*) (Fig. 3.7). *Organs* are cutans of organic matter and usually do not effervesce upon application of 30% H_2O_2.

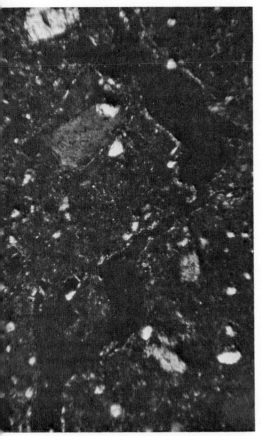
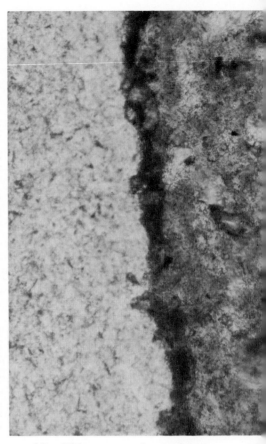

F.G. 3.6. Craze planes in the argillic horizon of a Haplargid from Arizona. Thin section under crossed polarizers. ×71.

FIG. 3.7. Thin mangan in the lower part of a Haplargid argillic horizon. Thin section under crossed polarizers. ×350.

Soluans (*solu*ble salt cut*ans*) consist of coatings of the crystalline salts such as carbonates, chlorides, and sulfates. By solubility tests or observations of their crystal form, a more precise identification is often possible. Some of the soluans are quite transient in the soil. In arid regions where irrigation water containing appreciable salt is used, they often form around root channels each growing season only to be leached from the profile by a preplant deep irrigation the next spring. Frequently, they have been described as "mycelialike" lime.

Silans (*silica* cut*ans*) appear as light-colored, low-luster coatings. Although their appearance could be confused with that of soluans, a quick check of their solubility in water or crystal form with a hand lens will usually be sufficient for identification.

Skeleton grains adhering to a cutanic surface can be referred to as *skeletans*. Such coatings have frequently been referred to as "silica flour" or "silica dust." They are frequently observed in the upper part of the argillic horizon.

Two of the most studied cutanic formations do not appear to separate clearly into classification groups. These are the *argillans* and *sesquans*. The argillans, various forms of which have been referred to as clay skins, clay films, Tonhautchen, or clay coatings often are composed not only of clay but also concentrations of iron and organic matter (Buol and Hole 1959). Thus, there is a problem in the separation of this type of cutan from the *sesquan* or sesquioxide cutan. Brewer has handled this area quite nicely, indicating that cutans where clay is in evidence should be referred to as argillans and modified as *ferri-argillans* or *organo-argillans* when iron or organic matter concentrations are indicated.

The name *sesquan* is thus reserved for those cutans that are nearly opaque in thin section and are almost completely consumed by dithionite extraction of the soil (Aguilera and Jackson 1953). Lacking the facilities of thin section and/or other laboratory observation, the waxy luster as observed in reflected light with the aid of a hand lens can be used as a sufficient clue. *Argillans* should have a waxy luster in reflected light and *sesquans* may have a glazed appearance but no waxy luster. The above examples, selected for their common occurrence, do not form a complete list of cutan classes based on composition composed by Brewer.

CUTAN CLASSIFICATION ACCORDING TO INTERNAL FABRIC AND BOUNDARY CONDITIONS. It is obvious that when a cutan is present on the wall of a void the boundary with the air space is abrupt. However, the boundary between the cutan and the soil material is of value in genetic interpretations of the cutan. It is difficult to obtain the magnification needed for the examination of internal boundaries by any method other than a thin section and light microscope. The genetic implications of the inner cutan boundary concern cutan formation by illuviation (illuviation cutan) or cutan formation resulting from differential pressures or a "troweling" effect (stress cutan), often referred to as a "pressure face." The illuviation cutan has an abrupt boundary with the other soil material and usually consists of several laminae (Fig. 3.8). In cross section the stress cutan has a diffuse inner boundary (Fig. 3.9).

The internal fabric of the cutan also has genetic implications. Internal fabric studies require transmitted light and a petrographic microscope. Opaque cutans such as mangans and sesquans of course do not lend themselves to internal fabric study. Where crystalline clay minerals are involved as in argillans, the degree of orientation can be observed and studied. Present indications are that strongly oriented internal fabric in argillans indicates well crystalline clay and lack of pedoturbation in the layer (Fig. 3.10). Poor orientation can result from either pedoturbation or poorly crystalline clay (Fig. 3.11) (Buol and Yesilsoy 1964; Khalifa and Buol 1968; Nettleton, Flach, and Brasher 1969).

Also in studies of the internal fabric of cutans, any layering or "vary-

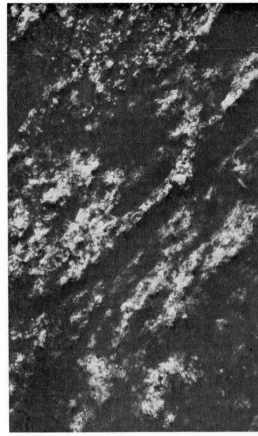

FIG. 3.8. Illuviation cutan in the argillic horizon of a Haplustalf from Arizona. Although primarily an argillan in composition, thin lamina of iron and organic material are present. A A narrow channel is observed near the cutan boundary with soil matrix. Thin section in plain light. ×350.

FIG. 3.9. Stress cutan in the argillic horizon of an Argiustoll from Arizona. The thin section, under crossed polarizers was oriented to show the anisotropic character of the material adjacent to the channel. ×350.

ing" of material should be noted. Crystalline clays interbedded with iron, manganese, or organic matter are often found in illuviation cutans (Fig. 3.8).

OTHER MICROPEDOLOGICAL FEATURES. Many other features found in the soil have been named by Brewer. He defines a pedotubule as

a pedological feature consisting of soil material (skeleton grains or skeleton grains plus plasma, as distinct from concentrations of fractions of the plasma) and having a tubular external form, either single tubes or branching systems of tubes; its external boundaries are relatively sharp. Tubular form, in this context, means that the feature as a unit, or its impression in the enclosing soil material, has a

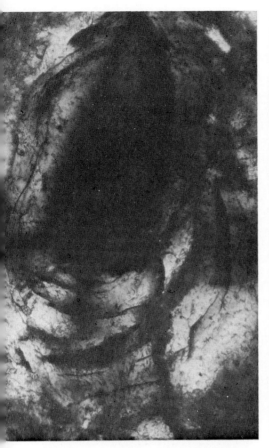

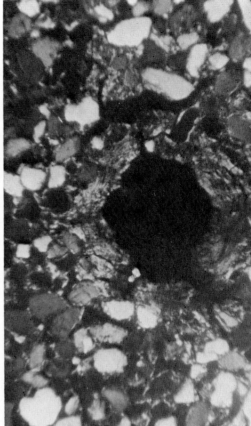

FIG. 3.10. Strongly oriented argillan. Two extinction bands, which were observed to move in an orderly fashion as the microscope stage was rotated, are visible. Thin section under crossed polarizers. ×350.

FIG. 3.11. Weakly oriented argillan coating a void in a sandy matrix. Mottled extinction rather than distinct bands are seen. Thin section under crossed polarizers. ×71.

relatively uniform cross-sectional size and shape, most commonly circular or elliptical; that is, the impression of the pedotubule conforms to the definition of channels.

Pedotubules are often interpreted as channels created by animals or plants and then back-filled with soil material, often from other horizons. In macro form they have been known as krotovina (Soil Survey Staff 1951). Subclasses of pedotubules can be recognized according to internal fabric, composition of the plasma, external form, and distinctness.

Brewer defines a glaebule as

a three-dimensional unit within the s-matrix of the soil material, and usually approximately prolate to equant in shape; its morphology (especially size, shape, and/

or internal fabric) is incompatible with its present occurrence being within a single void in the present soil material. It is recognized as a unit either because of a greater concentration of some constituent and/or a difference in fabric compared with the enclosing soil material, or because it has a distinct boundary with the enclosing soil material.

Within the general grouping of glaebules, Brewer identifies: *nodules*— "Glaebules with an undifferentiated internal fabric; in the context undifferentiated fabric includes recognizable rock and soil fabrics . . ."; *concretions*—"Glaebules with a generally concentric fabric about a center which may be a point, a line, or a plane . . ."; *papules*—"Glaebules composed dominantly of clay minerals with continuous and/or lamellar fabric; they have sharp external boundaries. Most commonly they are prolate to equant and somewhat rounded." He interpreted these named types as to their mode of formation. Nodules can form either in situ or by accretion. Most concretions are accretionary and papules are generally relics, often of argillans that have been disturbed through pedoturbation (Hole 1961).

    Fecal pellets or faunal excreta are recognized as a particular type of micropedological feature. They often appear to be more stable than pedotubules.[1]

EXAMINATION FOR
MICROMORPHOLOGICAL FEATURES.
Many micromorphological features require the use of thin sections and petrographic equipment. Identification and study, however, begin in the field with careful observation and recording of the location of the feature in the soil profile. A description of the total soil is necessary and careful descriptions of the micromorphological features should be made with the aid of a hand lens.

    Preparation of thin sections for microscopic observation can take place via several techniques (Bourbeau and Berger 1947; Brewer 1964; Buol and Fadness 1961; Grossman 1964). In most cases the procedure involves drying the sample, evacuating air from the pores, and filling them with some kind of resin. Drying is not desirable in soils of high organic matter content or those with expanding 2:1 lattice clays. To avoid this, freeze-drying techniques may be used. Also high molecular weight polymers such as Carbowax 6000 can be used in wet soils (Mitchell 1956).

    Once a thin section is prepared, observation techniques need to be used to quantify the observations. A practical method of doing this is the point count method as described by Anderson and Binnie (1961).

PURPOSE OF MICROPEDOLOGICAL STUDY.
Much of the micromorphological study has been designed to aid in the understanding of soil genesis. The identification of illuvial argillans has been made part of the criteria for the identification of argillic horizons (Soil Survey Staff 1960). Slickenside features are used to identify churning soils, that is, Vertisols and Vertic subgroups (Chapter 16).

---

1. Only the more commonly used micromorphological features have been discussed. Many other features, considered beyond the scope of this text, are discussed by Brewer 1964.

Micromorphology is also an important part of the study of soil-plant interactions. Soileau, Jackson, and McCracken (1964) found that artificial illuviation argillans mixed with iron oxides reduced the uptake of potassium by plants. Khalifa and Buol (1969) found that natural illuviation argillans acted to reduce P, K, and to a lesser extent N, uptake by plants.

Although it is not always possible to arrive at an indisputable conclusion as to the formation of a specific feature in the soil, the techniques of micropedological study are an aid toward that end. Careful observation and recording of micropedological features have become accepted and even expected parts of every careful soil study. Micromorphological studies add form to the substance determined from other types of analysis (Kubiena 1964). As the student will note in later chapters, the features described in this chapter play important roles as criteria upon which soil genesis theories are constructed.

LITERATURE CITED

Aguilera, N. H., and M. L. Jackson. 1953. Iron oxide removal from soils and clays. Soil Sci. Soc. Am. Proc. 17:359–64.

Anderson, D. M., and R. R. Binnie. 1961. Modal analysis of soils. Soil Sci. Soc. Am. Proc. 25:499–503.

Bourbeau, G. A., and K. C. Berger. 1947. Thin sections of soils and friable materials prepared by impregnation with plastic "Castolite." Soil Sci. Soc. Am. Proc. 12:409–12.

Brewer, R. 1964. Fabric and mineral analysis of soils. John Wiley & Sons, New York.

Buol, S. W., and D. M. Fadness. 1961. New method of impregnating fragile material for thin sectioning. Soil Sci. Soc. Am. Proc. 25:253.

Buol, S. W., and F. D. Hole. 1959. Some characteristics of clay skins on peds in the B horizon of a Gray-Brown Podzolic soil. Soil Sci. Soc. Am. Proc. 23:239–41.

Buol, S. W., and M. S. Yesilsoy. 1964. A genesis study of a Mohave sandy loam profile. Soil Sci. Soc. Am. Proc. 28:254–56.

Evans, D. D., and S. W. Buol. 1968. Micromorphological study of soil crusts. Soil Sci. Soc. Am. Proc. 32:19–22.

Grossman, R. B. 1964. Composite thin sections for estimation of clay-film volume. Soil Sci. Soc. Am. Proc. 28:132–33.

Hole, F. D. 1961. A classification of pedoturbations and some other processes and factors of soil formation in relation to isotropism and anisotropism. Soil Sci. 91:385–77.

Khalifa, E. M., and S. W. Buol. 1968. Studies of clay skins in a Cecil (Typic Hapludult) soil. I. Composition and genesis. Soil Sci. Soc. Am. Proc. 32:857–61.

———. 1969. Studies of clay skins in a Cecil (Typic Hapludult) soil. II. Effect on plant growth and nutrient uptake. Soil Sci. Soc. Am. Proc. 33:102–5.

Kubiena, W. L. 1964. The role and mission of micromorphology and microscopic biology in modern soil science, pp. 1–13. In A. Jongerius (ed.), Soil micromorphology. Elsevier, New York.

Mitchell, J. K. 1956. The fabric of natural clays and its relationship to engineering properties. Proc. Highway Res. Board 35:693.

Nettleton, W. D., K. W. Flach, and B. R. Brasher. 1969. Argillic horizons without clay skins. Soil Sci. Soc. Am. Proc. 33:121–25.

Soil Survey Staff. 1951. Soil survey manual. U.S. Dept. Agr. Handbook 18. U.S. Govt. Printing Office, Washington.

———. 1960. Soil classification, a comprehensive system—7th approximation. U.S. Dept. Agr. U.S. Govt. Printing Office, Washington.

Soileau, J. M., W. A. Jackson, and R. J. McCracken. 1964. Cutans (clay films) and potassium availability to plants. J. Soil Sci. 15:117–23.

Springer, M. E. 1958. Desert pavement and vesicular layer of some soils of the desert of the Lahontan Basin, Nevada. Soil Sci. Soc. Am. Proc. 22:63–66.

✣

Soil Composition and Characterization *

$\mathbf{A}$N EXPERIENCED pedologist can read a great deal from the morphology of a soil that he sees in the field. From the morphology he sees and describes, as discussed in the previous chapters, he can infer much about the properties of the soil and make many interpretations about its qualities. However, for modern soil science and for modern agriculture, quantitative data on the composition of soils are needed.

ROLE OF THE LABORATORY AND THE COMPOSITIONAL DATA IT PROVIDES. To make precise differentiations among soil groups, it has become necessary to rely on laboratory measurements of selected soil properties. For example, to make a distinction between the more highly weathered soils of the tropical regions and those in tropical and other areas which are less weathered and developed, it is necessary to obtain information about the chemical properties of these kinds of soils. The most helpful type of information in this case includes measurement of the "free" or extractable iron content, the cation exchange capacity, and the content of weatherable minerals.

Thus modern soil classification systems place a great deal of reliance on information about the quantitative compositions of soils. Certain kinds of soil properties are selected for use in the classification process, based on their assumed importance in understanding the genesis of the soil and on the number of other important properties which co-vary with changes in the property under consideration. Analytical methods are selected which offer the best means of measuring the compositional properties of interest.

DEVELOPMENT OF METHODS AND CRITERIA FOR ACQUISITION OF DATA. The soil compositional properties which are selected as a basis for soil classification reflect the current understanding and theories about the processes of soil formation. These theories determine what kinds of properties are thought to be of greatest value in interpreting the soil properties for various uses. This can be seen in a study of the history of the development of ideas about the types of soil compositional data thought to be important and the parallel improvements in methods of soil analysis. This development over the past fifty years makes a very fascinating story. Indeed, it is a reflection of the "knowledge explosion" which has taken place in soil science in the past few decades. These rapid changes in our ideas about which soil properties are the most important are best reflected in that field of soil science known as pedology, for soil classification is after all a mirror of our understanding of soils.

Prior to the 1930s, total or elemental analysis of the soil was widely practiced. A great deal of time and effort went into the collection of data on the elemental composition of soils; that is, the total amount of an element contained in a soil sample was determined by "wet chemical" analyses and the results expressed on an oxide basis. For example, the percentages of aluminum, silicon, iron, calcium, potassium, and other "basic" cations were all carefully measured on each soil sample collected for analysis. Weight percentages of each of these were reported in the oxide form. These weight percentages were then divided by the molecular weight of each of these compounds to give molecular percentages. Ratios of the molecular percent values (called "molecular ratios") were then calculated for each of the soil horizons to determine losses and gains and thus infer the nature of the soil-forming processes taking place. To give an example, molecular ratios of the silica (silicon dioxide) to the sesquioxides (oxides of iron and aluminum, Fe_2O_3 and Al_2O_3) were computed to determine if the latter two were accumulating with respect to the silica, an indication of "laterization." Also ratios of alkali and alkaline earth species (expressed as molecular oxides) to alumina and alumina plus silica were used as leaching factors (Jenny 1941).

These total analyses were very time-consuming and tedious to perform. In this type of determination, it is necessary to convert the insoluble forms of the elements of interest to soluble forms so they may be measured by chemical means. (Most of these elements are present in soils in the form of insoluble silicates.) The method commonly used is to fuse the soil sample with a flux, sodium carbonate being most commonly used for soil samples. Prior to the 1940s and the advent of colorimeters and spectrophotometers, the elements had to be precipitated and determined gravimetrically. As can be confirmed by those who have spent some time with these "classic" methods of quantitative analysis, this is a very arduous and time-consuming procedure. (Sodium carbonate fusion for total elemental analysis of soils is described by Kanehiro and Sherman 1965, and by Jackson 1958.) An example of the type of data collected under this system and of the types of interpretations that were made from them can be found in the 1935 *Atlas of American Agriculture* (Marbut 1935). In the section of this *Atlas* prepared by Marbut and his staff are excellent total elemental analysis data for the most important soils of the United States. These are tabulated by soil horizon, and molecular ratios have been carefully calculated. The soil scientist of today is somewhat inclined to "turn up his nose" at these "old-fashioned" data. But it should be kept in mind that such elemental and molecular ratio data are still very useful, as will be described later. Also, these early soil scientists—our intellectual ancestors—had good reasons (at that time) to perform these determinations so painstakingly and to interpret carefully the molecular ratios derived from them. Among the reasons for use of total elemental analysis techniques by these early-day soil scientists were:

1. The procedures and techniques were readily available from classical analytical chemistry.

2. It was not generally known or appreciated in those earlier days that the finer fractions of the soil—the clays—consist for the most part of crystalline minerals with a definite and regular internal arrangement of the ions of the elements. Nor was it understood that a given element might occur in several different structural positions and as part of several different mineralogical entities. The finer particle sizes were thought to consist of an amorphous noncrystalline mixture of the oxides and/or hydrous oxides of the elements.

3. In earlier days, the concept of nutrient element availability had not been established. That is, it was not appreciated that the amount of potassium available to plants at any one time was that held in exchangeable form by the clays, and that this amount is only 5 to 10% of the total potassium present in some soils. A common concept in that period was of the soil as a storage bin of nutrient elements, and one needed to take inventory of the total amounts of the various nutrient elements present in the various soil horizons.

In the 1930s came the realization that the soil clays (or most of them) were indeed crystalline and that the active or available portion of a nutrient element was more important from an agronomic point of view than the total amount present. In the 1940s came the development of new instruments and new procedures. These included X-ray diffraction units fitted with attachments suitable for the study of soil clay minerals, colorimeters, flame photometers, spectrophotometers, and new ideas on extraction solutions for removing from the soil the active portion of the nutrient elements.

These developments brought about rapid changes in the methods of characterizing the composition of soils. Thus, most soil laboratories now use these techniques and procedures to the almost complete exclusion of the total elemental analyses—regardless of the purpose of the analysis of the soil composition, be it for pedological (classification) or soil fertility evaluation purposes.

Let us now turn our attention to the different aspects of soil composition, the methods used for such characterization, and the inferences we can make when the data are at hand for a soil profile.

PRINCIPLES AND PROCEDURES IN USE
OF COMPOSITIONAL DATA. Those soil properties which result from soil genetic processes or affect soil genesis are selected as differentiating. This we call the *genetic thread* principle. If there is a choice between two

properties of apparently equal genetic significance, that one is selected for use which has the greatest agricultural significance. For those soil properties which cannot be readily and quantitatively described in the field, samples are taken for laboratory measurement of their composition. By empirical cut-and-try procedures, critical limits between sets of soils are established. For example, soils of the temperate subhumid region are compared with soils of the adjacent warm temperate humid region (as in the case of soils of northeastern United States versus those of southeastern United States). As hypothesized from soil genesis theory, the latter soils are found to be more highly leached and thus of lower base saturation, particularly with increasing depth in the soil profile. By empirical procedures, limits of base saturation (which must be measured in the laboratory) are set for each class, and samples are analyzed to determine in which class they fall. By this procedure of *empirical correlation,* we are insured against too heavy reliance on our hypotheses (which are generally incomplete) and we thereby avoid "prejudicing the future" (Cline 1963; Smith 1963). Another advantage of this procedure is that every class established has a real counterpart on the landscape. For the soil properties thus selected for use in classification and characterization and which cannot be quantified by field study and description of the soil profile, an operational definition is prescribed in terms of a certain laboratory method (Bridgman 1927; Smith 1963). For example, the clay percentage is defined in terms of a certain set of laboratory procedures used in its measurement. The procedure must be stipulated because, worldwide, soil clays vary greatly in their behavior in any given procedure.

Like the old recipe for rabbit stew which begins "first get the rabbit," the laboratory procedure for determining soil composition begins with directions for obtaining the soil samples. Selection of the pedons for sampling is a very important procedure, well worth taking a great deal of extra care. The time, effort, and expense invested in laboratory analyses of the profile samples will be wasted if the samples are not representative of the soil. The data resulting from the analyses likely will be used by a number of people for different purposes, and these people will be misled and their interpretations and extrapolations erroneous if the samples are not representative and carefully collected. Too, the soil sampled may be incorrectly classified if the samples are not properly collected, both as to appropriate pedon sampling site and care in taking samples from the horizons of the profile once the sampling procedure is started. The soil profile samples should be taken from a pedon determined by field studies and observations to be truly representative of the map unit and taxon under study. Our experiences and observations are that field soil surveyors and others working closely with a particular soil or group of soils tend to develop an idealized and exaggerated image of the kind of profile the soils have; that is, they are prone to develop a concept of the soil with better developed A horizons, thicker solum, and more fully expressed horizonation than the real soils possess. As a consequence, they tend to seek out a pedon with this rather atypical profile development for sampling for characterization purposes rather than one more representative of the central tendencies of the soil's morphology. This common type of bias must be avoided; it has already caused us much difficulty. Bryant (1964) developed and experimented with a system of sample site selection for sampling for detailed characterization and genesis studies. This program involves detailed field mapping and morphological studies on a grid basis of at least two blocks, each containing a hundred or more hectares (half a square mile or more). Preliminary samples of A and B horizons are taken at a number of points (about 15) for laboratory determination of key parameters. The results of the detailed morphological studies plus the laboratory analyses are then used as a basis for selecting pedon sites for detailed profile sampling. Dolman (1967) followed a similar procedure in characterizing Histosols of eastern North Carolina. Obviously, such detailed procedures cannot be used for routine partial soil characterization studies but should be considered for detailed studies of soil genesis and classification from which many important inferences are to be drawn. In general, for specialized soil genesis studies, duplicate profiles not more than a few miles apart have ordinarily been sampled. However, to establish central tendencies and ranges of properties of a taxon (or its "map unit" as mapped in a given area), a larger number of soil profile samples is required. Nelson (1961) determined that 15 or more soil profile samples are needed to obtain reasonable estimates of the true mean of soil series on the coastal plain of North Carolina.

Some of the basic principles of sampling soils have been developed and presented by Cline (1944, 1945). He points out that soil *volumes* are the populations from which samples are taken, not soil areas. Many of the details and mechanics of soil sampling for genesis and characterization studies are presented in the *Soil Survey Manual* (Soil Survey Staff 1951). We recommend careful reading of this reference before detailed soil profile sampling is undertaken. We emphasize these points about soil profile sampling, based on our experiences and the discussion in the *Soil Survey Manual:*

1. Uncultivated and virgin sites should be sought out for sampling, unless one of the major objectives is to characterize cultivated soils.

2. Roadcuts and ditchbanks are not satisfactory for sampling, unless they are freshly made and there has been no topsoil disturbance. The wetting-drying cycles, high degree of oxidation, plant root and animal activity, and contamination from dust or other atmospheric pollutants in these sites all tend to modify the soil structure and compositional properties to such an extent the samples would not be representative.

3. There is no satisfactory substitute for a well-located sampling pit, about the size of an average pedon (about 2 m long by 1 or 2 m wide) as a site for satisfactory sampling. Auger borings, probes, and tube sampling do not allow one to delineate and sample the various horizons satisfactorily as they occur in the pedon, and also have a high risk of introducing contamination.

4. At least one side of the pit should be kept undisturbed and un-

trampled as a sampling face, preferably the one which has the best lighting. This is especially important if a backhoe mounted on a tractor is used for opening the pit.

5. After that face of the pit which is to be sampled is cleaned, horizons should be delineated and a careful morphological description prepared, utilizing observations all around the pit to describe the pedon properly, including variation in horizon thickness and boundaries.

6. Thick horizons (greater than 10 to 15 cm in width) should not be sampled in those areas directly transitional to adjoining horizons.

7. If a large number of cores are to be taken and the soil is dry and hard, it is usually best to sample from the top downward. In this procedure, after a horizon is sampled, the remainder of the soil of that horizon in the sampling area is removed, exposing the next lower horizon for coring and bulk sampling. In this procedure, care must be taken to avoid contaminating the lower horizons with material from upper horizons. If taking cores is not a major consideration, then it is ordinarily best to sample from the bottom of the profile upward, after sampling the A subhorizons.

8. Bulk and core samples should be very carefully labeled, using tags or other identification both inside and outside the container.

9. For micromorphological work, blocks of undisturbed soil with orientation marked (as to which was the upper side, in place) can be collected from each horizon to be thus sampled and placed in containers, or a monolith, placed in a prepared wooden box, extending over several horizons may be collected and brought to the laboratory for binocular microscopic study and thin section preparation.

SOIL PROPERTIES MEASURED BY LABORATORY TECHNIQUES

PHYSICAL DETERMINATIONS

Particle Size Analysis (Mechanical Analysis). This analysis is the measurement of the percentages of the primary separates: sand, silt, and clay. At present, the diameter limits of these classes used in pedological studies are sand, 2 to 0.05 mm; silt, 50 to 2μ; and clay, $< 2\mu$ (see Fig. 2.2). This procedure has two aspects: *dispersion* of the soil, and *fractionation* into particle size classes. Dispersion is ordinarily achieved by shaking or stirring the soil sample with a sodium hexametaphosphate (Calgon) solution. Determination of the content of clay and silt ordinarily is by a sedimentation-pipette or sedimentation-hydrometer procedure; the sand percentage is determined by sieving. (A nest of sieves is used where sand and subfractions are desired.) (Day 1965; SCS-USDA 1967.) For soil genesis studies in which close estimates of the clay content of the various horizons of a soil profile are desired, the sedimentation-pipette procedure has been the method of choice. However, problems are encountered with dispersion of iron- and aluminum-rich soils, due to the cementation effect of the sesquioxides. In such cases, a factor of 2.5 multiplied into the water percentage held at 15 bars tension gives a better approximation to the clay content (Soil Survey Staff 1967).

It should be noted that in studies of clay movement in soil profiles, it is very useful and helpful to calculate ratios of the fine clay ($< 0.2\mu$ diameter) to the coarse clay (2 to 0.2μ), which requires use of centrifugation (Jackson 1956). This is based on the hypothesis that fine clay is more mobile in the illuviation process than is the coarse clay. Some of the inferences which may be drawn from particle size analysis are:

Within soil profiles—among horizons

1. Clay translocation and presence or absence of argillic horizons. The fine clay: coarse clay ratio is especially useful, as fine clay is present in larger amounts in illuvial horizons.

2. Amount of forest influence, based on increased clay translocation relative to prairie soils of similar parent material and age—in forest-prairie transition areas of the United States.

3. Type of soil developmental process.

4. Detection of lithologic discontinuities, based on shifts in sand subfraction percentages or abrupt shifts in silt and clay percentages in adjacent horizons.

Bulk Density. Bulk density of a soil is the weight of a known soil volume compared to the weight of an equal volume of water, or weight *per unit volume*. Or, expressed in still another way, bulk density is "the ratio of the mass to the bulk or macroscopic volume of soil particles plus pore spaces in a sample" (Blake 1965). It is ordinarily expressed in units of g/cc. Bulk density data are used for computing total porosity (assuming a soil particle density of 2.65 g/cc is the usual procedure) and volume percent of available water. Methods of obtaining bulk density data are the core, plastic- or paraffin-coated clod, excavation, and gamma radiation densitometry techniques (Blake 1965). The latter two methods are ordinarily not used at present in collecting data for soil classification studies. The gamma technique requires sophisticated equipment, and the excavation technique often cannot be used to relate to a single soil horizon as a relatively large vertical distance is required. A resin-coated clod technique has come into increased use recently. It is based on a comparison of the weight of a soil clod in air and in water (Brasher et al. 1966). The clod method is useful in collecting data from soil horizons too dense or brittle to sample with a coring device, and also allows calculation of shrink-swell potential or linear extensibility from the same clod specimens (Grossman et al. 1968). However, it should be noted that bulk density values from a coated clod method will ordinarily yield relatively high bulk density values because it does not include the interped void or "pore" space.

Some of the inferences which may be drawn and uses made of bulk density data in soil classification, characterization, and interpretation are:

1. Detection of pan layers in soil profiles, especially fragipans, and quantification of their degree of development.

2. Determining and quantifying the presence of significant amounts of volcanic ash and pumice in soil materials (bulk density of less than 0.85 g/cm^{-3} indicates volcanic ash in mineral soils low in organic matter).

3. Determining degree of weathering and alteration of C horizons

formed from igneous and metamorphic rocks (saprolite), as bulk density drops from values near 2.65 to less than 2 with increasing weathering and associated development of void spaces.

4. Evaluation of root impedance possibilities, as relationships have been established between high bulk density and lack of root penetration (Veihmeyer and Hendrickson 1948).

5. Evaluation of volume changes during soil genesis, as gains and losses from specific horizons, if profile uniformity and absence of lithologic discontinuities can be established.

Soil Moisture Parameters. Of interest and use in soil genesis and characterization is the percent (volume basis) of water held at 15 bars tension, as this is the "permanent wilting point" for plants. "Dry" soil for classification purposes refers to a soil with moisture content below this permanent wilting point (Soil Survey Staff 1960). Also as mentioned in the first part of this chapter, it has been determined that multiplication of the 15-bar water percentage by a factor of 2.5 yields a good estimate of the clay percentage in soils in which dispersion is a problem (Soil Survey Staff 1967). Of course, complete moisture sorption data, at several points on the curve from less than 1 bar to 15 bars, are useful in irrigation and moisture supply studies but are not at present used in pedological activities. In general, tension table and pressure chamber ("pressure cooker") techniques, using soil cores with field structure preserved, are employed for soil moisture retention measurements at low suction values (1 to 2 bars). Pressure plate and membrane apparatus, using crushed soil samples, is ordinarily employed for the higher tension measurements, especially the 15-bar limit (Richards 1965; SCS-USDA 1967). This is because the water retention at the dry end is chiefly controlled by the specific surface of the soil (Richards 1965).

For classification purposes, a soil is considered "usually dry" if its moisture content is below that at 15 bars more than half of the time it is not frozen; "usually moist" refers to the converse—that the soil moisture content is above that of the 15-bar percentage more than half the time the soil is not frozen (Soil Survey Staff 1960).

Shrink-Swell Capacity or Linear Extensibility. Certain soils have the capacity to swell significantly when moist and to shrink and crack when dry, related to a relatively high content of montmorillonite clay. As this is important not only because of the physical qualities of the soil surface (large, deep cracks in dry seasons) but also because it is a soil genetic process, it is of considerable importance in classification. This quality is quantified through the use of a Coefficient of Linear Extensibility (COLE) or a Potential Volume Change (PVC or Swell Index). The former is ordinarily used for pedological purposes; the latter is more commonly used for on-site evaluation of possible highway routes or potential building sites (Franzmeier and Ross 1968). PVC values are obtained by measurement in a specially designed instrument involving a strain gauge (Henry and Dragoo 1965). The Coefficient of Linear Extensibility (COLE) is defined as follows (Grossman et al. 1968; Soil Survey Staff 1967):

$$COLE = \frac{L_m}{L_d} - 1$$

L_m = length of moist sample, and
L_d = length of dry sample

The coefficient is actually calculated from differences in bulk density of plastic-coated clods (Brasher et al. 1966) when moist (1/3 bar, or 1/10 if coarse, sandy soil) and when oven dry (Grossman et al. 1968; SCS-USDA 1967):

$$COLE = \sqrt[3]{\frac{Db_d}{Db_m}} - 1$$

where Db_d = bulk density dry, and
Db_m = bulk density moist

Some uses made and inferences to be drawn from COLE data are:

1. If COLE exceeds 0.09, significant shrink-swell activity can be expected (Soil Survey Staff 1967).

2. If COLE exceeds 0.03, a significant amount of montmorillonitic clay is present (Grossman et al. 1968).

CHEMICAL DETERMINATIONS

Cation Exchange Capacity. The capacity of a soil to sorb or hold cations and to exchange species of these ions in reversible chemical reactions is a quality important for both soil fertility-nutrition studies and for soil genesis. Thus this type of data is widely used in soil classification considerations. However, its measurement is rather empirical, and several different analytical methods have been proposed which yield different results (Chapman 1965; Coleman and Thomas 1967; Jackson 1958). Among the factors contributing to differing cation exchange capacity (CEC) values by various methods are:

1. Variations in CEC, depending upon the pH at which the determination is made, due to differing reactivity of the various exchangers in soil systems—clay minerals, hydrous oxides, amorphous compounds, and organic materials.

2. Variation in results with chemical composition of the exchanging or displacing solution used. Certain species of ions are more readily displaced or exchanged than others, and certain species (potassium especially) may be actually trapped or fixed by some clay mineral species present in some soils.

Despite these difficulties and problems, CEC determinations yield numbers which are valuable in evaluating the capacity of the soil to retain cations, its degree of weathering, and general chemical reactivity. The two types of CEC determinations which have been more widely employed are the ammonium saturation-displacement method commonly conducted at pH 7, and the summation method in which all exchangeable cation species (including the acid-generating hydrogen and aluminum or exchange acidity) are added. As the commoner method of determining exchange acidity is performed at pH 8.2 buffered against fluctuation (Mehlich 1938), this ordinarily gives a CEC value higher than that with the ammonium saturation method. This is due to the increasing CEC, with higher pH displayed by organic matter, the clay mineral kaolinite especially, and amorphous compounds such as allophane. Soils rich in montmorillonite do not display this feature.

Because of this variation in CEC with pH which becomes quite large in the more highly weathered soils rich in kaolinite and hydrous oxides, attempts have been made to partition CEC into two components. These are the "permanent charge" and the "pH-dependent charge" components (Coleman, Weed, and McCracken 1959). The permanent charge component results from a net negative charge on the clay lattice due to the substitution of ions of lower charge (valence) for more highly charged ones during the formation of the clay mineral. The pH-dependent or pH-sensitive component results from increasing ionization, as the pH of the soil system increases, of certain functional groups of organic matter and of OH's associated with the broken edges of clay mineral plates and hydrous oxides of iron and aluminum. This produces apparent increased CEC as the pH is increased. In principle, the permanent charge of mineral soil samples can be measured by leaching the soil with an unbuffered salt solution, such as $1N$ KCl, and measuring the basic cations and aluminum thus displaced (Coleman, Weed, and McCracken 1959). In practice, however, it has been found that hydrous oxide coatings and aluminum (plus iron) in interlayer spaces of 2:1 layer clays actually "block" a part of the permanent negative charge by countering it with positive charge (Coleman and Thomas 1964, 1967). Because of this feature, it is not considered desirable to attempt to measure and use "permanent charge" CEC as a precise entity in soil classification activity. However, an approximation to it, obtained by summing the exchangeable bases and salt-displaceable (exchangeable) aluminum, *is* useful in establishing chemical properties diagnostic for the oxic horizon in highly weathered soils (Soil Survey Staff 1967). Some uses, inferences, and interpretations from CEC data are:

1. Inferences as to clay mineral species present in the soil. Clay minerals have been determined (Grim 1968) to have these ranges in CEC (in meq/100 g soil) as measured by the ammonium acetate method at pH 7: kaolinite, 3 to 15; smectite group (including montmorillonite), 80 to 150; illite (clay mica), 10 to 40; vermiculite (noninterlayered), 100 to 150; chlorite, 10 to 40.

2. Relative degree of weathering of the soil. Low CEC's are correlated with disappearance or absence of primary weatherable minerals and accumulation of secondary clay minerals of low CEC as a result of the weathering process. High CEC's tend to be associated with the less weath-

ered soils, with weatherable primary minerals as a plant nutrient reserve. An arbitrary breaking point between high and low CEC's is suggested as 10 me/100 g whole soil in subsoils (B horizons) with more than 30 to 40% clay.

3. Agronomic and forest nutrition significance. High CEC's of mineral soils indicate a high plant nutrient storage capacity. However, if such a soil is acid, it is likely to contain large quantities of exchangeable aluminum, with serious acidity problems.

4. Engineering practice. Mineral soils (relatively low in organic matter) with very high CEC's (greater than 20 to 25 me/100 g) are likely to contain significant amounts of montmorillonite, with associated high shrink-swell potential and high linear extensibility.

5. CEC data are used as the base in computing "percentage base saturation," a widely used pedologic and nutritional quality of soils, as discussed in the following section. It is important to know what method is used in the CEC determination.

Exchangeable Cations and Exchange Acidity. Those positively charged ion species countering the negative charge on soil clays and organic matter can be placed into two groups: exchangeable bases and exchangeable acid-generating cation species. Those commoner species of the former group, and hence of importance in soil classification and genesis, are calcium, magnesium, sodium, and potassium. Those of the acidic group are hydrogen and aluminum. Ordinarily, both are present in acid soils and are referred to collectively as "exchange acidity." This exchange acidity is caused almost completely by aluminum ions (Coleman and Thomas 1967), though "exchangeable hydrogen" has been used as synonymous with exchange acidity by many who have written on this subject.

Methods for determining exchangeable bases involve displacement of the ions from the soil colloids and their measurement in the displaced solutions. They are measured by flame and atomic absorption photometry, or by titration techniques (Heald 1965; Jackson 1958; Prince 1965; Rich 1965; SCS-USDA 1967).

In laboratory studies of exchange acidity, it has become common practice—in soil classification and genesis studies, at least—to measure total exchangeable acidity and to make a separate determination of exchangeable Al. The method commonly employed for exchange acidity is Mehlich's barium chloride-triethanolamine extraction buffered at pH 8.2 (Mehlich 1938). For determination of exchangeable Al, the soil is leached with an unbuffered salt solution (such as $1N$ KCl) and aluminum in the leachate is measured by titration or spectrophotometric techniques (Coleman, Weed, and McCracken 1959; Lin and Coleman 1960; McLean 1965).

Some inferences from and uses of data on relative abundance of species of exchangeable bases are:

1. Soil properties are adversely affected (such as increasing dispersion) if the proportion of exchangeable sodium to other bases and the CEC increases. Thus one of the main diagnostic features of the natric horizon is

a percentage sodium saturation (of the CEC) greater than 15. (If an under-lying C horizon has greater than 15% sodium saturation, a natric horizon is recognized if it contains more exchangeable Na + Mg than exchangeable Ca + H.)

2. The ratio of exchangeable calcium to magnesium is an indicator of relative weathering and degree of development. In humid and sub-humid regions, exchangeable Mg increases with respect to increasing ex-changeable Ca with increasing soil age and degree of development.

Other inferences of importance to plant nutrition can be drawn with respect to the proportions among exchangeable bases, but these are beyond the scope of this discussion.

Some uses of and inferences from exchange acidity data, with particular reference to partitioning into exchangeable Al and H are:

1. Exchange acidity increases with increasing leaching and weathering in the more humid environments.

2. If exchangeable Al occupies more than approximately 60% of the CEC, toxic levels of Al in soil solution result (Evans 1968; Nye et al. 1961). Thus the exchangeable Al content is of importance and widely used in plant nutrition and soil classification-genesis studies as well.

3. Exchangeable H is present in *significant* amounts only when large amounts of soluble acids are present and ordinarily is accompanied by pH values less than 4. Sources of large quantities of soluble acids are:

a. H_2SO_4 produced by oxidation of sulfides, as in a drained coastal marsh soil (the "cat clays") or spoil from mining operations.

b. Large amounts of very acid litter, as under conifers and Ericaceous shrubs growing on sandy soils generally in cool humid climates, as in the "mor" or organic layer above a spodic horizon.

c. Locally, following high rates of application of ammonium fer-tilizers to sandy soils.

A fourth possibility of high exchangeable H levels is in organic soils, where this ion may be dominant if the mucks and peats are very acid. How-ever, there is increasing evidence that this acidity may actually arise from slowly exchangeable Al held or complexed by the organic matter (Coleman and Thomas 1967).

Small amounts of true exchangeable H may persist at pH levels of 6 or 7, especially if the soil is dominated by 2:1 clay, but such levels are so small in the pH range above 4 to 4.5 as to be insignificant.

The greatest significance and use of exchangeable base-exchange acidity data comes from calculating the proportion of each occupying the CEC— the percentage base saturation.

pH Measurements. The definition and concepts of pH as a measure of soil acidity and alkalinity were presented in Chapter 2, Field Measurement and Interpretation of pH Values. We refer here to laboratory measurements of pH and the inferences for classification-genesis that may be drawn from the resulting data. Methods of making pH measurements in the laboratory are described in references on soil analysis (for example, SCS-USDA 1967).

In general, for mineral soils a measurement of soil pH using a 1:1 soil:water ratio (pH_w) is desirable and useful, although not as helpful in understanding the chemistry of the soil as measuring the exchange acidity, exchangeable aluminum, and exchangeable bases. From the pH_w of mineral soils we can make these general inferences:

1. pH < 3.5, associated with significant pH drop after wetting-drying cycles: acid sulfates present. If the sample is from a coastal marsh, the sample represents "cat clays" or acid sulfate soils with very serious agricultural problems. If the sample is from mine spoil or from pits, sulfides present in a buried geologic formation have been oxidized. Repeated wetting and drying, followed by pH measurement, is the diagnostic test here. If there is significant pH drop from the original measurement on moist soil after the wetting-drying, the serious problem of acid sulfates is present (Fleming and Alexander 1961; Moormann 1963).

2. pH < 4.5: a significant amount of exchangeable hydrogen is probably present in addition to exchangeable aluminum. Sources of this hydrogen could be:

    a. Dissociation from strong acid functional groups in the organic fraction (likely to occur significantly only in intergrades to Histosols, as in histic epipedons).

    b. Free acid produced by oxidation of sulfur and sulfides to sulfates, as in "cat clays" and mine spoils, but not as severe as in (1).

    c. Exchangeable hydrogen from hydrolysis of fertilizer salts in Ap horizons. This can be checked by leaching the sample with water and remeasuring the pH, which will be significantly (one-half to one pH unit or more) higher than the initial measurement if such salts are present.

3. pH 4.5 to 5.8 in mineral soils: sufficient exchangeable aluminum is present to affect plant growth significantly, and the percentage base saturation is low (Kamprath 1967).

4. pH 4 to 5.2 in Histosols or in a histic epipedon (soils relatively rich in organic matter): significant exchangeable aluminum and hydrogen are present to affect plant growth. If the content of mineral clay is low and the layer is not adjacent to a mineral layer, the acidity is probably due to exchangeable hydrogen, predominantly. It is our experience and observation that exchangeable Al is not a significant factor in Histosols above approximately pH 5 to 5.2, and sufficient Ca is present for growth of most temperate-region row crops. This is because of the different nature of the exchangers (organic functional groups) and complexing of Al in the Histosols.

5. pH 5.8 to 6.5: acidity is present (apparently from hydroxy-Al and organic functional groups), ordinarily hydronium, in amounts sufficient to affect acidity-sensitive crops such as alfalfa. The soil is 70 to 90% base saturated, depending on type of clay minerals present.

6. pH 6.5 to 8: the soil is essentially fully base saturated; no exchangeable Al is present; free $CaCO_3$ may be present only if well protected inside soil aggregates with restricted diffusion rates.

7. pH 8 to 8.5: the soil is fully base saturated and free $CaCO_3$ is present in the system; exchangeable cation population is largely $Ca + Mg$.

8. pH 8.5 to 10: the soil contains large amounts of soluble salts and conductivity is high; appreciable exchangeable Na is present but may not be enough to qualify as natric horizon.

9. pH > 10: the soil is highly sodium saturated—"alkali" soil.

For Histosols and histic epipedons, measurements of pH are best done at soil:water ratios of 1:5 or 1:10 in order to give meaningful results. Measurements of soil pH in $1N$ KCl (pH_s) are often useful, particularly if compared with the pH_w. If the pH_s is one-half or more units less than pH_w in the range below pH 6, significant amounts of exchangeable Al or complexed slowly exchangeable Al are present. (The drop is due to the hydrolysis of the Al displaced by the K.) If pH_s is greater than pH_w, the soil has a net positive charge due to dominance of the exchange complex by hydrous iron oxides. This is a good indication of "acric" great groups in the Oxisol order. (The rise in pH is due to displacement of OH's by the Cl.)

Percentage Base Saturation (PBS). The cation exchange capacity calculated by summing the exchangeable bases and exchange acidity measured by the Mehlich method is the base commonly used in computing PBS for soil classification studies (Coleman and Thomas 1967; SCS-USDA 1967). This can be expressed as follows:

$$PBS = \frac{\Sigma \text{ exchangeable bases} \times 100}{\Sigma \text{ exchangeable bases} + \text{ exchange acidity}}$$

Other methods include use of CEC determined by ammonium saturation method as a base, or measuring total exchangeable bases in a single determination and using a CEC measurement to compute PBS.

Some uses made of percentage base saturation data include:

1. Degree of leaching—the PBS of the subsoil of B and upper C horizons is especially diagnostic of the extent to which exchangeable basic cations have been removed from the soil and replaced by exchange acidity. Therefore, this characteristic is extensively used in soil classification, soil fertility, and mineral nutrition studies. Two of the soil orders are separated from each other by differences in PBS of the subsoil. Those soils in regions of higher rainfall, warmer temperatures, and an older landscape surface have been observed to have PBS less than 35 in their B horizons (soils with argillic horizons only), or PBS decreasing from B to C horizons. Higher PBS values are found in those soils of the humid temperate regions and on younger landforms of the subtropics and tropics.

2. Percentage base saturation greater than 50% is required for the mollic epipedon, the main diagnostic horizon for one of the soil orders. Thus PBS is used as a differentiating characteristic for three of the soil orders.

Free (Extractable) Iron. That portion of the total iron in a soil occurring as hydrous oxides and uncombined with layer silicate structures and which is reductant-soluble is designated free iron. It is in the form of clay particle coatings, or as discrete particles, or possibly in interlayer positions. It is of interest in soil genesis-classification studies because of its increasing concentration with increasing weathering, and its effect on soil colors. This iron is active in phosphate fixation. It contributes to greater soil aggregate stability.

A commonly used procedure for its measurement involves reducing the iron with sodium dithionite, chelating it with sodium citrate in a sodium bicarbonate buffer, following which it is measured colorimetrically (Mehra and Jackson 1960). A procedure different in certain respects is used in some laboratories (Kilmer 1960; SCS-USDA 1967).

Some uses and inferences of free iron data are:

1. In soils with weatherable iron-bearing minerals, the percentage of free iron increases with increasing weathering and soil age.

2. Free iron decreases with increasingly poor natural drainage, which can be quantified with measurement of percentage free iron as can be seen in the data in Table 4.1 of Rich and Obenshain (1956) on Virginia soils.

3. The content of extractable Fe is also used, together with extractable carbon and aluminum, to define and recognize spodic horizons with their illuvial accumulation of iron and/ or aluminum and humus.

Saturation Extract Conductivity. In semiarid and arid regions, there is appreciable accumulation of soluble salts in the soil solution in certain landscape positions. This undesirable quality can be quantified with measurement of the conductivity of the saturation extract. This saturation extract is obtained by preparing a saturated paste of soil, then vacuum filtering to obtain the saturated extract (U.S. Salinity Laboratory Staff, 1954). The amount of soluble salts in this saturation extract can then be measured with a Wheatstone bridge-conductivity cell combination. The results are expressed in units of mmho/cm at 25 C.

In soil classification activity, a conductivity of saturation extract of 2 mmho/cm (25 C) or greater in the upper part of the profile is used as one of the diagnostic criteria for recognition of the order Aridosols. Plant re-

TABLE 4.1 ❧ **Content of free iron oxide in some soils of Virginia**

Soil Series	Natural Drainage Class	Free Fe, Whole Soil
		(%)
Matapeake	well drained	1.72
Bertie	somewhat poorly drained	1.03
Othello	poorly drained	0.39
Portsmouth	very poorly drained	0.03

Source: Rich and Obenshain 1956.

sponse to varying levels of conductivity have been described in detail by the U.S. Salinity Laboratory Staff (1954). A conductivity > 4 mmho/cm (25 C) is used to define a saline soil (U.S. Salinity Laboratory Staff 1954).

ORGANIC MATTER AND ITS COMPONENTS. Principal components of soil organic matter of interest in general soil formation-classification activities are carbon and nitrogen. Total organic matter content is commonly measured indirectly in soil classification-characterization studies, determining the organic carbon percentage and calculating percent organic matter by use of a factor. In the past, a factor of 1.724 has been multiplied by the organic C percentage to obtain organic matter percentage. However, as Broadbent (1965) and Allison (1965) point out, modern data indicate the ratio of organic matter to organic C is variable in different soils, and that a value in the range of 1.8 to 2.0 (approximately 1.9) is more appropriate for surface soils.

Organic carbon is most commonly determined by either dry combustion (heating in furnace and measuring CO_2 involved) or wet combustion (measuring degree of reduction of a strong oxidizing agent) (Allison 1965; Broadbent 1965). As the dry combustion-CO_2 evolution technique is quite quantitative (Allison 1965), it is considered as the primary standard. Organic C data are obtained in this way when precise, quantitative information is needed for soil classification-genesis studies. The wet combustion procedure most commonly used is the Walkley-Black method, in which the soil is digested in an excess of chromic acid, with titration of the unused oxidant (Allison 1965). Since some assumptions and approximations are involved in this procedure, it is less accurate and precise than dry combustion, but it requires less time and less complicated laboratory apparatus, and thus is commonly used where semiquantitative organic C data are acceptable.

In the Comprehensive Soil Classification System (Soil Survey Staff 1960, 1967), it has been found more desirable to express organic matter in terms of kilograms of carbon per square meter of soil surface area. This makes a more useful and discriminating diagnostic criterion. For such purposes, the weight percentages of organic carbon should be multiplied by the bulk density (to give volume percentage) and by the thickness of each horizon involved (SCS-USDA 1967).

Nitrogen in soils is commonly determined by the macro-Kjeldahl method (Bremner 1965), where such data are desired for soil genesis and classification purposes. A complete review of nitrogen analysis has been prepared by Bremner (1965); a summary of the most commonly used method is presented in *Soil Survey Investigations Report 1* (SCS-USDA 1967).

Some uses of and inferences from carbon and nitrogen data are:

1. Organic carbon content of the various horizons is a useful indication of the extent of accumulation of organic matter under differing environmental conditions. It is a key diagnostic criterion of those soils in which organic matter accumulation in the soil is a dominant pedogenic

process, as in the soils of the grasslands. Thus organic carbon content is a key criterion of the major diagnostic horizons mollic and umbric epipedons. Organic carbon accumulation per square meter is used as a diagnostic criterion in certain of the suborders and great groups of the comprehensive system, as in the case of certain soils in tropical areas that have relatively high amounts of organic matter in the soil profile (Soil Survey Staff 1960, 1967).

2. The content or organic carbon (organic matter) is used to differentiate organic soils (Histosols) from mineral soils.

3. The ratio of organic carbon to nitrogen, C/N, is a useful indication of the degree of decomposition of organic matter in soils, as it narrows with increasing modification. Well-decomposed soil humus has a C/N of approximately 12 or 13 in humid temperate soils, whereas straw has a C/N of approximately 40, for example. However, C/N is not used as a major diagnostic criterion in the Comprehensive Soil Classification System.

MINERALOGICAL COMPOSITION. For ease of better understanding and discussion, it seems convenient to separate soil minerals into two broad groups: (1) skeletal minerals—primary minerals, grains, microcrystalline aggregates, and fragments—mostly inherited from the parent rock; and (2) clay minerals and amorphous fine-grained (clay size) compounds. The former group comprises the sand and silt fractions for the most part, whereas the latter are primarily in the clay fraction. However, it should be noted that primary minerals are found in the clay fractions of some soils, and that the clay minerals may be found in the silt and sand fractions of certain soils.

Skeletal Mineralogy. The primary minerals of this group mostly consist of sand and silt-size grains, each of which is a single mineral. The microcrystalline aggregates placed in this group include volcanic ash (a mixture of various primary minerals) and chert (microcrystalline silica). Fragments include small chips of rocks, containing various minerals, found in sand and silt sizes. The weatherability of these minerals and aggregates is discussed in the following chapter on weathering and soil formation.

These skeletal minerals and aggregates are best identified and percentages established by use of a polarizing microscope, after appropriate sample preparation. X-ray diffraction and infrared techniques have been used to some extent for study of these coarser fractions, but they as yet lack the precision that can be obtained by use of optical mineralogy methods with the use of a light microscope fitted with polarizers. However, such studies are quite time-consuming and tedious—one reason that soil mineralogical data are sparse! Procedures of sample preparation suitable for soil petrographic analysis are presented in standard references and texts (for example, Brewer 1964; Milner 1952) and are summarized in *Soil Survey Investigations Report No. 1* (SCS-USDA 1967). Techniques and criteria used to identify mineral grains can be found in mineralogy texts and in Milner (1952); notes on mineral identification of particular use and importance in soil mineralogy have been prepared by Cady (1965).

Some uses of and inferences from data on skeletal mineralogy obtained by petrographic techniques are:

1. Uniformity of parent material for the various horizons of a profile—or lack of uniformity.

2. Presence of lithologic discontinuities (based on shifts in mineral species percentages from horizon to horizon).

3. Nutrient status and soil fertility reserve of soils (based on content of weatherable minerals which release plant nutrients as they weather).

4. Degree of weathering, based on ratio of weatherable to nonweatherable minerals.

5. Probable course of clay mineral formation and soil development, based on types of minerals present.

6. Recognition of presence of certain types of primary minerals or aggregates which impart unique and distinctive properties to the soil. An example is volcanic ash which, if present in sufficient amounts in sand and silt fractions, imparts lower bulk density, higher moisture retention, and higher phosphate fixation.

7. Because of the above-listed important inferences, mineralogy is extensively used as a diagnostic criterion in the Comprehensive Soil Classification System. This is especially true for soil families, for which types of minerals present are important criteria (Soil Survey Staff 1967).

Clay Minerals and Amorphous Clays. Layer aluminosilicate clays constitute the main portion of the clay fraction of most soils. Also present in significant amounts in certain soils are materials such as hydrous iron oxides and gibbsite in more highly weathered soils, and allophane (amorphous aluminosilicate) which is especially abundant in soils formed from volcanic ash in humid regions. Structure and composition of these clay minerals are described in a number of texts and reference books such as the recent one by Grim (1968). These materials are best determined by a combination of X-ray diffraction, differential thermal analysis (DTA), specific surface, and electron microscopic techniques. These techniques, together with methods for preparing the soil samples for analysis, are described by authors of several chapters in *Methods of Soil Analysis,* published by the American Society of Agronomy (Barshad 1965; Kittrick 1965; Kunze 1965; Mortland and Kemper 1965; Whittig 1965). For preliminary, semiquantitative data, X-ray diffraction alone or in combination with DTA is reasonably satisfactory. Even with these techniques and instrumentation available, quantification of amounts of clay minerals in soil samples still remains more nearly an art than a science. This is because of the empirical nature of sample preparation, poor crystallinity, and variable chemical composition of soil clay minerals (the clay mineral species in soils do not behave in the well-defined manner of reference or museum samples of standard clay minerals), and diffraction is not in proportion to amounts present, as when small islands of a given species are present or when high "scattering" factors are present. Problems associated with estimation of percentages of soil clay minerals are reviewed and some techniques suggested in papers by Jackson (1964), Jackson and Mackenzie (1964), and Whittig (1965).

Some inferences from clay mineral data include:

1. Physical qualities, as shrink-swell potential, plasticity, moisture retention, permeability.

2. Cation exchange characteristics.

3. Potassium reserve and release rates; K^+ and NH_4^+ fixation potentials.

4. Weathering stage—intensity of weathering factors to which soils have been submitted.

Because of these many effects of clay minerals on soil properties, clay mineral content is used as a classification criterion. Most intensive use of clay mineral data in the Comprehensive Soil Classification System is as one of a set of soil family differentiating characteristics. Clayey soils are subdivided according to types of soil clay minerals present in this system.

LITERATURE CITED

Allison, L. E. 1965. Organic carbon, pp. 1367–78. In C. A. Black (ed.), Methods of soil analysis, Agron. 9. Am. Soc. Agron., Madison, Wis.

Barshad, I. 1965. Thermal analysis techniques for mineral identification and mineralogical composition, pp. 699–742. In C. A. Black (ed.), Methods of soil analysis, Agron. 9. Am. Soc. Agron., Madison, Wis.

Blake, G. R. 1965. Bulk density, pp. 374–90. In C. A. Black (ed.), Methods of soil analysis, Agron. 9. Am. Soc. Agron., Madison, Wis.

Brasher, B. R., D. P. Franzmeier, V. Valassis, and S. E. Davidson. 1966. Use of Saran resin to coat natural soil clods for bulk-density and water-retention measurements. Soil Sci. 101:108.

Bremner, J. 1965. Total nitrogen, pp. 1149–1341. In C. A. Black (ed.), Methods of soil analysis, Agron. 9. Am. Soc. Agron., Madison, Wis.

Brewer, R. 1964. Fabric and mineral analysis of soils. John Wiley & Sons, New York.

Bridgman, P. W. 1927. The logic of modern physics. Macmillan, New York.

Broadbent, F. E. 1965. Organic matter, pp. 1397–1400. In C. A. Black (ed.), Methods of soil analysis, Agron. 9. Am. Soc. Agron., Madison, Wis.

Bryant, J. P. 1964. Soils of the Carolina Bays and interbay areas in Scotland County, North Carolina. Ph.D. thesis, N.C. State Univ. (Order 24-4915) Univ. Microfilms. Ann Arbor, Mich. (Dissertation Abstr.)

Cady, J. G. 1965. Petrographic microscope techniques, pp. 604–31. In C. A. Black (ed.), Methods of soil analysis, Agron. 9. Am. Soc. Agron., Madison, Wis.

Chapman, H. D. 1965. Cation-exchange capacity, pp. 891–901. In C. A. Black (ed.), Methods of soil analysis, Agron. 9. Am. Soc. Agron., Madison, Wis.

Cline, M. G. 1944. Principles of soil sampling. Soil Sci. 58:275–88.

———. 1945. Methods of collecting and preparing soil samples. Soil Sci. 59:3–5.

———. 1963. Logic of the new system of classification. Soil Sci. 96:17–22.

Coleman, N. T., and G. W. Thomas. 1964. Buffer curves of acid clays as affected by the presence of ferric iron and aluminum. Soil Sci. Soc. Am. Proc. 28:187–90.

———. 1967. The basic chemistry of soil acidity. In R. W. Pearson and F. Adams (eds.), Soil acidity and liming, Agron. Monograph 12:1–41. Am. Soc. Agron., Madison, Wis.

Coleman, N. T., S. B. Weed, and R. J. McCracken. 1959. Cation-exchange capacity and exchangeable cations in Piedmont soils of North Carolina. Soil Sci. Soc. Am. Proc. 23:146–49.

Day, P. R. 1965. Particle fractionation and particle-size analysis, pp. 545–67. In C. A. Black (ed.), Methods of soil analysis, Agron. 9. Am. Soc. Agron., Madison, Wis.

Dolman, J. D. 1967. Genesis, morphology and classification of organic soils in the Tidewater Region of North Carolina. Ph.D. thesis, N.C. State Univ. (Order 67-11,276) Univ. Microfilms. Ann Arbor, Mich. (Dissertation Abstr. 28B:764B)

Evans, C. E. 1968. Ion exchange relationships of aluminum and calcium in soils as influenced by organic matter. Ph.D. thesis, N.C. State Univ. (Order 68-14, 651) Univ. Microfilms. Ann Arbor, Mich. (Dissertation Abstr. 29:1233-B)

Fleming, J. F., and L. T. Alexander. 1961. Sulfur acidity in South Carolina tidal marsh soils. Soil Sci. Soc. Am. Proc. 25:94–95.

Franzmeier, D. P., and S. J. Ross, Jr. 1968. Soil swelling: Laboratory measurement and relation to other soil properties. Soil Sci. Soc. Am. Proc. 32:573–77.

Grim, R. E. 1968. Clay mineralogy, 2nd ed. McGraw-Hill, New York.

Grossman, R. B., B. R. Brasher, D. P. Franzmeier, and J. L. Walker. 1968. Linear extensibility as calculated from natural-clod bulk density measurements. Soil Sci. Soc. Am. Proc. 32:570–73.

Heald, W. R. 1965. Calcium and magnesium, pp. 999–1010. In C. A. Black (ed.), Methods of soil analysis, Agron. 9. Am. Soc. Agron., Madison, Wis.

Henry, E. F., and Mary C. Dragoo. 1965. Guide to use of the FHA soil PVC meter. FHA-595. Federal Housing Admin., Washington.

Jackson, M. L. 1956. Soil chemical analysis—advanced course. Published by the author, Dept. of Soils, Univ. Wis., Madison.

———. 1958. Soil chemical analysis. Prentice-Hall, Englewood Cliffs, N. J.

———. 1964. Soil clay mineralogical analysis, pp. 245–94. In C. I. Rich and G. W. Kunze (eds.), Soil clay mineralogy. Univ. N.C. Press, Chapel Hill.

Jackson, M. L., and R. C. Mackenzie. 1964. Chemical analysis in the quantitative mineralogical examination of clays, pp. 313–25. In C. I. Rich and G. W. Kunze (eds.), Soil clay mineralogy. Univ. N.C. Press, Chapel Hill.

Jenny, H. 1941. Factors of soil formation. McGraw-Hill, New York.

Kamprath, E. J. 1967. Soil acidity and response to liming. Tech. Bull. 4, Intern. Soil Testing Series, Soil Sci. Dept., N.C. State Univ., Raleigh.

Kanehiro, Y., and G. D. Sherman. 1965. Fusion with sodium carbonate for total elemental analysis, pp. 952–58. In C. A. Black (ed.), Methods of soil analysis, Agron. 9. Am. Soc. Agron., Madison, Wis.

Kilmer, V. J. 1960. The estimation of free iron oxides in soils. Soil Sci. Soc. Am. Proc. 24:420–21.

Kittrick, J. A. 1965. Electron microscope techniques, pp. 632–52. In C. A. Black (ed.), Methods of soil analysis, Agron. 9. Am. Soc. Agron., Madison, Wis.

Kunze, G. W. 1965. Pretreatment for mineralogical analysis, pp. 568–77. In C. A. Black (ed.), Methods of soil analysis, Agron. 9. Am. Soc. Agron., Madison, Wis.

Lin, C., and N. T. Coleman. 1960. The measurement of exchangeable aluminum in soils and clays. Soil Sci. Soc. Am. Proc. 24:444–46.

Marbut, C. F. 1935. Soils of the United States. In USDA Atlas of American agriculture, part 3. Advance sheets, no. 8.

McLean, E. O. 1965. Aluminum, pp. 978–98. In C. A. Black (ed.), Methods of soil analysis, Agron. 9. Am. Soc. Agron., Madison, Wis.

Mehlich, A. 1938. Use of triethanolamine acetate-barium hydroxide buffer for the determination of some base exchange properties and lime requirement of soil. Soil Sci. Soc. Am. Proc. 3:162–66.

Mehra, O. P., and M. L. Jackson. 1960. Iron oxide removal from soils and clays by a dithonite-citrate system buffered with sodium bicarbonate, pp. 317–27. In Clays and clay minerals, Proc. 7th Natl. Conf., Monograph 5, Earth Science Series. Pergamon Press, New York.

Milner, H. B. 1952. Sedimentary petrography, 4th ed. T. Murby & Co., London.

Moormann, F. R. 1963. Acid sulfate soils (cat-clays) of the tropics. Soil Sci. 95: 271–75.

Mortland, M. M., and W. D. Kemper. 1965. Specific surface, pp. 532–44. In C. A. Black (ed.), Methods of soil analysis, Agron. 9. Am. Soc. Agron., Madison, Wis.

Nelson, L. A. 1961. Evaluation of properties and yield potentials of some important Atlantic coastal plain soils. Ph.D. thesis, N.C. State Univ. (Order 61-4634) Univ. Microfilms. Ann Arbor, Mich. (Dissertation Abstr. 22:1764)

Nye, P., Doris Craig, N. T. Coleman, and J. L. Ragland. 1961. Ion exchange equilibria involving aluminum. Soil Sci. Soc. Am. Proc. 25:14–17.

Prince, A. B. 1965. Absorption spectrophotometry, pp. 866–78. In C. A. Black (ed.), Methods of soil analysis, Agron. 9. Am. Soc. Agron., Madison, Wis.

Rich, C. I. 1965. Elemental analysis by flame photometry, pp. 849–65. In C. A. Black (ed.), Methods of soil analysis, Agron. 9. Am. Soc. Agron., Madison, Wis.

Rich, C. I., and S. S. Obenshain. 1956. Distribution of free iron oxides in four coastal plain soils of Virginia. Sixth Congr. Intern. Soil Sci. Soc. 2:447–51.

Richards, L. A. 1965. Physical condition of water in soil, pp. 128–52. In C. A. Black (ed.), Methods of soil analysis, Agron. 9. Am. Soc. Agron., Madison, Wis.

SCS-USDA. 1967. Soil survey investigations report no. 1. Soil survey laboratory methods and procedures for collecting soil samples. U.S. Govt. Printing Office, Washington.

Smith, G. D. 1963. Objectives and basic assumptions of the new soil classification system. Soil Sci. 96:6–16.

Soil Survey Staff. 1951. Soil survey manual. U.S. Dept. Agr. Handbook 18. U.S. Govt. Printing Office, Washington.

———. 1960. Soil classification, a comprehensive system—7th approximation. U.S. Dept. Agr. U.S. Govt. Printing Office, Washington.

———. 1967. Supplement to Soil classification, a comprehensive system—7th approximation. U.S. Dept. Agr. U.S. Govt. Printing Office, Washington.

U.S. Salinity Laboratory Staff. 1954. Diagnosis and improvement of saline and alkali soils. U.S. Dept. Agr. Handbook 60. U.S. Govt. Printing Office, Washington.

Veihmeyer, F. J., and A. H. Hendrickson. 1948. Soil density and root penetration. Soil Sci. 65:487–93.

Whittig, L. D. 1965. X-ray diffraction techniques for mineral identification and mineralogical composition, pp. 671–98. In C. A. Black (ed.), Methods of soil analysis, Agron. 9. Am. Soc. Agron., Madison, Wis.

Weathering and Soil Formation   ⚘

WEATHERING refers to the chemical and physical disintegration and decomposition of rock, and the minerals contained in them that are not at equilibrium under the temperature, pressure, and moisture conditions of the atmosphere-lithosphere interface.

Weathering of the initial materials precedes soil formation in hard rocks and accompanies it in soft rocks and soil materials. It is a continuing reaction during soil development, to the point where no more reactants are available. Weathering proceeds both below the solum and within the solum itself. Thus we find it useful to distinguish between geochemical weathering and pedochemical weathering, a distinction proposed by Jackson and Sherman (1953). *Geochemical weathering* is that taking place below the soil solum (in C horizons) and that which would take place were the soil solum not there. *Pedochemical weathering* is the disintegration and chemical modification of minerals taking place within the soil A and B horizons, with all the associated biological and other soil-forming processes.

Our aims in this chapter are to define and describe these weathering processes, describe the relative stability and persistence of common soil minerals as weathering proceeds, and indicate the general conditions for the synthesis of new soil minerals formed from the wreckage and debris of the weathered minerals.

GEOCHEMICAL WEATHERING. The weathering reactions that take place as part of geochemical weathering are oxidation, reduction, combinations of these in alternating cycles, hydration, solution, and hydrolysis.

OXIDATION. Oxidation is an important geochemical reaction occurring in well-aerated rock and soil materials where oxygen supply is high and biological demand for it is low. The most important specific reaction is that of ferrous to ferric ion:

$$Fe^{++} \longrightarrow Fe^{+++} + e^{-}$$
where e^{-} = electron transfer.

Oxidation of iron is a disintegrative weathering process in those minerals containing the ferrous ion as part of their structure. The change in size and charge of this element as it is converted to ferric form causes mineral structures to break apart. Examples are those species of biotite mica containing appreciable amounts of ferrous iron, the mineral glauconite (commonest mineral in "greensand") (Cloos, Fripiat, and Vielvoye 1961) which is rich in ferrous iron, and the iron-bearing species of the hornblendes and pyroxenes in the ferromagnesian primary mineral group. The iron released by this disintegration of primary minerals unites with hydroxyl (OH) and/ or oxygen to form iron minerals, as described later. Manganese is likewise released into "free" form by the oxidation of primary minerals, but its chemistry is more complex, due to the number of oxidation states it possesses.

REDUCTION. Reduction in the geochemical environment occurs where the material is water saturated (such as below the water table level), oxygen supply is low, and biological oxygen demand is high. The effect is to reduce the iron to the highly mobile ferrous form (Cate 1964). In this form it may be lost from the system if there is net downward and outward movement of the groundwater. If the ferrous iron persists in the system, it reacts to form sulfides and related compounds. These impart the characteristic green and blue-green colors to many reduced soil materials. If the iron remains in hydrous ferrous oxide form (lepidocrocite) in the soil material, the characteristic orange and yellow mottling results. This phenomenon is associated with relatively high contents of organic matter (Bloomfield 1952; Brown 1953; Jeffery 1960; Marel 1951).

OXIDATION-REDUCTION. A commoner feature in soil C horizons and other initial materials of soils is fluctuation from oxidizing to reducing conditions, often cyclic in response to variations in weather through the year. Or, the soil initial material may be in a reducing environment during the geochemical weathering stage, then be introduced into an oxidizing environment as it becomes immediate parent material for soils (the C horizons). This change may come about by lowering of the land surface by erosion, uplift of coastal areas, lowering of regional water tables due to landscape development processes, climatic change, or lowering and raising of the water table by man as in paddy rice culture. To understand the changes in oxidation state and consequent mineral form and with these fluctuations from oxidizing to reducing conditions and vice versa, it is useful to refer to charts which have recently been prepared on this subject (Collins 1968; Garrels and Christ 1965). In using the example in Figure 5.1, one looks at the pH and the Eh (redox potential) to determine the form of Fe and Mn under a variety of conditions. This type of diagram serves as a very helpful geochemical model. If the Eh is not known, one can approximate the intensity of these redox conditions. Some generalizations can be drawn from this chart to illustrate its utility. One such generalization is that with increasing acidity, ferrous iron becomes increasingly

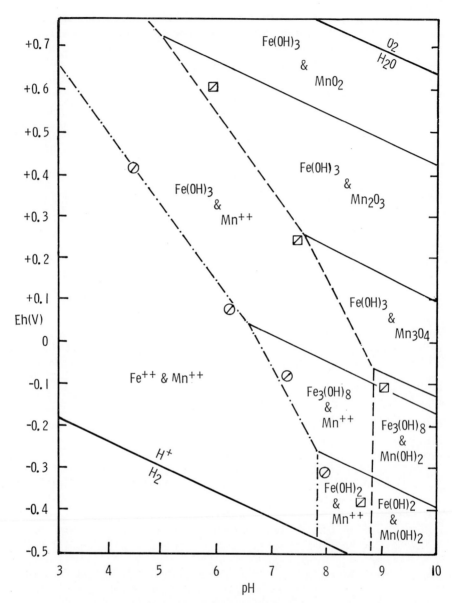

FIG. 5.1. Stability fields of iron and manganese related to Eh and pH in 0.01N chloride solutions. (After J. F. Collins.)

stable in more oxidizing conditions. That is, in very acid soil material systems, we may expect to find ferrous iron, even though the system is slightly to moderately oxidized. Manganese follows a pattern similar to iron, but is in reduced form in higher redox conditions at a given pH than is iron. Manganese remains in reduced form at higher pH's than iron.

(This model explains the deeper occurrences of manganese than iron in the C horizons of our soils which have a redox gradient, that is, which become more oxidizing with depth.)

HYDRATION. Hydration refers to the association of water molecules or of hydroxyl groups with minerals, often without actual decomposition or modification of the mineral itself. It occurs primarily on surfaces and edges of mineral grains but in cases of simple salts may pervade the entire structure, with some changes in properties. An example of the latter case is hydration of the mineral anhydrite to form gypsum:

$$CaSO_4 + H_2O \longrightarrow CaSO_4 \cdot 2H_2O \text{ (gypsum)}$$

Commoner are the sorption of water molecules on mineral surfaces and the association of hydroxyls and water in the coordination sphere of aluminum and silica at the broken edges of minerals such as the layer silicates (micas, for example). This sorbed water provides a bridge or entryway for hydronium (hydrated hydrogen) ions to attack the structure. The association of water or hydroxyls with Al and Si at broken edges is but the first step in hydrolysis.

HYDROLYSIS. Hydrolysis refers to the attack of the small, highly charged hydrogen ion (which together with its hydration shell is referred to as *hydronium*) on the crsytal structures. The result is replacement of the basic ions by the hydrogen, with consequent collapse and disintegration of the structure. A simplified example is the following equation for hydrolysis of orthoclase feldspar.

$$KAlSi_3O_8 + H^+ \longrightarrow HAlSi_3O_8 + K^+$$

The resulting "silicic acid" is not really of importance because it is very short-lived in soil, if present at all. Actually the silica and aluminum reorganize the oxygen and hydroxyl to form the amorphous mineral allophane or the crystalline halloysite, $Al_2Si_2O_5(OH)_4$.

Another example of hydrolysis is the attack of hydrogen (actually hydronium) on the interlayer potassium of micas to produce the clay mineral illite (by partial K removal) or vermiculite (by complete K removal). This process is illustrated in diagrams and discussion presented in Rich (1964) and in Rich and Black (1964), based on studies of K removal from interlayers.

In general, hydrolysis is the most important chemical weathering process and results in the complete disintegration or drastic modification of weatherable primary minerals.

SOLUTION. Solution refers to the dissolving of simple salts such as carbonates and chlorides which occur as mineral grains in some soil initial materials. An example is the dissolving of calcium carbonate contained in a calcareous glacial till or loess deposit:

$$CaCO_3 + 2H^+ \longrightarrow H_2CO_3 + Ca^{++}$$

PEDOCHEMICAL WEATHERING. Certain weathering reactions take place in the soil solum almost exclusively or at least react in their greatest intensity there. Such have been included in the term *pedochemical weathering* proposed by Jackson and Sherman (1953). These are reactions which the pedologist can claim for his own. However, they also occur to some extent under geochemical conditions in the soil initial material, and thus were discussed in the previous section. In this section, we will describe these reactions as they occur in the soil solum.

OXIDATION-REDUCTION CYCLES. Alternation between reducing and oxidizing conditions is responsible for the release of iron and manganese from primary minerals and their localization into mottles and concretions in the soil solum. Of particular interest and importance in silicate clay destruction in soils is the alternation between strong reducing and oxidizing conditions in poorly drained soils, especially those of coastal lowland areas. This process, described by Cate and Sukhai (1964) and by Patrick and Wyatt (1964), consists of replacement of exchangeable Al^{+++} by exchangeable Fe^{+2} upon onset of reducing conditions. With the return of oxidizing conditions, this exchangeable ferrous iron is displaced and aluminum appears from the clay lattice to occupy the exchange sites. This appearance of Al^{+++} causes some destruction and disintegration of the silicate clay structure. Nettleton (1966) suggested this mechanism as an explanation for the apparent clay destruction in poorly drained North Carolina coastal plain soils. This set of reactions may be an important pedochemical weathering process in wet reduced soils which periodically become dry and oxidized, but its full extent and importance is not yet fully assessed.

SHUTTLING OF ALUMINUM FROM CLAY LATTICES
TO HYDROUS OXIDES VIA EXCHANGE SITES. This pedochemical weathering mechanism is responsible for destruction of clay (especially montmorillonitic clay) in the solum, under some conditions. Essentially the process works like this: Assume that soil clays are initially saturated with exchangeable Ca^{++} and Mg^{++}, and that these are displaced by H^+ in acid weathering. The H^+ causes instability, bringing out Al^{+++} from the clay lattice, with subsequent disintegration of part of the lattice. Hydrolysis of this Al^{+++} results in additional H^+ ions which cause further weathering of the clay (Coleman 1962; Coleman, Ragland, and Craig 1960). This process is apparently responsible for the decomposition of montmorillonite present in soil sola as an inheritance from initial materials containing this mineral in those environmental areas (high rainfall and temperature) where this mineral is unstable and not in equilibrium.

POTASSIUM REMOVAL FROM MICAS. This weathering process, described in the previous section on geochemical weathering, is particularly important in those soil sola in which there is a high hydronium supply from biological sources and a plentiful source of clay micas originating from the initial material. Removal of a small to moderate amount of the potassium from the interlayers of the mica does not cause great distortion or loss of alignment of the silica-alumina packets. As a result, potassium added to the system may be trapped in the vacated "holes." Exchange capacity is somewhat increased. This is characteristic of the mineral commonly referred to as illite. But with removal of more than approximately 50% of the interlayer K, sheet alignment is lost, and strain and distortion of the lattice takes place. Therefore, added K is not easily trapped or fixed, and the remaining interlayer K becomes more available (White 1962; White, Anderson, and Hensel 1959). With complete removal of K from interlayer planes, vermiculite and montmorillonite clay mineral types are produced.

ALUMINUM INTERLAYERING
OF 2:1 CLAY MINERALS. An important pedogenic mineral modification in acid soils is the precipitation of "islands" of hydroxy-Al in interlayer spaces of vermiculite and, to a lesser extent, in montmorillonite clays. The Al-interlayered clay is referred to as "2:1-2:2 intergrade." This weathering modification is one that is essentially peculiar to soil sola. As a result, the cation exchange capacity of the clay is partially blocked and neutralized. This interlayer Al^{+++} contributes to acidity, though it is slowly or difficultly exchangeable, and consequently the contribution to acidity is difficult to assess.

STABILITY INDICES AND
WEATHERING SEQUENCES. Soil minerals can be arranged in an order of stability, or conversely their weatherability, which makes a useful model. Such a model can be used to determine the general "degree of weathering" of a given soil, to predict the native nutrient (soil fertility) reserve of soils, to generalize about the behavior of soils (as physical properties in relation to clay mineral types present), to assess the effects of various environmental conditions on the path of soil formation, and to account for the effect and contribution of minerals present in the soil initial material. In view of the great differences in specific surface and consequent reactivity, it is desirable to separate soil mineral particles into two size classes when we discuss their weatherability: clay size and sand-silt size.

A "stability series" proposed by Goldich (1938) well illustrates the weatherability of the commoner soil primary minerals and generally coincides with empirical observations on stability. The Goldich sequence in order of increasing stability from top to bottom is shown in Figure 5.2.

Those geochemists and geologists familiar with the "reaction series" of rocks at higher temperatures will recognize this as the converse of that

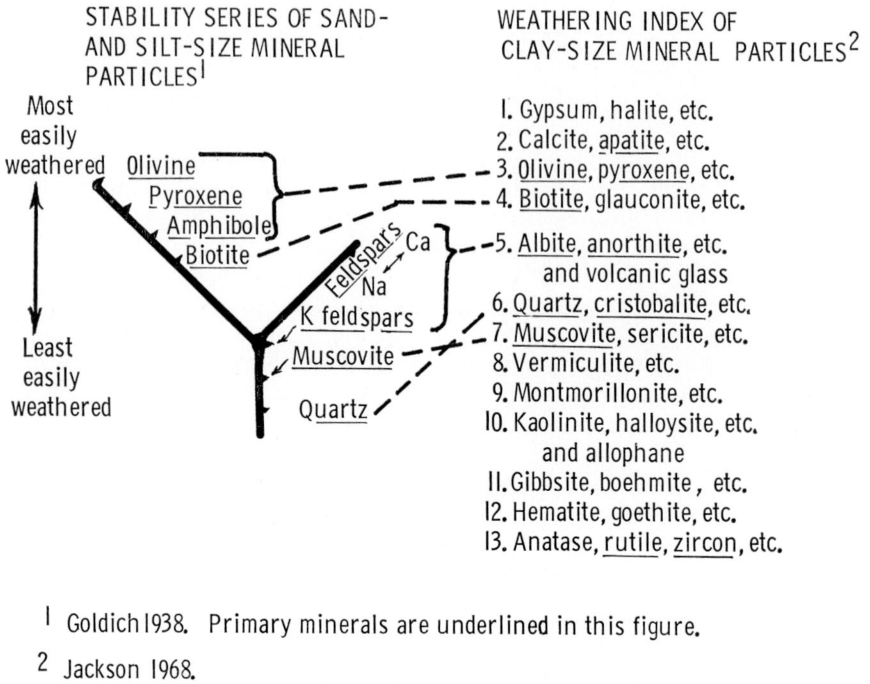

STABILITY SERIES OF SAND-
AND SILT-SIZE MINERAL
PARTICLES[1]

WEATHERING INDEX OF
CLAY-SIZE MINERAL PARTICLES[2]

Most
easily
weathered

Least
easily
weathered

Olivine
Pyroxene
Amphibole
Biotite
Feldspars Ca
Na
K feldspars
Muscovite
Quartz

1. Gypsum, halite, etc.
2. Calcite, apatite, etc.
3. Olivine, pyroxene, etc.
4. Biotite, glauconite, etc.
5. Albite, anorthite, etc.
   and volcanic glass
6. Quartz, cristobalite, etc.
7. Muscovite, sericite, etc.
8. Vermiculite, etc.
9. Montmorillonite, etc.
10. Kaolinite, halloysite, etc.
    and allophane
11. Gibbsite, boehmite, etc.
12. Hematite, goethite, etc.
13. Anatase, rutile, zircon, etc.

[1] Goldich 1938. Primary minerals are underlined in this figure.

[2] Jackson 1968.

FIG. 5.2. A comparison between the stability series of sand-
and silt-size mineral particles and the weathering index
series of clay-size mineral particles. The first series consists
of primary minerals arranged, from top to bottom, in
the order of their crystallization from molten material, and
also in the order of decreasing ease of weathering. The
second series consists of a condensed version of the first in
which the positions of muscovite and quartz have been
interchanged because of the greater stability in soils of
clay-size mica. At the top and in most of the lower part of
this series are secondary minerals.

ranking. That is, the least stable minerals are those that crystallize from
a "melt" at the highest temperatures. Apparently this greater instability
is related to their greater disequilibrium with the environment in the
lithosphere-atmosphere interface, the pedosphere.

In the left hand or "basic" branch of this series, there is increasing
silica tetrahedral linkage with increasing stability from top to bottom.
That is, the least stable mineral (olivine) is composed of single "unhooked"
silica tetrahedra. The structure is held together by bonds to the easily
hydrolyzable magnesium and the oxidizable iron. In quartz, the most stable
mineral, there is complete silica tetrahedral linkage. All oxygen atoms are
shared with more than one silicon. Also, there is a decrease in the content
(percentage) of easily hydrolyzable bases from the least to the most stable
minerals.

In the right hand or feldspar branch, there is decreasing distortion of the lattice from calcic to the potassic feldspars. The bivalent calcium does not fit well into the feldspar chain structure, though it does satisfy the charge imbalance from considerable substitution of aluminum for silicon. But the large monovalent potassium well suits the role of satisfying the smaller charge imbalance from lesser Al-for-Si proxying and fits nicely into holes in the feldspar chain. Thus orthoclase is more stable than plagioclase.

A useful and convenient model for weatherability of clay-size minerals is the weathering sequence proposed by Jackson and associates (Jackson et al. 1948), with later modifications (Jackson 1968). This sequence is composed of 13 stages, as illustrated in Figure 5.2.

A diagrammatic comparison of this clay-size weathering sequence with the sand-silt stability series is also presented in Figure 5.2.

Quartz changes position from the coarser to finer sizes because of greatly increased solubility with the increased specific surface associated with the smaller particle size. Muscovite mica is relatively more stable in clay size because of the stabilizing effect of the silica-alumina layers. In general, the concept is that in any soil the clay will be found to contain a modal distribution of two or three adjacent minerals in the weathering sequence, reflecting the weathering stage of the soil. Over time, with impact of the climatic factors, the clay fraction of a soil is assumed to be translated through the stages of the sequence. As in the sand-silt size Goldich stability series, we can note here the effect of chemical composition and mineral internal structure on resistance to weathering. Minerals with high contents of hydrolyzable bases are high in the sequence (weather easily). The silica-alumina layers of the resistant layer silicates have a stabilizing effect due to the silica tetrahedral linkage. The 1:1 layer structure is more resistant than the 2:1 layer minerals because of the absence in the former of hydrolyzable bases or oxidizable iron substituting for Al. The simple Al and Fe hydrous oxides are more stable yet, because of their very low solubility and high metal-to-hydroxyl or metal-to-oxygen bonding.

In using this weathering sequence as a model, one should be aware of several pitfalls. One is the effect of previous climates in polygenetic soils or paleosols. Another is the contribution of the mineralogy of the soil initial material, for a soil rich in kaolinite may have inherited this mineral from its parent material. There are problems with the placement of the 2:1-2:2 intergrade, because the aluminum interlayering apparently contributes to higher stability. We need to keep in mind also that there is no single path or main road of weathering. High intensity rainfall with rapid rates of removal of weathering products may produce gibbsite quite quickly and readily, in contrast to the millions of years required under other conditions. The concentration of soil solutions in a soil in a climate with a pronounced dry season may produce quite different clay mineral suites than a soil under the same amount of rainfall well distributed through the year.

SYNTHESIS OF MINERALS IN SOILS. Among the basic principles to consider in pedosynthesis of secondary or clay minerals are the importance of ionic concentrations and ionic equilibria in the soil systems, solubility products of the chemical entities involved, Eh-pH conditions, and the kinetics or time rate of the various coupled weathering-synthesis reactions, including the rate of removal of weathering products such as bases and silica.

Principles of ionic concentrations and of ionic equilibria in mineral synthesis have recently been presented (Garrels and Christ 1965) in such a manner as to provide a quantitative basis and model. These new interpretations impress upon us the point that it is the concentration of ions and their equilibrium reactions in a local volume of the soil that control the kind of minerals formed, not solely the external climate or environmental conditions. In a detailed study in Virginia of clay minerals formed in proximity to different minerals occurring in the same rocks, it was found that the chemical composition of the weathering mineral controlled the kind of clay mineral formed in that local area (Barnhisel and Rich 1967).

As pointed out in an earlier section, the Eh-pH status of a soil system controls the kind of iron and manganese minerals formed. For example, if the soil system is acid and the Eh low (redox potential negative or slightly on the positive side), then lepidocrocite, iron pyrites, and other ferrous-iron minerals form from the iron released by weathering. Otherwise goethite is formed which then "ages" to hematite with time if a high positive Eh (strong oxidizing condition) is maintained.

Solubility of weathering minerals, rate of aging of gels to form crystalline soil minerals, and the rate of removal of weathering products such as silica and bases all relate to the time required to form new minerals and the type of minerals formed. If weathering products are quickly removed from the locale of weathering minerals, then the chemical reactions are rapidly driven far toward production of minerals commonly associated with advanced stages of weathering, such as gibbsite. On the other hand, if weathering products are not removed, due to slow or no water movement in the soil, for example, then the ionic concentrations of silica and magnesium are such that montmorillonite is formed.

GENERALIZED CONDITIONS FOR FORMATION AND PERSISTENCE OF COMMON SOIL CLAY-SIZE MINERALS

SMECTITES (MONTMORILLONITE AND RELATIVES). A relatively high ionic concentration of silica and of magnesium is required for synthesis of montmorillonites. Generally these conditions are met in the vicinity of decomposing silicate minerals rich in magnesium (and iron). High silica concentration is maintained by slow movement or stagnation of soil water. Montmorillonite is unstable under conditions of high hydronium concentration and rapid leaching. However, it often occurs in dense clay layers such that leaching is slow, and thus it persists under conditions of high weathering intensity when inherited from the parent material.

VERMICULITE. This mineral is formed under conditions of moderate hydronium concentration such that potassium (and magnesium) are completely removed from interlayers. Mica must be present in the initial material. Concentration of Si must be high. But concentration of Al in solution must be low, or it will be precipitated in interlayers to form 2:1-2:2 intergrade.

ILLITE. This mineral forms where mica is present in the initial material, under conditions of moderate to low hydronium concentration necessary for partial stripping of K from interlayers. Moderate to relatively high concentrations of Si and Al are required for stability. Moderate to high hydronium concentration causes instability and disappearance, as it is converted to vermiculite.

2:1-2:2 INTERGRADE OF
Al-INTERLAYERED VERMICULITES. This mineral is synthesized under conditions of moderate to high hydronium concentration and medium to high Al and Si concentration. The interlayer spaces serve as a sink for the Al in solution, an "antigibbsite" effect (Jackson 1963). Mica or montmorillonite is required in the initial material as a precursor.

KAOLINITE. This mineral is synthesized under conditions of approximately equal concentrations of Si and Al, with high hydronium concentration and essentially an absence of Mg and other bases. Formation is aided by the presence of layer silicates as "templates" or patterns for the 1:1 sheet structure.

HALLOYSITE. This mineral forms where concentration of Al approximately equals that of Si, as a result of rapid weathering of feldspars (or of organization of an amorphous alumino-silicate such as allophane into more crystalline structure over time). This requires high hydronium concentration and zero or low concentration of bases.

GOETHITE. This mineral requires relatively high Eh (positive redox potential) and medium hydronium concentration. It forms from rapid disintegration of ferromagnesian minerals or condenses from hydrated amorphous gel over time. With persistence of high oxidizing conditions (Eh greater than $+$ 100 MV) and moderate pH, it proceeds slowly to the formation of hematite.

HEMATITE. This mineral forms under conditions of high Eh and moderate to high pH with increasing loss of hydroxyls from more hydrous goethite, or forms directly from disintegration of ferromagnesian minerals under conditions of high Eh-pH.

GIBBSITE. This mineral forms and persists under conditions of low Si concentration and high hydronium concentration, with absence or low concentration of bases. This may form by aging of aluminous allophanes or gels.

ALLOPHANE. This amorphous mineral forms under conditions of medium to high hydronium concentration in highly moist or wet systems, by rapid weathering of noncrystalline volcanic ash, or to a limited extent by rapid weathering of feldspars.

SUMMARY. Except for true solution and ion removal from an area, weathering to the soil scientist is a process that alters the material in and below the soil. Chemical weathering processes tend to reduce the initial material to the lowest energy level stable at ambient conditions. It produces new minerals in the initial material of the soil.

LITERATURE CITED

Barnhisel, R. I., and C. I. Rich. 1967. Clay mineral formation in different rock types of a weathering boulder conglomerate. Soil Sci. Soc. Am. Proc. 31:627–31.

Bloomfield, C. 1952. The distribution of iron and aluminum oxides in gley soils. J. Soil Sci. 3:167–71.

Brown, G. 1953. The occurrence of lepidocrocite in some British soils. J. Soil Sci. 4:220–28.

Cate, R. B., Jr. 1964. New data on the chemistry of submerged soils: Possible relationship to bauxite genesis. Econ. Geol. 59:161–62.

Cate, R. B., Jr., and A. P. Sukhai. 1964. A study of aluminum in rice soils. Soil Sci. 98:85–93.

Cloos, P., J. J. Fripiat, and L. Vielvoye. 1961. Mineralogical and chemical characteristics of a glauconitic soil of the Hageland region (Belgium). Soil Sci. 91:55–65.

Coleman, N. T. 1962. Decomposition of clays and the fate of aluminum. Econ. Geol. 57:1207–18.

Coleman, N. T., J. L. Ragland, and Doris Craig. 1960. An unexpected reaction between Al-Clay or Al-soil and CaCl₂. Soil Sci. Soc. Am. Proc. 24:419–20.

Collins, J. F. 1968. The Eh-pH environment and iron-manganese equilibria in soils. Ph.D. thesis, N.C. State Univ. (Order 69-7652) Univ. Microfilms. Ann Arbor, Mich. (Dissertation Abstr. 29:3575-B)

Garrels, R. M., and C. L. Christ. 1965. Solutions, minerals, and equilibria. Harper and Row, New York.

Goldich, S. S. 1938. A study in rock-weathering. J. Geol. 46:17–58.

Jackson, M. L. 1963. Aluminum bonding in soils: A unifying principle in soil science. Soil Sci. Soc. Am. Proc. 27:1–10.

———. 1968. Weathering of primary and secondary minerals in soils. Trans. 9th Intern. Congr. Soil Sci. 4:281–92.

Jackson, M. L., and G. D. Sherman. 1953. Chemical weathering of minerals in soils. Adv. Agron. 5:219–318.

Jackson, M. L., S. A. Tyler, A. L. Willis, G. A. Bourbeau, and R. P. Pennington. 1948. Weathering sequence in clay-size minerals in soils and sediments. I. Fundamental generalizations. J. Phys. Colloid. Chem. 52:1237–60.

Jeffery, J. W. O. 1960. Iron and the Eh of waterlogged soils with particular reference to paddy. J. Soil Sci. 11:140–48.

Marel, H. W. van der. 1951. Gamma ferric oxide in sediments. J. Sediment. Petrol. 21:12–21.

Nettleton, W. D. 1966. Pedogenesis of certain Aquultic and Aquic Normudultic soils of the North Carolina coastal plain. Ph.D. thesis, N.C. State Univ. (Order 66-12,899) Univ. Microfilms. Ann Arbor, Mich. (Dissertation Abstr. 27:1674-B)

Patrick, W. H., Jr., and R. Wyatt. 1964. Soil nitrogen loss as a result of alternate submergence and drying. Soil Sci. Soc. Am. Proc. 28:647–53.

Rich, C. I. 1964. Effect of cation size and pH on potassium exchange in Nason soil. Soil Sci. 98:100–106.

Rich, C. I., and W. R. Black. 1964. Potassium exchange as affected by cation size, pH, and mineral structure. Soil Sci. 97:384–90.

White, J. L. 1962. X-ray diffraction studies on weathering of muscovite. Soil Sci. 93:16–21.

White, J. L., J. U. Anderson, and D. R. Hensel. 1959. Applications of mineralogical techniques to soil genesis studies. Silicates Ind. 24:301–5.

❧

Pedogenic Processes: Internal, Soil-Building Processes

THE SEQUENCES of mineral weathering and pedochemistry discussed in the preceding chapter are combined with various physical phenomena to constitute processes of soil formation. This chapter is concerned with fundamental pedogenic concepts and brief definitions of numerous special pedological terms, some of which have overlapping meanings, as will be indicated. Different schools of pedologic thought place emphasis on different aspects of those complexes of reactions called processes of soil formation. For example, some soil scientists regard silication (Vilenskii 1957), and others illuviation (Stobbe and Wright 1959), of sesquioxides as the essential feature of podzolization. No attempt is made here to differentiate between such competing points of view.

THE GENERAL NATURE OF THE PEDOGENIC PROCESS. A process of soil formation is a complex of or sequence of events, including both complicated reactions and comparatively simple rearrangements of matter, that intimately affects the soil in which it operates. Numerous events may take place simultaneously or in sequence to mutually reinforce or contradict each other (Rode 1962; Simonson 1959). For example, calcification and podzolization (Table 6.1) operate concurrently in certain Boralfs (Gray Wooded soils). A given process may tend to maintain the soil in its current condition or it may tend to change the soil. Some processes, such as crystal growth in a void in the interior of a large boulder and movement of fluids inside a large tree root or in a hibernating rodent in a soil, are remote from the active part of the soil and hence, even though they take place within the soil, may be excluded from the long list of soil-forming processes. On the other hand, crystal growth at the surface of the buried boulder, exchange of fluids between soil and root hairs, and exchange of gases between dormant rodents and the soil atmosphere are intimate parts of the processes of soil genesis. Geochemical weathering of minerals (Chapter 5) is a geologic process in the formation of initial material of soil and continues in the soil profile where

TABLE 6.1 ✖ **Some processes of soil formation that are complexes of subprocesses and reactions**

Term	Fourfold Categori- zation*	Brief Definition
1a. Eluviation	3	Movement of material out of a portion of a soil profile as in an albic horizon
1b. Illuviation	3	Movement of material into a portion of soil profile as in an argillic or spodic horizon
2a. Leaching (depletion)	2	General term for washing out or eluviating soluble materials from the solum
2b. Enrichment	1	General term for addition of material to a soil body
3a. Erosion, surficial	2	Removal of material from the surface layer of a soil
3b. Cumulization	1	Aeolian and hydrologic additions of mineral particles to the surface of a soil solum
4a. Decalcification	3	Reactions that remove calcium carbonate from one or more soil horizons
4b. Calcification	3	Processes including accumulation of calcium carbonate in Cca and possibly other horizons of a soil
5a. Salinization	3	The accumulation of soluble salts such as sulfates and chlorides of calcium, magnesium, sodium, and potassium in salty (salic) horizons
5b. Desalinization	3	The removal of soluble salts from salic soil horizons
6a. Alkalization (solonization)	3	The accumulation of sodium ions on the exchange sites in a soil
6b. Dealkalization (solodization)	3	The leaching of sodium ions and salts from natric horizons
7a. Lessivage	3	The mechanical migration of small mineral particles from the A to the B horizons of a soil, producing in B horizons relative enrichment in clay (argillic horizons)
7b. Pedoturbation	3	Biologic, physical (freeze-thaw and wet-dry cycles) churning and cycling of soil materials, thereby homogenizing the solum in varying degrees
8a. Podzolization (silication)	3,4	The chemical migration of aluminum and iron and/or organic matter, resulting in the concentration of silica (i.e., silication) in the layer eluviated
8b. Laterization (desilication, feralization, ferritization, allitization)	3,4	The chemical migration of silica out of the soil solum and thus the concentration of sesquioxides in the solum (goethite, gibbsite, etc.), with or without formation of ironstone (laterite; hardened plinthite) and concretions
9a. Decomposition	4	The breakdown of mineral and organic materials
9b. Synthesis	4	The formation of new particles of mineral and organic species
10a. Melanization	1,3	The darkening of light-colored mineral initial unconsolidated materials by admixture of organic matter (as in a dark A1 or mollic or umbric horizon)
10b. Leucinization	3	The paling of soil horizons by disappearance of dark organic materials either through transformation to light-colored ones or through removal from the horizons

TABLE 6.1. *(continued)*

Term	Fourfold Categorization*	Brief Definition
11a. Littering	1	The accumulation on the mineral soil surface of organic litter and associated humus to a depth of less than 30 cm
11b. Humification	4	The transformation of raw organic material into humus
11c. Paludization	4	Processes regarded by some workers as geogenic rather than pedogenic, including the accumulation of deep (>30 cm) deposits of organic matter as in mucks and peats (Histosols)
11d. Ripening	4	Chemical, biological, and physical changes in organic soil after air penetrates the organic deposit, making it possible for microbial activity to flourish
11e. Mineralization	4	The release of oxide solids through decomposition of organic matter
12a. Braunification, Rubifaction, Ferrugination	3,4	Release of iron from primary minerals and the dispersion of particles of iron oxide in increasing amounts; their progressive oxidation or hydration, giving the soil mass brownish, reddish brown, and red colors, respectively
12b. Gleization	3,4	The reduction of iron under anaerobic "waterlogged" soil conditions, with the production of bluish to greenish gray matrix colors, with or without yellowish brown, brown, and black mottles, and ferric and manganiferous concretions

   * The four categories (see text) are: 1. additions to a soil body; 2. losses from a soil body; 3. translocation within a soil body; 4. transformation of material within a soil body.

it is considered to be a process of soil formation, that is, pedochemical weathering (Chapter 5). Pedogenic processes include gains and losses of materials from a soil body in accordance with the degradational, aggradational, or intermediate geomorphic character of the site as well as translocations within a soil body.

A soil, like a house, is put together or dismantled by specific processes. In view of the more disorderly and infinitely more complicated course of soil formation, one might better compare soil formation to the collapse and gradual disintegration, under influence of numerous organisms, of a combined chemical storehouse and natural history museum. Marbut (1935) suggested that nobody has ever seen a mature soil form in toto. Yet we do observe some processes in operation such as the cracking of clayey soils during dry periods and the incorporation of plant debris into soil by earthworms and ants.

The possible number of pedogenic events and combinations and interactions between them in soils is staggering. Although laboratory experiments can demonstrate that specific processes can produce specific soil features, the actual course of events within undisturbed soil will probably never be fully known. Some soil features are ephemeral, others enduring. A relatively permanent krotovina may be produced by rodent activity and burrow filling by soil in a single season. Sesquioxide concretions and nodules may last for millenia in soils. On the other hand, earthworm casts may be produced and dispersed in an A horizon in a matter of days or hours.

Work accomplished by geologic agencies in forming initial materials has influenced their mineralogy and texture (Simonson 1959). Soils may develop much more rapidly in deep, loose clastic deposits than in solid bedrock. In each principal kind of soil, discussed in later chapters, processes of soil formation occur in unique proportions, intensities, and sequences. Resulting soil features may differ not only in kind but also in durability ("pedologic half-life"). For example, carbonate concretions are abundant and lasting in Aridisols (Desert soils), persist in small numbers under the clay pan of some Albaqualfs (Planosols) in humid temperate regions, but are absent in associated Hapludalfs (Gray-Brown Podzolics) that lack the clay pan. The ensembles of pedogenic processes that characterize principal ecosystems and particular kinds of tesseras[1] (Jenny 1958, 1961) are known by pedological terms such as podzolization and calcification (Kellogg 1936) (Table 6.1). Fundamental pedogenic processes include both generalizations and specific subdivisions of these.

GENERAL FUNDAMENTAL PEDOGENIC
PROCESSES AND CONDITIONS. General fundamental processes and conditions (see Table 6.1) provide a framework for later consideration of more specific reactions and processes.

Two overlapping trends in soil development are horizonation and haploidization.[2] Horizonation (Hole 1961) includes the proanisotropic processes and conditions by which initial materials are differentiated into soil profiles with many horizons (as in D profiles of Northcote 1965). We think of the Spodosol (Podzol) profile (0-A2-Bhir-C) as being a horizonated soil (Fig. 6.1). Pronounced horizonation of some young red tropical soils is chiefly mineralogical and may not be visible to the observer of a freshly exposed profile. Haploidization includes proisotropic processes and conditions by which horizonation is inhibited or decelerated or by which horizons are mixed or disturbed. We think of a Vertisol (Grumusol; Black Cotton soil) profile with only A-C horizonation as being haploid or simple (U soil profiles of Northcote 1965). A Vertisol may be 10,000 years old (Thorp 1965), whereas a Spodosol may be no older than 2,250 years (Franzmeier and Whiteside 1963). It may be concluded that haploidization is at least as important a general process or condition as horizonation.[3]

RELATIVELY SPECIFIC FUNDAMENTAL
PEDOGENIC PROCESSES. Processes of soil formation include (1) additions of organic and mineral materials to the soil as solids, liquids, and

---

1. Jenny (1965) defines tessera as "the operational unit which we collect in the field, examine and analyze." It is usually smaller than a pedon. An eco-tessera samples both soil and vegetation. A soil tessera samples only the soil part of the ecosystem.
2. The term is from *haploidy,* the condition of being haploid, that is, simple in appearance or arrangement.
3. Horizonation and haploidization express both environmental inheritance and pedogenesis. In stratified pseudoprofiles, lithogenetic horizonation is more prominent than pedogenetic. South polar Aridisols (Antarctic Desert soils) exhibit the dominance of environmental aridity and frigidity over pedoturbation in a haploid soil.

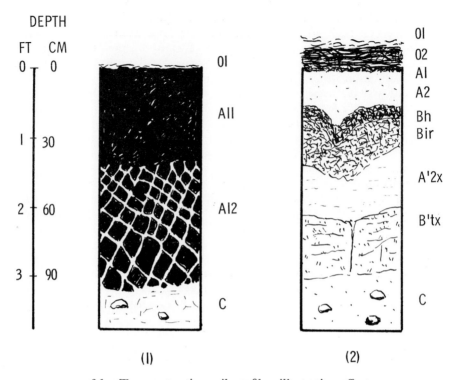

FIG. 6.1. Two contrasting soil profiles, illustrating effects, respectively, of haploidization (1) and horizonation (2). Soil profile number one is that of a Vertisol in which argillipedoturbation has homogenized the solum. Soil profile number two is of a bisequal Alfic Spodosol in which pedoturbation has been at a minimum, allowing for differentiation of numerous distinct soil horizons in the solum.

gases, (2) losses of these from the soil, (3) translocations of materials from one point to another within the soil, and (4) transformation of mineral and organic substances within the soil (Simonson 1959).

The list of terms in Table 6.1 begins with those that related to item 3, above. *Eluviation* and *illuviation* (words analogous to emigration and immigration) are phases of translocation distinguished as a sequence of movement from one part (usually a horizon) to another in a soil profile. More than half of the terms relate to translocation within a soil body. Two aspects of eluviation are mobilization and translocation. Illuviation involves the same translocation processes and their interruption by immobilization of material in a soil horizon.

Leaching is analogous to eluviation by solution but connotes removal from the entire solum, although it is common to speak of a leached horizon. It is a primary process, prerequisite in many soils to translocation of colloids. Calcium is prominent in Polynov's (1937) list of relative mobility indices of some soil constitutents, as based on analyses of igneous rocks and of dissolved loads of rivers: Cl⁻, 100; $SO_4^=$, 57; Ca^{++}, 3.00; Na^+, 2.40; Mg^{++}, 1.30; K^+, 1.25; SiO_2, 0.20; Fe_2O_3, 0.04; Al_2O_3, 0.02. On the basis of Clarke's (1908) geochemical data, these same units can be expressed as to relative abundance (by weight) in the earth's crust: Cl⁻, tr; $SO_4^=$, tr; Ca, 7; Na, 5; Mg, 2; K, 5; SiO_2, 100; Fe_2O_3, 13; Al_2O_3, 25. Trapping of Mg and K and Al_2O_3 in soil clay mineral lattices, and Fe_2O_3 in nodules, concretions, and ironstone deposits may account in considerable part for the low mobility of these constituents. Depth of leaching of carbonates in soil profiles is of first importance in landscapes originally blanketed with calcareous materials such as limestone, dolomitic sandstone, loess, and glacial drift.

Enrichment may be with respect to horizons but is usually taken in the sense of the entire soil receiving material from surrounding pedons as in depressional parts of the landscape or by air from remote areas. Under a humid temperate climate, such soils in areas of slightly leached initial material are typically enriched in plant nutrients and carbonate by laterally moving water from surrounding areas. In highly leached terrains without calcareous or fertile initial materials, the depressional soils are typically not enriched, but rather are the most acid and leached of the entire landscape.

Surficial erosion refers to lateral removal of surface layers of soil as by raindrop splash, run-off waters, wind, solifluction, creep, and other processes of mass-wasting.

Cumulization (Fr. *cumulus:* to heap) is offered as a term to express the accumulation of mineral material onto the surface of the soil by either air or water. In fact, this may be considered a geogenic rather than a pedogenic process. The effects of this process are most evident in depressional areas where material eroded from soils upslope has accumulated.

Decalcification is specifically used for the eluviation of carbonates within a soil body. The process may lead to the complete removal of carbonates from the entire profile, as is common in more humid areas, or be accompanied by calcification which is the accumulation of carbonates commonly observed in more arid regions. The general reaction involved in carbonate movement is as follows: $CaCO_3 + H_2O + CO_2 \rightleftharpoons Ca(HCO_3)_2$. Decalcification can be thought to occur when $H_2O + CO_2$ are present and the reaction moves to the right with the formation of the soluble bicarbonate. *Calcification* occurs when either CO_2 or H_2O is removed from the system and the reaction moves to the left.

Desalinization is most frequently used with reference to the removal by leaching of soluble salts from horizons or total soil profiles that have previously contained enough soluble salt so that plant growth was impaired. Therefore, it is a process that can be active only after soluble salts have accumulated, that is, salinization.

Salinization operates chiefly in subhumid, semiarid, and arid regions, and some coastal humid regions, wherever depressions are enriched in salts faster than they are leached. Solubilities in grams per 100 ml of pure water of common compounds (at 0 C unless otherwise indicated) are listed by

Hodgman et al. (1962): K_2CO_3, 112; $CaCl_2$, 59.5; $MgCl_2$, 54.3 (20 C); NaCl, 35.7; KCl, 27.6; $MgSO_4$, 26.0; Ca (HCO_3), 16.2; $FeSO_4$, 15.7; K_2SO_4, 12.0 (25 C); Na_2SO_4, 4.8; $CaSO_4$, 0.2; $MgCO_3$, 0.01; $CaCO_3$, 0.001 (25 C); FeS, 0.006 (18 C). Salt accumulation is preferential in depressional soil high in content of clay and low in permeability, with reduced leaching. Sulfates and chlorides are the predominant salts. Nitrates and borates occur rarely.

Alkalization involves the accumulation of sodium ions on the exchange sites of the clay. All cations in solution engage in a reversible reaction with the exchange sites on the clay and organic matter particles. This reaction can be represented by the following formula where X represents clay or organic matter exchange sites: $CaMg2Na$ $X \rightleftharpoons Ca^{++} + Mg^{++} + 2Na^{+} + X^{-6} + 3CO_3^{=} \rightleftharpoons Na_2CO_3 + MgCo_3 + CaCO_3$. From this equation and consideration of the carbonate solubilities (Na_2CO_3 is more than 100 times as soluble as Ca or Mg carbonate), it is seen that much of the Ca^{++} and Mg^{++} may be precipitated from solution before Na^{+} is precipitated. Thus, the concentration of Na^{+} left in solution to react with the exchange sites is high. The precipitation of the Ca^{++} and Mg^{++} in carbonate takes place first as the soil dries. The Na^{+} then attaches itself to the clay and organic matter exchange sites.

Dealkalization refers to the removal of the Na^{+} ions from the exchange sites. This process also involves much dispersion of clay. Dispersion occurs when the Na^{+} ion becomes hydrated. Much of the dispersion can be eliminated if Ca^{++} and/or Mg^{++} ions are concentrated in the water used to leach the alkali soil as they can replace the Na^{+} on the exchange complex.

Lessivage, the washing in suspension of fine clay and lesser amounts of coarse clay and fine silt down cracks and other voids in a soil body, is reflected in (1) depletion of A horizons of clay, (2) enrichment of the B horizon in clay content relative to the C and/or A horizon, (3) higher fine clay: total clay ratio in the B horizon than in the A horizon, and (4) presence of argillans in the B and C horizons. The mobile clay involved may be a product of weathering in the A horizon or may be of eolian origin added to the soil during development (Buol and Hole 1961; Khalifa and Buol 1968).

Pedoturbation is the process of mixing in the soil. A certain amount of mixing takes place in all soils. Seven kinds of pedoturbation are recognized. *Faunal pedoturbation* is soil mixing by animals such as ants, earthworms, moles, rodents, and man himself; *floral pedoturbation* is mixing by plants as in tree tipping that forms pits and mounds; *congellipedoturbation* is mixing by freeze-thaw cycles as in patterned ground of tundra and alpine landscapes; *argillipedoturbation* is mixing of materials in the solum by mass wasting movements of expansible clays; *aeropedoturbation* is mixing by movement of gases in the soil, during and after rains; *aquapedoturbation* is mixing by upwelling currents of water within the solum; *crystalpedoturbation* is mixing by growth of crystals, such as of halite (NaCl); *seismipedoturbation* is mixing by vibrations, notably earthquake tremors. Pedoturbation by biotic and physical agents can destroy argillans, embed fragments of them in the soil matrix, or even move clay back up into the A

TABLE 6.2 ❧ **Solubility of 2- to 5-** μ **quartz and amorphous silica**

Temperature	Quartz	Amorphous Silica
(° C)	*(ppm)*	*(ppm)*
5	≃6	≃60
25	≃11	≃120–140
50	≃25–30	≃200
75	≃50	≃300

SOURCE: Siever 1962.

horizon. Microscopic fragments of argillans have been observed in thin sections of material from large ant mounds (Baxter and Hole 1967).

Podzolization is defined as the process by which sesquioxides are translocated in a soil profile (Stobbe and Wright 1959) and can be explained in part by the solubilities of ferrous and ferric iron. The soluble ferrous iron forms at the sites of eluviation and the insoluble ferric iron forms at the point of illuviation. The role of chelation will be discussed in Chapter 20.

Laterization defined in a general way in Table 6.1 refers to processes that remove silica from the soil. High temperatures and extreme leaching favor rapid desilication and accumulation of iron (ferritization) immobilized in ferric oxide forms under oxidizing conditions. As compared to mid- and high-latitude regions, intertropical zones include areas of high temperatures and extreme leaching that favor rapid desilication and accumulation of iron (ferritization) immobilized in ferric oxide forms under oxidizing conditions. This may be particularly true, on a small scale, in desert varnish which consists of iron and manganese oxide stains, including residues from desilication of surfaces of stones (Hooke, Yang, and Weiblem 1969). The rate at which solubilities (in ppm) of 2- to 5-μ quartz and amorphous silica increase with rise in temperature were estimated by Siever (1962) (Table 6.2).

Decomposition and *synthesis* of minerals have been discussed in the preceding chapter. No attempt is made here to make a comparable exposition of transformation of organic materials, even if this were possible. A general scheme of organic matter decomposition is in Table 6.3.

Oosting (Edelman 1950) reported that in some forest soils in the Netherlands, the color of the root zone depends on the vegetation: "Under oak trees the soil is dark-brown, under beeches a bit more reddish; under fir trees orange-brown and under birches more yellowish." These observations

TABLE 6.3 ❧ **Some steps in the decomposition of organic matter**

Organic Compounds in Plant Tissue*		Hydrolysis Reactions		Decomposition Compounds	Further Decomposition	Resistant Complex
Starch, cellulose, hemicellulose, pectin, uronic acid	+	HOH	=	Simple sugars	Polymerization → Humus (chain building) or (humification)	
Protein	+	HOH	=	Amino acids		
Lignin, waxes, resins	+	HOH	=	Phenols		

* Ease of decomposition decreases from top to bottom of the list. Components, listed in order of increasing degree of polymerization are: fulvic acid (yellow), humic acid (brown), humic acid (black). These three acids are sometimes considered to be characteristic, respectively, of mor, moder, and mull.

suggest that different organic compounds have been synthesized at the different sites.

Melanization and *leucinization* refer to changes in color value in soil, whether caused by addition or losses, respectively, in the content of organic matter (the common case), or by transformations from dark-colored (melanized) to light-colored (leucinized) organic compounds or vice versa.

The word *littering* is offered here for the accumulation of vegetable and associated faunal debris (O1 horizon) on the mineral soil surface. Their further decomposition or *humification* is often expressed in an O2 horizon.

Paludization is the accumulation of a thicker mass of organic materials in a poorly drained site where preservation under anaerobic conditions has allowed a net gain through time. This process is geogenic in that it is an accumulation of initial soil material. Subsequent pedogenic chemical, physical, and biologic changes in organic soils (Histosols) constitute the complex process of *ripening*. Release of mineral components of organic matter through decomposition is the complex process of *mineralization*. The vast stores of nutrients in the standing crop of a forest ecosystem and in soil organic matter in grassland ecosystems are unavailable to plants except through mineralization by fire or by the slower oxidation of decay.

The general trend of reddening of subsoils of uplands along transects from the polar to equatorial regions is an expression of dispersion through the soil and progressive oxidation of iron. Three processes, *braunification*, *rubifaction*, and *ferrugination* (development of brown, reddish brown, and reddish soil colors), are self-explanatory as to soil appearance, although imperfectly understood as to the exact mechanisms involved. *Gleization* in poorly drained soils involves the reduction of iron, its segregation into mottles and concretions, or its removal by leaching from the gleyed horizon. Formation of iron sulfide (FeS) is common in gleyed soils, which upon drainage may become acid from formation of H_2SO_4 through oxidation.

Among other elements in soils are manganese (more mobile than Fe) and titanium (less mobile than Fe). Manganese oxide is commonly found as black coatings on surfaces of joints in the C horizon below the iron oxide-stained B horizon in Hapludalfs. Titanium oxides are often found intimately associated with ferric iron oxide. The naming of processes involving specific ions or ecological niches could become almost endless. At present this seems of little value in as much as their use would not have general application.

A SIMPLIFIED MODEL OF
SOIL SOLUM DEVELOPMENT.
Consideration of a simplified theoretical model can give us an intimation of the complexities of soil genesis. The model presented here is of an open system with respect to water and some soluble products of weathering of feldspar and biotite, but a closed one in regard to other materials. Organic matter is not considered in the calculations. We assume that the soil has developed from a column of homogeneous material originally like the present C horizon.

Coarse sand grains of resistant minerals are used as "index minerals."[4] Because of their large size, the coarse sand grains of the index minerals (hereafter referred to as IM) are assumed not to have been affected during pedogenesis. They have moved only insofar as collapse or expansion of entire horizons have raised, lowered, and separated them. We can calculate gains and losses of mineral matter for the first three horizons by comparing the weight of gains per cc in a horizon with their weight in the C horizon. Increase or decrease of populations of IM grains per unit volume in the solum indicates, respectively, collapse or expansion of the soil material as compared to the C horizon.

The horizon designations of our model and corresponding data are in Table 6.4.

In the model (Table 6.4) we have designated four horizons of equal thickness to simplify calculations. The bulk density of each horizon is determined so that all measurements can henceforth be made on a weight basis. By optical and particle size distribution methods the index mineral, nonclay, and clay contents are determined for each horizon (columns 5, 7, 8, respectively). The IM factor (column 6) is computed by setting the IM content in each horizon (column 5) equal to the IM content of the C horizon, that is, the assumed composition for the entire profile prior to soil formation. By multiplying the C horizon values of present total weight, clay, and nonclay contents (columns 4, 7, and 8) by the IM factor (column 6), we obtain a value for the original contents in each horizon (columns 9, 10, and 11). To determine the nonclay content change in each horizon (column 12), columns 8 and 11 are compared. Note that in the example 20 grams of nonclay are lost from the profile, most of it from the upper horizons. A similar comparison of columns 7 and 10 calculates the change in clay content. In the example the nonclay loss equals the clay gain. When in actual calculations the clay formed is less than nonclay loss, erosion or leaching losses can be calculated. Also, in the example column 14 (sum of columns 12 and 13) estimates the net clay translocation within the profile, assuming that the clay is formed at the site of the nonclay loss.

Real soils, unlike the theoretical model just demonstrated, are open to both gains and losses of materials. Although the model used has been prepared for simplicity, it illustrates profile trends that are common in actual examples.

---

4. For reports on the use of index minerals in investigations of actual soils, see Barshad (1964), Graham (1950), and Haseman and Marshall (1945).

LITERATURE CITED

Barshad, I. 1964. Chemistry of soil development, pp. 1–70. In F. E. Bear (ed.), Chemistry of the soil. Reinhold, New York.

Baxter, F. P., and F. D. Hole. 1967. Ant *(Formica cinerea)* pedoturbation in a prairie soil. Soil Sci. Soc. Am. Proc. 31:425–28.

Buol, S. W., and F. D. Hole. 1961. Clay skin genesis in Wisconsin soils. Soil Sci. Soc. Am. Proc. 25:377–79.

TABLE 6.4 ❧ A simplified model of soil solum development using an index mineral calculation

(1) Horizon	(2) Thickness of 1-cm² Column	(3) Bulk Density	(4) Present Total Weight Oven-dry	(5) IM Weight	(6) IM Factor	(7) Present Clay Oven-dry Weight	(8) Present Nonclay Oven-dry Weight	(9) Original Total, Oven-dry Weight	(10) Original Clay Oven-dry Weight	(11) Original Nonclay Oven-dry Weight	(12) Change in Nonclay Content	(13) Change in Clay Content	(14) Net Clay Translocation
	(cm)	(g/cc)	(g)	(g)		(g)	(g)	(g)	(g)	(g)	(g)	(g)	(g)
A	20	1.0	20	3.2	1.6	5	15	40	8	32	−17	− 3	−20
B2t	20	1.5	30	1.2	.6	20	10	15	3	12	− 2	+17	+15
B3t	20	1.25	25	1.6	.8	10	15	20	4	16	− 1	+ 6	+ 5
C	20	1.25	25	2.0	1.0	5	20	25	5	20	0	0	0
			100	8.0		40	60	100	20	80	−20	+20	0 loss

Clarke, F. W. 1908. The data of geochemistry. U.S. Geolog. Survey Bull. 330.

Edelman, C. H. 1950. Soils of the Netherlands. North-Holland Pub. Co., Amsterdam.

Franzmeier, D. P., and E. P. Whiteside. 1963. A chronosequence of Podzols in northern Michigan. I. Ecology and description of pedons. Mich. State Agr. Exp. Sta. Quart. Bull. 46:2–20.

Graham, E. R. 1950. The plagioclase feldspars as an index to soil weathering. Soil Sci. Soc. Am. Proc. 14:300–302.

Haseman, J. F., and C. E. Marshall. 1945. The use of heavy minerals in studies of the origin and development of soils. Missouri Agr. Exp. Sta. Res. Bull. 387.

Hodgman, C. D., R. C. Weast, R. S. Shankland, and S. M. Selby. 1962. Handbook of chemistry and physics. Chem. Rubber Pub. Co., Cleveland, Ohio.

Hole, F. D. 1961. A classification of pedoturbations and some other processes and factors of soil formation in relation to isotropism and anisotropism. Soil Sci. 91:375–77.

Hooke, R. L., H. Yang, and P. W. Weiblem. 1969. Desert varnish: An electron probe study. J. Geol. 77:275–88.

Jenny, H. 1958. Role of the plant factor in the pedogenic functions. Ecology 39:5–16.

———. 1961. Derivation of state factor equations of soils and ecosystems. Soil Sci. Soc. Am. Proc. 25:385–88.

———. 1965. Tessera and pedon. Soil Survey Horizons 6:8–9.

Kellogg, C. E. 1936. Development and significance of the great soil groups of the United States. U.S. Dept. Agr. Misc. Publ. 229.

Khalifa, E. M., and S. W. Buol. 1968. Studies of clay skins in a Cecil (Typic Hapludult) soil. I. Composition and genesis. Soil Sci. Soc. Am. Proc. 32:857–61.

Marbut, C. F. 1935. Soils: Their genesis and classification. Pub. 1951 by Soil Sci. Soc. Am., Madison, Wis.

Northcote, K. H. 1965. A factual key for the recognition of Australian soils, 2nd ed. C.S.I.R.O. Australia, Div. of Soils, Div. Rep. 2/65.

Polynov, B. B. 1937. The cycle of weathering. (Transl. by A. Muir.) Murby & Co., London.

Rode, A. A. 1962. Soil Science (Pochvovedeniye). (Transl. by A. Gourevich.) Israel Prog. for Sci. Trans., Jerusalem. Available U.S. Dept. Commerce, Washington.

Siever, R. 1962. Silica solubility, 0–200° C and the diagenesis of siliceous sediments. J. Geol. 70:127–50.

Simonson, R. W. 1959. Outline of a generalized theory of soil genesis. Soil Sci. Soc. Am. Proc. 23:152–56.

Stobbe, P. C., and J. R. Wright. 1959. Modern concepts of the genesis of Podzols. Soil Sci. Soc. Am. Proc. 23:161–64.

Thorp, J. 1965. The nature of the pedological record in the Quaternary. Soil Sci. 99:1–8.

Vilenskii, D. G. 1957. Soil science, 3rd enlarged ed. State Teachers' College Pub. House, Ministry of Culture, USSR, Moscow. (Transl. by A. Birron and Z. S. Cole.) Israel Prog. for Sci. Trans., Jerusalem, 1960. Available U.S. Dept. Commerce, Washington.

❧

Soil Environment: External Factors
of Soil Formation

IT HAS BEEN pointed out that the character and development of soils are not controlled by genes but rather by external factors (Crowther 1953). A study and classification of these factors are great helps in understanding the soils.

It is much less possible to observe soil formation than it is to observe growth of plants and animals. Catastrophic physical phenomena which are easily observable, such as volcanic eruptions, spectacular erosion by wind and water, and earthquake disruptions, are geologic but not pedologic. Soil formation is so hidden from view or so slow as to elude observation except for a limited number of phenomena, particularly at or near the surface. We can, however, observe differences in factors of soil formation and correlate these with soil differences. Thus, it is that there is so much interest among soil geneticists in the environment of a soil. A zoologist who is studying an elephant species carefully studies both the animal and its environment and can depend on the phenomena of inheritance of genetically determined characteristics in his investigations of this animal. Many students of soil bodies consider the environment to be a more dependable index of the soil than the "parent material," which may be changing to soil at the interface between bedrock and C horizon, in the case of soils developed in situ; and may be changing from initial to secondary mineral at points throughout the solum as in the case of soils formed from unconsolidated geologic deposits rich in silicates.

DEFINITIONS OF FACTORS
OF SOIL FORMATION.
A factor of soil formation is an agent, force, condition, or relationship, or combination of these, which influences, has influenced, or may influence a parent material of a soil, with the potential of changing it. Examples of factors of soil formation are radiant solar energy or a condition of saturation of the soil with water. A detailed list of factors of soil formation would be very long indeed. Therefore, a few factors have been selected by soil geneticists, from Dokuchaev to the pres-

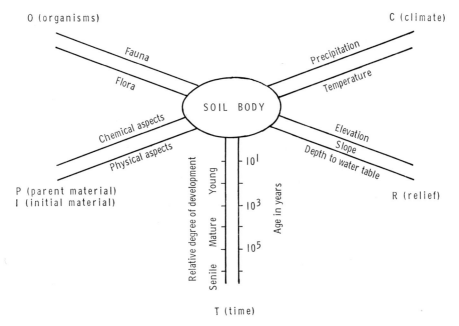

FIG. 7.1. "Flower" diagram of factors of soil formation.

ent, to serve as a basis for organizing pedological investigations and data gathered.

Five general factors of soil formation are indicated in the diagram in Figure 7.1: parent material, relief, climate, organisms, and time. Four of these (all but relief) were included in 1898 by Dokuchaev in his equation of the factors (Jenny 1961):[1]

$$S = f(cl,o,p,)t^0 \tag{7.1}$$

where S represents soil; cl, climate of a region; o the organisms (both plants and animals); p the "geologic substratum"; and t^0 the relative age of soil (young, mature, or senile). He considered the factors to be interdependent. Sibirtsev emphasized the importance of climate as a factor of soil formation. He grouped soils under three headings: zonal (soils corresponding in their distribution to climatic belts), semizonal (soils, usually referred to as azonal, which are much more weakly developed than zonal soils), and intrazonal (soils greatly influenced by local conditions, such as excess of water, salts, or carbonates).

Joffe (1936) emphasized mass and energy as factors of soil formation. Marbut (1935) stressed environment. Glinka (1927) referred to forces not

---

1. From "Significance of N. M. Sibirtsev's Work in the Founding of Soil Science" as written by Professor S. S. Sobolev as an introduction to N. M. Sibirtsev. Selected Works (Izbrannye sochineniya), vol. 1. Soil Science (Pochvovedeniye). Published in Moscow, 1951. Transl. from Russian by N. Kaner and published by IPST, Jerusalem, 1966 for USDA and NSF. Available from U.S. Dept. Commerce, Washington, D.C.

subject to quantitative measurement. Some soil scientists applied to soil genesis the concept of cause and effect as developed by John Stuart Mill (1925). Baldwin, Kellogg, and Thorp (1938) used the three-fold grouping of Sibirtsev. Jenny (1941) considered the five factors named by Dokuchaev as independent variables. Crocker (1952) considered water table (w) to be a dependent factor of soil formation, as did Russell and Rhoades (1956).

Rode (1961) wrote of eight factors of soil formation, the additional three being gravity, water (surface, soil, and ground), and man. The effects of many other phenomena such as daily earth tidal movements are considered to be inconsequential in soils, and so are not listed among factors of soil formation.

V. R. Williams (1949) stressed the role of biological processes in soil genesis, most prominently expressed in the Chernozem.

Space is included in Figure 7.2 as an element in parent material, relief, climate, and organisms. John Stuart Mill (1925) stated that all phenomena of nature exist in two distinct relationships to one another: (1) simultaneity (spatial, geometric relationship), and (2) succession (causation and effect). The spatial relationship in soil involves size of pores, which is related to the kinds of organisms living in the soil. A preponderance of boulders and other coarse fragments in a soil favors development of profiles which are deeper than in nonstony soils, because percolation is funneled by coarse

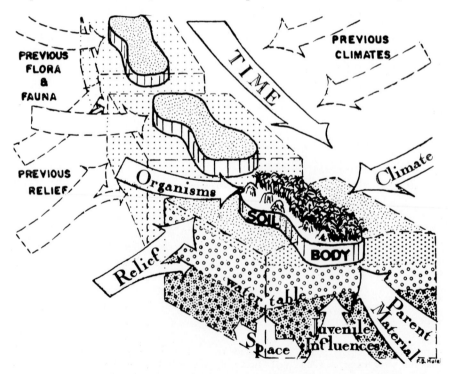

FIG. 7.2. "Freight train" diagram of factors of soil formation.

fragments through a smaller mass of soil material. A one-inch-thick layer of glacial drift over a hard, polished granite surface allows for less soil profile development during postglacial time than do ten feet of glacial drift over granite. The horizontal dimension of a topographic sequence of soils can be so limited on a particular hillside that there is not room for all common members of a toposequence to develop. Each ecological niche at the land surface has a characteristic soil. Exact location in space of each soil in terms of latitude and longitude is helpful.

Juvenile influences, which include slight emanations of heat, vapor, and liquids from deep-seated sources in the bedrocks of the earth, are part of both parent material and soil climate. Severe alteration of igneous rocks along joints, and the upward flow of heat from rock to soil in the winter, are examples of phenomena related to juvenile influences. Soils around geysers and hot springs receive unusual amounts of heat from the earth. Succession of states in soils is treated in Chapter 12.

STATE FACTOR EQUATIONS. Jenny (1961) outlined the genesis of equations which concisely express "generalizations of soil behavior related to genetic factors." The equation of Dokuchaev quoted above conceives of the factors simply as soil formers. Jenny's equations of 1941 and 1958 describe the relationship between a soil property and "state factors," which are groups of factors as follows: environmental climate, "species germules" of organisms (not their actual growth), topography (including hydrologic features such as the water table), parent material (defined as the state of soil at soil formation time zero), time (age of soil, absolute period of soil formation), and additional, unspecified factors.

$$S = f\ (cl,o,r,p,t,\ .\ .\ .) \tag{7.2}$$

The state factors define the state of the soil system. But what is the soil system? We can define it on the basis of an arbitrarily chosen volume. This could be the solum of a soil body or a pedon in it (Soil Survey Staff 1960). Or it could be an entire ecosystem of a component tessera (Jenny 1958). The second choice avoids the impossible task of trying to separate living material as nonsoil from inanimate material, true soil, in an ecosystem where they interpenetrate inextricably. The equation is most usefully interpreted to relate soil properties to factors in the entire ecosystem.

The ecosystem is an open system with influxes and outfluxes of energy and matter (Jenny 1961). Influx of energy includes solar radiation, heat transfer, and entropy transfer from the outside heat reservoir. Outflux of energy is in the form of heat radiation and reflection of light. Influx of matter involves gases entering the ecosystem by diffusion or mass flow (wind); water in liquid and solid forms entering the ecosystem from above, below, or sides; solids dispersed and dissolved in water; solids dispersed and moved in air (wind); organisms that immigrate into the ecosystem. All these things may also move out of the ecosystem and then constitute outflux. Solifluction and other forms of mass movement are special forms of outflux. Measurements of inflow and outflow in ecosystems can be made with lysimeters and biotrons over limited periods of time. Valid estimates of fluxes in natural ecosystems over extended periods of time can be obtained only by careful analysis of the dynamics of an ecosystem.

If l is a symbol for any property of the ecosystem, then (l_{in}, l_{out}) may signify a flux or transport of matter and energy along gradients or potential differences against resistance to permeation through an interface or boundary plain in the ecosystem. Rain falling on a smooth granite surface results in no flux into the granite because of resistance to permeation (m) at the interface. Rain on loose sand initiates flux into the sand because of a low value for m. Jenny's equation

$$\text{flux} = -\frac{(P \text{ outside} - P \text{ inside})}{\Delta x} m \tag{7.3}$$

refers to the difference between outer and inner potential or gradient, the thickness of the boundary plane (Δx) and its resistance to permeation (m). The flux is in the opposite direction to the gradient, hence the negative sign.

Delineation of an entire ecosystem or a component tessera as separate from the surrounding environment makes possible a distinction between external flux potential (P_x), such as external climate (cl), and internal ecosystem state-dependent potentials, such as the climate of the vegetation or of the soil (cl'). The biotic factor (o) is another external potential. It includes all species active and in dormant forms as seeds and spores that are present in the ecosystem at or just after time zero and that can migrate into or be carried into the ecosystem at any time during its history. But the actual array of species present in an ecosystem (after time zero) is an internal state-dependent potential (o'). Deposits of dust by windstorms, illuvium by floods, and increments of fertilizers by farmers are among many other external flux potentials that impinge upon ecosystems and the soils in them.

Each ecosystem has been acted upon for a certain period of time by the external flux potential (P_x). The "initial mineral and organic matrix of the soil portion of the ecosystem" (Jenny 1961) is the parent material (p), defined in terms of a standard state of pressure and temperature. New materials and states derived from these through the influence of soil formation are state dependent (p'). The configuration of the system (r) comprises the topographic features, such as slope and aspect, and hydrologic features, such as water table. Changes in configuration during soil genesis are state dependent (r').

Jenny's general state factor equation

$$l, s, v, a = f(L_0, P_x, t) \tag{7.4}$$

states that a property (l) of the ecosystem, a soil property (s), a vegetation property (v), or an animal property (a) is a function of the state of the larger ecosystem (L) at time zero (L_0), the external flux-potential (P_x) and the age of the system (t).

In a given landscape at a given moment, the relative effectiveness of the classic five factors of soil formation (cl,o,r,p,t) is distinctive. A factor that has a narrow range throughout the landscape is ineffective in producing variation in a given soil property from place to place. A factor that has a wide range in the landscape may be ineffective in part of the range and effective in another. The following mathematical expression from Jenny (1961)

$$\int_{c}^{d} \frac{\partial N}{\partial F} \, dF \tag{7.5}$$

states that a factor (F) has a range $(c \rightarrow d)$ in a landscape and that it relates to a property $(N =$ average % nitrogen in surface soil) as may be shown by a curve on a graph. Where the curve has a slope of 0, the factor is ineffective. Where the curve has a steep slope, the factor is effective. Jenny has applied this to soil properties in ecosystems in India (Jenny and Raychaudhuri 1960).

This discussion of factors of soil formation may well include a consideration of the relevance of Whitehead's concept of "scientific organism." In this context a soil body or soil association can be considered to be a kind of scientific organism with a degree of dynamics of its own that is not simply imposed upon the soil by external factors (P_x). For example, a tropical, black, self-mulching soil may be self-liming, self-land-forming, and may act as a complex of valves which alternately admit and shut off the flow of percolating water. To the extent that a soil develops its own dynamic, it participates as a factor in its evolution and exerts an influence on the environment, modifying microclimate, microtopography, and vegetative cover, and altering the course and speed of the geomorphic erosion-sedimentation cycle.

SOIL SEQUENCES. Jenny has recognized the interdependence of the state factors of soil formation but has looked for situations in which, because all factors but one are "ineffective" in the landscape, the influence of the one variable factor is revealed. Sequences of soils can be sought which are dominated by single factors. There are climo-, bio-, topo-, litho- and chronosequences of soils, as will be illustrated in the following chapters.

A study of 95 soil individuals from California, highly stratified as to genetic factors, showed notable correlations of soil N content with parent material and clay mineralogy; and of soil C content with climate and flora. A greenhouse pot experiment, in which barley was grown in these soils, demonstrated that genetic factors correlated with crop yields (Jenny, Salem, and Wallis 1968).

**RELATING FACTORS OF SOIL
FORMATION TO PEDOGENIC PROCESSES.** Environmental factors set limits and directions for soil development just as these factors determine

the kinds of houses that Eskimos, North American pioneers, and African pygmies build. But the actual steps by which soils (or houses) are formed constitute pedogenic processes. Jenny's references to fluxes into and out of and within soils are in fact references to processes of soil formation. Factors and processes external to the ecosystem influence and are influenced by parallel factors and processes within the ecosystem, both in the vegetation and in the soil.

The processes of soil formation enumerated in Table 6.1 may be expressed in terms already justified in this chapter. All pedogenic processes are uneven in operation in some degree. In the regions of highest rainfall, leaching does not proceed at a constant rate throughout the year. Congellipedoturbation may operate only once a year, and significant movement of colloids and their deposition to form clay skins may take place only once in many years. Yet the effects of such uneven processes may endure for long periods.

LITERATURE CITED

Baldwin, M., C. E. Kellogg, and J. Thorp. 1938. Soil classification, pp. 979–1001. In Soils and men, yearbook of agriculture. U.S. Dept. Agr. U.S. Govt. Printing Office, Washington.

Crocker, R. L. 1952. Soil genesis and the pedogenic factors. Quart. Rev. Biol. 27:139–68.

Crowther, E. M. 1953. The sceptical soil chemist. J. Soil Sci. 4:107–22.

Glinka, K. D. 1927. The great soil groups of the world and their development. (Transl. by C. F. Marbut.) Edwards, Ann Arbor, Mich.

Jenny, H. 1941. Factors of soil formation. McGraw-Hill, New York.

———. 1958. Role of the plant factor in the pedogenic functions. Ecology 39:5–16.

———. 1961. Derivation of state factor equations of soil and ecosystems. Soil Sci. Soc. Am. Proc. 25:385–88.

Jenny, H., and S. P. Raychaudhuri. 1960. Effect of climate and cultivation on nitrogen and organic matter reserves in Indian soils. Indian Council Agr. Res., New Delhi, India.

Jenny, H., A. E. Salem, and J. R. Wallis. 1968. Interplay of soil organic matter and soil fertility with state factors and soil properties, pp. 5–37. In Study week on organic matter and soil fertility, Pontificiae Academiae Scientarius Scripta Varia. John Wiley & Sons, New York.

Joffe, J. S. 1936. Pedology. Rutgers Univ. Press, New Brunswick, N. J.

Marbut, C. F. 1935. Soils: Their genesis and classification. Publ. 1951 by Soil Sci. Soc. Am., Madison, Wis.

Mill, J. S. 1925. A system of logic, 8th ed. Longmans, Green and Co., London.

Rode, A. A. 1961. The soil-forming process and soil evolution. Edited by V. S. Volynskaya and K. V. Krynochkina. (Transl. by J. S. Joffe.) Israel Prog. for Sci. Trans., Jerusalem. Available U.S. Dept. Commerce, Washington.

Russell, J. S., and H. F. Rhoades. 1956. Water table as a factor in soil formation. Soil Sci. 82:319–28.

Soil Survey Staff. 1960. Soil classification, a comprehensive system—7th approximation. U.S. Dept. Agr. U.S. Govt. Printing Office, Washington.

Williams, V. R. 1949. Basic soil science for agriculture (Pochvovedeniye Zemledelie
s osnavami pochvoedeniya), Moscow. (Transl. from Russian by N. Kaner.)
Israel Prog. for Sci. Trans., Jerusalem, 1968. Available U.S. Dept. Commerce,
Springfield, Va.

✿

Parent Material: Initial Material of the Solum

$\mathcal{L}$

P ARENT MATERIAL was cited as a significant soil-forming factor by the Russian pioneers in pedology (Dokuchaev 1883). Indeed, many of the earliest approaches to soil survey and classification were based on the geology and composition of the soil-forming material (Richthofen 1886; Thaer 1809, 1810, 1812). In these early studies, soils were often designated as "granite soils" or "glacial soils" and similar terms indicating the geologic origin and composition of the initial material. Following this pioneering work, extensive research on the effects of parent material on soil properties was carried out by the Russian soil scientist and geochemist Polynov (1930), who pointed out the controlling effects on soil properties exerted by the parent material. In the previous chapter, we indicated how Professor Jenny (1941) has conducted a systematic analysis and survey of the relationships between important soil properties and the parent materials from which the soils formed, as one of the soil-forming factors. In his book, Jenny found it desirable to formulate a clear definition of parent material as an *independent* soil-forming factor, with accompanying mathematical expression, as described in Chapter 7. Thus he defined parent material as "the state of the soil system at time zero of soil formation," that is, the physical body of soil and its associated chemical and mineralogical properties at the starting point of effect of a particular set of other soil-forming factors (environment and landscape position). We can see that a previous soil or a previously weathered rock mass (saprolite) could be "parent material" in this definition and concept.

This is an important point for us to keep in mind—that the present, modern soil as we see and study it owes its properties to (1) the composition of the surficial layer present when the current array of environmental factors started their effect and (2) the modifications resulting from the effect of these environmental factors over time.

In general, the younger the soil the more influence and relationship there is of the soil parent material. As weathering and pedogenic processes proceed, the imprint of the initial material is less and less. In extremely weathered and old soils, there is relatively little influence of the initial material unless it was of extreme composition such as a sterile quartz sand. What we need to sort out, then, is the nature of the initial material and its likely persisting influence on the soil, the kind and amount of "pre-weathering" of the initial material before it started to serve as parent material of the soil we have under study, and the effect of the dynamic environmental or active soil-forming factors on the parent material of this soil. These types of studies are difficult to make because of the complex environmental history (climatic and vegetational changes in the recent geologic past) of so many of our soils, and because of problems in separating parent material influences on soil properties from other effects (of natural drainage, climate, vegetation). This is one of the many problems in soil science—problems which make it such a challenging and fascinating field! In the discussion that follows, we will first consider the generalized effects various parent material can have on soils. Then we will consider attempts to separate parent material effects from those of other soil-forming factors.

INFLUENCES OF ROCK TYPES
ON SOIL PROPERTIES. In this discussion, we have organized the generalizations somewhat according to classical subdivisions of rock types: sedimentary, igneous, and combinations of mineralogically similar metamorphic and igneous rocks. We have then subdivided within these classes. Reference to some of the standard manuals or texts on rock classification will aid in following the discussion if the reader is unfamiliar with basic properties of the rock types (for example, Travis 1955).

SEDIMENTARY ROCKS. Unconsolidated glacial and loessial deposits are important soil parent materials in the temperate regions, especially in North America and northern Europe.

Glacial till tends to reflect the lithology and composition of the material over which the glaciers passed. In the North American Midwest, loamy textures are predominant as the glaciers passed over residuum from limestone and shales. In New England, the northern lake states of the United States, and northern Europe, the tills tend to be of a sandy loam texture because the glaciers have passed over sandstones and granites, yielding a coarser, quartzose, more siliceous and acidic glacial deposit. Illite (or clay mica) is an important clay mineral in these deposits and is accompanied by moderate to dominant amounts of montmorillonite and small amounts of kaolin, vermiculite, and chlorite. If the glacial till is derived primarily from limestone and shale residuum (as in the American Midwest), montmorillonite may be the dominant clay mineral, and the pH and base saturation are high, with little modification due to pedogenesis because these deposits are on young landscapes under a mild climate. If the till is derived from sandstones and granites, the pH and base saturation tend to be low, and this is reflected in the acidity and low-base status of the soils.

Loess is generally thought to be wind-blown silt derived from glacial floodplains, though on the Moscow Plain silts previously thought to be loess or loesslike may be water deposited. These are extensive and important soil parent materials in the North American Midwest and in western Europe. Before modification by pedochemical weathering, the loessial parent materials contain large amounts of silt, about 10 to 20% clay (content of clay in part depends upon distance of the deposit from the loess source), have a high content of weatherable minerals, show high base saturation, or are calcareous, and their clay mineralogy is ordinarily dominated by montmorillonite with varying amounts of illite (clay mica or "hydromica") and perhaps some vermiculite. As these parent materials are on relatively young landscapes in a mild climate, they have been only slightly modified by pedochemical weathering. Consequently, soils formed from these initial materials are silty, high in nutrient reserve, and generally have excellent physical properties.

Unconsolidated coastal plain sediments are important soil parent materials along the Gulf and Atlantic coasts of North America (for example, 40% of North Carolina is coastal plain) and elsewhere outside the glaciated zone where the coast is not precipitous, as in parts of northern South America and Central America. These are in greater part of marine origin at lower elevations nearer the coast, and of alluvial-colluvial-deltaic origin farther inland. In general, these are secondary sediments derived from older, more highly weathered landscapes. They tend to occur in regions of high rainfall (though there are exceptions—as the dry coast of Peru) and temperature. Thus, the initial materials are likely to be acid, low (to moderate) in content of weatherable minerals, and of variable texture. They range from sand to clay, depending on the environment of deposition. The alluvial-colluvial-deltaic coastal plain sediments tend to be rich in kaolinite, acid, and of low nutrient reserve. Those initial materials derived by marine (sea) deposition tend to be clayey and to contain appreciable amounts of montmorillonite, if lagoonal deposits. If the initial materials were deposited as off-shore bars or beaches, they tend to be coarse and acid.

Limestones and dolomites are greater than 50% carbonates by definition, and the balance of the rock is composed of silt and clay and/or quartz and/or iron and other "contaminants." Soil forms in the residuum left from the dissolution of the carbonates during weathering. Thus, the kind of soil formed is related to the dominant kind of "impurities" in the limestone. If the limestone is a "dirty" one (rich in clay), then clayey and impermeable soils are the result. Due to the low rate of leaching through these dense soils, they ordinarily are not highly leached and are high in pH and base saturation. If the limestone is rich in sand and chert (granules of microcrystalline or amorphous silica), the soils tend to be coarse loamy, gravelly, acid, and low in base status. If the limestone is rich in iron impurities such as hematite, a red soil results, generally acid in reaction if in a humid climate.

Sandstones (quartzose sandstones or orthoquartzites) by definition contain more than 50% sand-size particles predominantly of quartz. The cements are variously silica, iron, and carbonates, which together with the "impurities" (such as feldspar or mica content) have great influence on the kind of soil formed from sandstones. In general, soils formed from these rocks are of coarse texture (especially in surface horizons) and highly permeable. They tend to be low in base status, nutrient reserve, and pH, especially if formed in humid climates in which the high permeability promotes acid leaching. The soils tend to be deep, unless they are forming in residuum from sandstones cemented with silica, in which case they are shallow because of the slow rate of dissolution of this type of cement. Those soils formed from sandstone with some iron cement tend to be reddish in color. If the feldspar content is greater than 25%, the rock is described as an arkosic sandstone, or as arkose if the feldspar content is very high. Soils formed in residuum from rocks such as these tend to be clayey because of the feldspar weathering to clay, and high in nutrient reserve because of nutrient release from the feldspars.

Shales are laminated or fissile (thinly bedded or layered), somewhat indurated or hardened rocks called clay shale, claystones, or mudstones if comprised of predominantly clay, or siltstones if predominantly of silt-size particles. In general, their mineral composition is of layer silicates, feldspars, quartz, small amounts of mica, and sometimes calcium carbonate. The layer silicates tend to be illite, except in very acid shales that ordinarily are rich in kaolin clay. Soils formed in residuum from clay shales are generally fine textured (clayey), relatively impermeable, and consequently little leached and with shallow sola. They are of high base status and pH, unless formed from the black or gray acid shales occurring in connection with coal or lignite deposits. Illite and montmorillonite generally are the main clay minerals in soils formed from these shales except for the acid shales which tend to give rise to kaolinitic soils—the kaolinite inherited from the parent rock. Soils formed from the siltstones tend to be of medium (silty or loamy) texture, with medium to high nutrient reserve.

LIGHT-COLORED SILICEOUS CRYSTALLINE ROCKS. These rocks include the more "acidic" quartzose igneous and metamorphic rocks. For our discussion, they are separated into broad groups according to general mineralogical composition.

Granites and granite gneiss average approximately 25% quartz, 65% or less orthoclase (potassium) feldspar, with lesser amounts of mica (ordinarily predominantly muscovite), and small amounts of hornblende. These rocks may show slight differences in weathering pattern due to differences in structure—the gneisses being banded, with mineral segregation in the bands. Basically, though, they tend to produce the same kinds of soils. Soils formed from saprolite derived from these rocks by geochemical weathering tend to be coarse (coarse loamy), especially in surficial horizons. They tend to be friable and permeable soils, generally acid and of low base status because of the high quartz content of the parent rock and the acid leaching resulting from the coarse textures. Mineral nutrient reserve tends to be low in these soils, except in cool climates. The soils tend to be yellow or a yellowish brown in color because of the low iron content of the parent rocks. Clay mineralogy in these soils tends to be highly kaolinitic in warmer, more humid climates, and vermiculite-illite-montmorillonite in the cooler and/or more arid climates.

Schists are foliated (thin platy), metamorphic rocks rich in mica (or chlorite or sericite), with varying amounts of quartz, and with very small amounts of other weatherable minerals. Because of lower quartz content, soils formed in saprolite from mica schists tend to be silty and less coarse than those formed from granitic saprolite. They tend to have a high potassium reserve in the micas, except in older, more humid landscapes where there has been long-term weathering. The clay minerals in these soils are predominantly illite (clay mica) and vermiculite, except in older and/or more highly weathered soils in which kaolinite is predominant, and in soils with dry seasons in which significant montmorillonite is present. Soils from sericite schist saprolite are likely to be very silty but acid, and rich in exchangeable aluminum because of its derivation from the breakdown of the aluminous sericite. Soils formed in residuum from chloritic schists are likely to be clayey, plastic, and rich in montmorillonite, and possibly contain excessive amounts of magnesium.

DARK-COLORED FERROMAGNESIAN
(MAFIC OR BASIC) ROCKS. In this group of parent rocks, we include andesites, diorites, basalt, and hornblende gneiss. All are rich in iron- and magnesium-bearing minerals and calcic plagioclase feldspars that weather rapidly (Chapter 5), yielding a good deal of clay and free iron. These minerals also keep the base status high, as long as they persist. Quartz content is very low in these rocks, consequently relatively small amounts of sand are concentrated in the soils. Soils resulting from the saprolite produced by geochemical weathering of these rock types tend to be rich in clay, with very little quartz sand to produce the sandy surface soil textures found in soils originating from the granitic rocks. Thus, surface horizons generally are loam or clay loam. These soils tend to be dark red or dark brown due to the high free-iron content. Base status and pH are relatively high, and exchangeable aluminum levels are low or nonexistent. The soil clay minerals tend to be kaolinite and halloysite if the soil is well drained, but montmorillonitic if the soil is poorly drained or is in a region with a distinct dry season.

VOLCANIC ASH. Volcanic ash parent materials are composed of noncrystalline, tiny, glass fragments, bits of the easily weatherable feldspars and ferromagnesian minerals, and varying amounts of quartz. Most of the volcanic ash deposits are andesitic (moderately basic in composition), especially those of the volcanic zones around the Pacific Basin. Those which are highly siliceous are very vesicular (apparently derived from "frothy" lava) and are commonly called "pumice." They occur in limited small areas in northwestern United States, in Central America, and in northwestern South America. Very iron- and magnesium-rich volcanic debris (ferromagnesian or strongly basic) likewise occurs in very small areas, usually as granular basaltic sand. Neither of these latter two groups tends to form soils with the same kind of properties as the more generally occurring volcanic ash, except under rather special limited conditions. Volcanic ash parent materials are very extensive in those parts of the world with current or recent volcanic activity, especially the regions around the Pacific Ocean. These parent materials thus occupy large areas in western South America, Central America, the Pacific Northwest of the United States, Japan, New Zealand, Indonesia, the Philippines, and other parts of the Pacific Basin. Some areas of ash also occur in the Antilles, in the Caribbean.

Volcanic ash imparts rather distinctive properties to soils over a wide range of climatic conditions. One of the main features soils inherit from their volcanic ash parent materials is allophane, an amorphous aluminosilicate with which abundant organic matter is complexed in the upper solum. However, soils formed from volcanic ash in arid regions or those with very pronounced dry seasons tend to form montmorillonite and are not as black, hence generally are not included with this group. These soils have received various names: Ando (previous United States designation), Andepts (present United States classification), Kuroboku (Japanese), Alvisols (New Zealand), and Trumao (Chile).

Common features of soils rich in allophane and formed from volcanic glass have been summarized by Wright (1964) as follows:

1. Generally thick soil profiles, friable in upper part, and ordinarily with distinct stratification.

2. Presence of intensely dark humic compounds in the topsoils, comparatively resistant to microbial decomposition.

3. Prominent yellowish brown subsoil colors, with marked "smeary" or "greasy" feel when squeezed between the fingers. (Note the color may range from brown to red, and the "smeary" consistency is most pronounced in the more humid climates, according to Swindale 1969.)

4. Very low bulk density—very light and porous with a fluffy feel.

5. High water-holding capacity.

6. Weak structural aggregates, with porous peds lacking cutans.

7. Almost complete lack of stickiness or plasticity when moist; on drying, the particles are often slow to rewet and may float on water.

8. High cation exchange capacity, extremely difficult to disperse for particle size distribution analysis.

It should also be noted that these soils tend to have high phosphate retention.

LITHOSEQUENCES. In the functional factorial analyses of Jenny (1941), a lithosequence is defined as a set of soils with property differences due solely to differences in parent material, with all other soil-forming factors constant. Expressed mathematically, the function is shown as:

$$S = f(pm)_{cl, o, r, t,} \ldots$$

Such a sequence or array of soils is difficult to recognize and establish because of the problems in establishing that all soils in the set have property differences due solely to parent material differences without some effects from environmental differences or from differences in local landscape position. However, several sets of soils have been defined as approaching this condition such that we can analyze the effects of parent material compositional differences, primarily on young and relatively simple landscapes, such as in recently glaciated regions. One such study is that reported from Manitoba, Canada, by Ehrlich, Rice, and Ellis (1955). They found that "the composition of parent materials had a profound effect on the type of profile formed." These are soils formed in glacial sediments of Mankato age (late Wisconsin, the last major period of glaciation) and which varied in particle size, calcium carbonate content, and mineralogical composition. They found that these differences controlled soil properties to the extent that the soils are placed in different orders (highest categorical level of classification).

In New York State, Cline (1953) has reported major differences in soil properties directly related to differences in composition and physical properties of the parent materials. These differences are illustrated in Figure 8.1. This set is probably not a true lithosequence, but approaches one sufficiently to illustrate our point.

An array of soils forming a near-lithosequence on the piedmont of North Carolina, in humid southeastern United States, is illustrated in Figure 8.2. This array has been developed from information reported by McCaleb (1959), McCaleb and Lee (1956), and Nyun and McCaleb (1955). As can be seen in this figure, profound differences in soil properties result from the differences in parent material. The Cecil soil, formed from saprolite from the more siliceous mica gneiss and granite gneiss, has a relatively thick and sandy A horizon because of the residual accumulation of the quartz sand present in large amounts in the parent rocks. The Davidson soil, formed from the saprolite from ferromagnesian or "basic" rocks, has loamier and more clay-rich A horizons because of the lower content of quartz and higher content of easily weatherable minerals in the parent rocks. The Davidson B horizon is redder and richer in free iron in com-

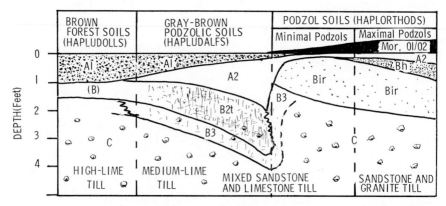

FIG. 8.1. Brown Forest-Podzol sequence in New York State. (After Cline 1953.)

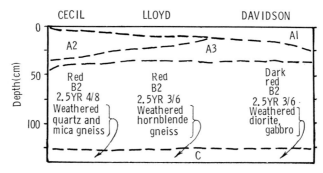

FIG. 8.2. Lithosequence on the piedmont of North Carolina (Cecil-Davidson series).

parison to the Cecil because of the higher iron content of the minerals in the "basic" rocks, such as diorite and gabbro, from which it forms. Likewise, the Davidson soil is higher in pH and much lower in exchangeable aluminum because of the higher content of calcium- and magnesium-containing minerals in the parent rock, thereby affecting the soil properties.

LITERATURE CITED

Cline, M. G. 1953. Major kinds of profiles and their relationships in New York. Soil Sci. Soc. Am. Proc. 17:123–27.

Dokuchaev, V. V. 1883. Russian Chernozem (Russkii Chernozem). In Collected writings (Sochineniya), vol. 3, Acad. of Sci. USSR, Moscow.

Ehrlich, W. A., H. M. Rice, and J. H. Ellis. 1955. Influence of the composition of parent materials on soil formation in Manitoba. Can. J. Agr. Sci. 35:407–21.

Jenny, H. 1941. Factors of soil formation. McGraw-Hill, New York.

McCaleb, S. B. 1959. The genesis of the Red-Yellow Podzolic soils. Soil Sci. Soc. Am. Proc. 23:164–68.

McCaleb, S. B., and W. D. Lee. 1956. Soils of North Carolina. I. Factors of soil formation and distribution of great soil groups. Soil Sci. 82:419–31.

Nyun, M. A., and S. B. McCaleb. 1955. The Reddish Brown Lateritic soils of the North Carolina piedmont region: Davidson and Hiwassee series. Soil Sci. 80:27–41.

Polynov, B. B. 1930. Das Muttergestein als Faktor der Bodenbildung und als Kriterium fur die Bodenklassification. Soil Res. 2:165–80.

Richthofen, F. F. von. 1886. Führer für Forschungsreisende. Berlin.

Swindale, L. D. 1969. The properties of volcanic ash soils, pp. B.10.1–B.10.9. In Panel on volcanic ash soils in Latin America. Training and Research Center of Inter-American Inst. of Agr. Sci., Turrialba, Costa Rica.

Thaer, A. D. 1809, 1810, 1812. Grundsätze der rationellen Landwirtschaft, vol. 1–4. (Cited in J. S. Joffe, 1949. Pedology. Pedology Publ., New Brunswick, N.J.)

Travis, R. B. 1955. Classification of rocks. Colo. School of Mines Quart. 50, no. 1.

Wright, A. C. S. 1964. The "Andosols" or "Humic Allophane" soils of South America, pp. 9–22. Report on the meeting on the Classification and Correlation of Soils from Volcanic Ash, Tokyo, Japan. World soil resources report 14, FAO, Rome, Italy. 1965.

❧

Relief and Landscape Factors of the
Soil and Its Environment

GENERALLY ACCEPTED definitions of relief and topography were not available when Jenny (1941) chose to use these terms synonymously. The Soil Survey Staff of the USDA considered *relief* as the elevations or inequalities of the land surface considered collectively, and *topography* to be those features shown on a topographic map (Soil Survey Staff 1951). For the purpose of this discussion, the term relief will be used essentially as defined by the Soil Survey Staff. Also, slope will be used in its restricted sense to refer to the angle which any part of the earth's surface makes with a horizontal datum (Wilson 1968).

SLOPE–SOIL PROPERTY RELATIONSHIPS. Specific statements about the relationships of slope to soil properties can be made only within specified geographic areas. This is probably due to variations in intensity and nature of the other soil-forming factors. Within specific geographic regions, the following soil properties are commonly found to be relief related: (1) depth of the solum; (2) thickness and organic matter content of the A horizon; (3) relative wetness of the profile; (4) color of the profile; (5) degree of horizon differentiation; (6) soil reaction; (7) soluble salt content; (8) kind and degree of "pan" development; (9) temperature; (10) character of the initial material.

The most obvious relationship of a soil property to relief probably occurs in humid regions where soils on nearly level relief tend to have thicker sola than those on slopes. This can usually be attributed to either the slow "geologic" erosion of soil material from the soil surface or the lack of percolating water due to losses by runoff, or both, occurring on the slope. Norton and Smith (1930) found that depth of solum varied from 8 to 10 inches on slopes of 14% and 8%, respectively, to 25 inches on 1% slopes in soils developed from loess. General relationships usually are not valid with respect to specific morphological features from one physiographic region to another. For example, A2 horizons have been reported to be thicker in Podzols formed in depressions in Norway than on upland areas (Lag

1951). The opposite effect was found in Ultisols on the coastal plain in North Carolina (Daniels, Gamble, and Nelson 1967) where A2 horizons were thicker on well-drained upland sites and thinner in poorly drained depressions.

Doubtless many other examples of slope-soil relationships could be cited; an individual should attempt to establish an understanding of these and constantly be aware of such possibilities when working in a given area.

RELIEF AS AN INDEPENDENT FACTOR. While it is true that relief severely modifies the other factors of soil formation, it can be considered as an independent factor of soil formation.

Jenny (1958) describes the concept of this factor of soil formation by means of an imaginary greenhouse tank in which a mass of sand is arranged in the form of two parallel ridges separated by a trough in the sand. A "dry" climate is maintained for three centuries and a salty water table is maintained so that the upper limit of capillary water just reaches the floor of the trough. Seeds from plants with various degrees of salt tolerance are made available to all parts of the surface of the mass of sand. After three centuries, a plant-free (but not seed-free) playa encrusted with salt exists in the trough, salt-tolerant plants grow on the lower slopes of the ridges, and progressively less salt-tolerant plants grow at higher sites on the ridges.

RELIEF AS AN INTERDEPENDENT FACTOR. General areal inter-actions of slope and soil properties are difficult to find because of the interplay of relief with the ambient characteristics of the other soil-forming factors. Although specific soil properties often become well recognized with relief factors in any given area, these relationships often do not hold in other areas of different climate, initial material, biological environment, or age of the soil.

RELIEF RELATED TO CLIMATE. Relief influences the distribution of climatic forces and agencies of soil material. Variation in aspect and elevation influences the distribution of energy, meteoric water, plant nutrients, and vegetation by varying (1) conditions for organic activity, such as exposure of soil flora and fauna to direct sunlight, and mixing of mineral soil and organic matter by animals; (2) the exposure of the soil to wind; (3) the exposure of the soil to precipitation, including snowfall and snowdrifts; (4) conditions for natural drainage, including depth to water table; (5) conditions for runoff and erosion, positive and negative; and (6) conditions for accumulation and removal of deposits by wind.

Microrelief on the Caspian plain (Rode 1959) has redistributed precipitation so that Chernozems have formed in depressions, Chestnut soils on the microslopes, and Solonchak and Solonetz soils on the microridges. For other examples of the role of relief in the modification of climate, the reader is referred to Chapter 18.

RELIEF RELATED TO ORGANISMS. The manipulation of climate by relief is readily seen expressed in the vegetation of north- and south-facing slopes.

The effect of aspect on soil development and vegetation was studied on a small hill on the south side of the Snake River in an arid region of south central Idaho (Klemmedson 1964) (Table 9.1). The hill stood about 200 feet high and was 1,500 feet long and 800 feet wide. It was a deposit of Crowsnest outwash gravel, exposed for at least 18,000 years. Lithology is quartzite, prophyry, and granite. The vegetation is sagebrush grass, disturbed by uniform grazing and occasional fires. Data were collected along a north-south transect and an oval transect along a contour at an average slope of 37%. In 1962 the following comparisons were made between north and south slopes. It should be noted that the organic matter content of the top 10 cm of soil is greater on the north slope than on the south slope even though the production of herbage was reversed. This is generally considered to be the result of higher soil temperatures on south-facing slopes, causing more rapid organic mineralization. The greater intensity of insolation, however, places the plants under a greater moisture stress and reduces their production of roots.

Losche (1967) found in the mountains of Virginia that the upper part of the south-facing slopes was composed of *Pinus rigida* and *Quercus coccinea,* whereas *Quercus coccinea* predominated on the lower part of the slope. On the north-facing slope the dominant species were *Quercus rubra, Quercus coccinea, Quercus prinus,* and *Carya* spp. As part of the same study in the western part of North Carolina he found that on the lower part of the north-facing slope, *Liriodenron tulipifera* and *Quercus rubra* dominated, and *Quercus prinus, Quercus rubra,* and *Carya* spp. composed the major part of the canopy on the upper slope. The south-facing slope was dominated by *Quercus coccinea* and *Pinus rigida.* Upon analyzing the litter under each site, he found that more basic cations were recycled and accumulated on the forest floor of the north-facing slope than on the south-facing slope. More clay and more pronounced argillic horizons were found on the south-facing slopes than on the north-facing slopes. The depth of organic litter accumulation was greater on the north-facing slope than on the south-facing slope.

TABLE 9.1 ❧ **Some characteristics of tesseras on north and south slopes of an isolated hill in Idaho**

Item	North Slope	South Slope
% gravel by vol. in 0–10 cm soil	5	20 (max. of 40% somewhat downslope)
% organic carbon in 0–10 cm soil	12	6
Water retention capacity in 0–10 cm soil, in kg/0.1 m³	6.2	3.2
Dominant vegetation	Perennial grasses (*Poa, Stipa*)	Annuals (*Bromus, Salsola*)
Wt. of dry herbage in g/m²	50	100
Relative root wt. per unit volume of soil, on the basis of 100% for the north slope	100	37

SOURCE: Klemmedson 1964.

RELIEF RELATED TO INITIAL MATERIAL. In many cases, relief prior to and during soil formation is related to the nature of the initial soil material. In broad river deltas, crests of natural levees near the stream channels have coarser material than the areas beyond the levees that are very nearly level and have the finer textured initial material (Russell 1967).

In steeper topography where the valleys below the mountain ranges are characterized by broad alluvial-colluvial fans, the initial material near the mountain range contains more coarse and angular material than areas farther away from the mountain range (Birot 1960).

RELIEF RELATED TO TIME. Relief changes with time. Therefore, the age of soil is determined to a large extent by the stability of the surface. Soil erosion is often discussed with the inference that it is detrimental because of effects of washing or blowing soil material caused by the mismanagement of the land by man. In addition to man-made or accelerated erosion is the natural or geologic erosion, which can be thought of as a continuous process but actually occurs sporadically. Coexistent with erosion is deposition. In the case of water erosion, the removal of material is usually from the sloping positions of the landscape, and the deposition takes place in the valleys and depressions. Wind erosion may be independent of slope direction. Wind erosion is severe on slopes facing predominant strong winds during seasons of the year when the surface is relatively devoid of vegetation.

If one conceives of the deepening of soil as taking place by the extension of weathering into the geologic substratum as measured from the air-soil interface, it is easy to see that erroneous relationships of soil depth as related to slope position and weathering intensities are possible because erosion and deposition of soil material is primarily at the soil-air interface. The absolute position of the soil-air interface is seen to change during the course of soil formation. Such changes are very difficult to measure.

In addition to losses or gains of soil material at the soil-air interface related to slope there is also creep or solifluction of the entire soil volume downslope (Birot 1960). This can often be seen by observing the bowed shape of trees growing in such situations. In general, the upper part of the soil profile tends to move faster than the lower part and the trees are tilted downslope. A growing tree then attempts to correct for this and to return to a vertical posture, with the result that it bends with convex side in a downslope direction (Fig. 9.1).

Geomorphology is the science that deals with the form or relief features of the earth's surface, classifies these features, and attempts to explain their origin and history. Because the formation of soil takes place at the earth's surface, the processes of soil formation are influenced by geomorphic changes. Changes in landform occur through time. It is not possible to express these stages of landscape development in terms of years. The rate of development depends on several factors but mainly on the resistance of the existing features to the type of environmental forces attempting to make the changes and on the intensity of the environmental forces involved.

FIG. 9.1.   Aspen trees growing in a curved shape to
recover a vertical posture as the soil solum creeps down
slope near Flagstaff, Arizona.

Therefore, changes that are considered normal in one environment are
abnormal in other areas. If more resistant layers occur in the stratigraphic
section, downwasting of the interfluves does not take place where such
surfaces provide protective cappings (Fig. 9.2). In this respect, specific
changes in geomorphology must be viewed on a regional basis much the
same as the influences relief has on soil development must be viewed on
a regional basis.

**GROUNDWATER TABLE LEVELS, WATER TABLE
FLUCTUATION, LATERAL WATER MOVEMENT,
AND THEIR EFFECTS ON SOIL PROPERTIES.** Water table is defined
as "the upper surface of groundwater or that level below which the soil
is saturated with water; locus of points in soil water at which the hydraulic
pressure is equal to atmospheric pressure" (Terminology Committee 1965).

In humid and subhumid regions, the groundwater table usually has a
relief which is like, but of lesser amplitude than, the relief of the land
surface; that is, it is closer to the soil surface in depressions than on high
points of the landscape. It is quite often the case that, seasonally at least,
the groundwater table invades the soil profile. This is more commonly true
in depressions than in convex positions.

A saturated condition in the soil alters many physical and chemical
reactions in the soil: (1) anaerobic or reducing reactions predominate

because of a limitation of the oxygen supply and may retard root growth; (2) percolation, or downward movement of water, is restricted, and instead, the lateral movement of water within the saturated zone tends to predominate; and (3) the specific heat of the soil is increased and the soil tends to be colder than surrounding nonsaturated soils.

Whereas the presence of a groundwater table is common in humid and subhumid regions and its presence can be observed in the properties of the soil profile, the water table is usually not of much significance in arid and semiarid regions where it is normally several feet below the soil surface. However, it does become extremely important in the drier parts of the world where it may occur in the soil profile.

Because the water table does have relief, a gradient tending to produce lateral water movement is established. Although this lateral movement is less rapid than surface water, because of the restrictions of the solid particles, it is present and serves to carry solutes and suspended material. In humid regions this may result in the accumulation of salts, oxides, or other suspendible or soluble material in landscape depressions. Such concentrations are usually minor because groundwater is usually removed from the area via springs and rivers.

In areas of less precipitation and especially when the potential evapotranspiration exceeds the precipitation, the presence of a groundwater table in the soil leads to the accumulation of soluble salts. Where there

FIG. 9.2. Downwasting landscape where hilltops are protected by plinthite layers in the Llanos of Colombia, S.A.

is no mechanism for liquid water to escape from a depression and the only water loss is by evaporation, salts dissolved in the water are left, upon evaporation, on or near the soil surface (see Chapter 6).

The water table depth in any individual soil is usually subject to seasonal fluctuations. A perched water table may seasonally develop (Carlisle, Knox, and Grossman 1957). This usually means that during at least part of the year the soil is free of the saturated condition and oxidation processes can proceed, and part of the year the soil environment is in a state of chemical reduction. This type of situation normally leads to the development of a mottled coloration. Gray bodies with chromas of two or less develop in the part of the profile subject to prolonged saturation. However, low chroma mottling may not always be a feature of saturated conditions. High pH and/or well-oxidized water will not develop the gley colors because the iron is maintained in a ferric form even though water saturated. Often, in soil material that is saturated and the iron reduced, one can observe tubes of iron oxide lining channels, to which oxygen gained access either via roots or, during periods of soil drying, by diffusion from the atmosphere.

TOPOSEQUENCES AND SOIL CATENAS. A soil catena has been defined as a topo-drainage sequence (Bushnell 1942, who quotes Milne). The catena is an often-used concept that has basis in fact mainly in those areas of the world where the groundwater table is present within the soil profile of the soils formed in depressions. The terms "well-drained, moderately well-drained, imperfectly drained, and poorly drained" have been used to describe soil individuals associated in toposequence where drainage related to water table is involved (Soil Survey Staff 1951).

The soils related in a catena are seen to have properties that can be related to their position on the landscape (Ruhe 1969). The reasons for these relationships, however, may not easily be seen. It may be because of microclimatic relationships, water table relationships, vegetative relationships, erosion-deposition, or a combination of these. Relationships of soil properties to topographic positions do occur and can be observed readily (Fig. 9.3). They should be studied and used in soil mapping. For a full understanding of the genetic causes however, many facts of the relationship of the relief factor to the other soil-forming factors must be examined with regard to their role in the local relief.

CHARACTERIZATION OF SOIL
INDIVIDUALS WITH RELIEF. A landscape may be characterized by the following information: (1) drainage density (total length of streams in the basin/area of the basin); (2) relief index (relief of the basin/main stream length); (3) soil body dimensions (internal measurements of width and length; external measurements of pattern length and width); (4) number of soil bodies per unit area; (5) soil profile characteristics and proportionate extent and catenal arrangement of kinds of soil bodies.

Hole (1953) suggested terminology for describing soils as three-dimensional bodies, including soil body patterns and landscape positions, as well

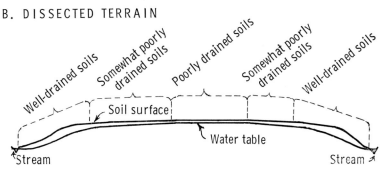

FIG. 9.3. Schematic sketches of catena relationships.

as drainage conditions (see Chapter 25). Each kind of soil can be said to occupy an ecological niche in a landscape. Areal slope spectra can be determined for soil series. The relative position of the niche is variable from one major landscape to another. Such a relief-soil relationship is seen in Table 9.2.

The most readily observable influence relief has on soil development is summed up briefly in the statement *water runs downhill*. Reference is made here to surface water, or that free water moving over the soil surface. Moving surface water almost always carries some solid particles with it and produces erosion or change in the relief. Also, it tends to cause the sloping areas to be drier, from the standpoint of infiltrated water, than level areas. As a result of the runoff from the sloping areas, the depressional areas and valleys receive more surface water than surrounding upland soil. This in effect results in greater leaching of those soils formed at the base of the slope than those formed on upland areas and a minimum of leaching on the steep slopes for any given climate.

The role of relief is finally viewed as a real factor in the development of soil. Its exact role is, however, difficult to evaluate with examples of a generalized nature. Relief functions differently in various environmental situations. It serves to alter the influence of parent material and time by the erosional and depositional changes that occur because of it. It serves to alter climate via its control on runoff, groundwater table, and aspect,

TABLE 9.2 & **Slope spectra for several soil series in selected counties of Wisconsin**

County	Soil Series	Proportionate Extent (%) on Slopes of the Ranges Approximating Those Indicated				
	(slope range)	0–2%	2–6%	6–15%	15–30%	30–60%
Richland	All soils	11	5	36	33	15
	Dubuque	0	3	55	42	0
LaCrosse	All soils	18	14	9	31	28
	Dubuque	0	3	3	80	14
Barron	All soils	23	36	32	2	7
	Omega	11	75	14	0	0
Bayfield	All soils	21	49	25	5	0
	Omega	27	73	0	0	0

and because of the alterations in climate, the vegetative factor is also altered by relief. The role of relief can be evaluated only in a regional or local area. Its role as a soil-forming factor changes from area to area. It should also be thought of as being itself affected by pedogenesis.

LITERATURE CITED

Birot, P. 1960. The cycle of erosion in different climates. (Transl. from French, 1968, by C. I. Jackson and K. M. Clayton.) Univ. Calif. Press, Berkeley.

Bushnell, T. M. 1942. Some aspects of the soil catena concept. Soil Sci. Soc. Am. Proc. 7:466–76.

Carlisle, F. J., E. G. Knox, and R. B. Grossman. 1957. Fragipan horizons in New York soils. I. General characteristics and distribution. Soil Sci. Soc. Am. Proc. 21:320–21.

Daniels, R. B., E. E. Gamble, and L. A. Nelson. 1967. Relation between A2 horizon characteristics and drainage in some fine loamy Ultisols. Soil Sci. 104:364–69.

Hole, F. D. 1953. Suggested terminology for describing soils as three-dimensional bodies. Soil Sci. Soc. Am. Proc. 17:131–35.

Jenny, H. 1941. Factors of soil formation. McGraw-Hill, New York.

———. 1958. Role of the plant factor in the pedogenic functions. Ecology 39:5–16.

Klemmedson, J. O. 1964. Topofunction of soils and vegetation in a range landscape, pp. 176–89. Am. Soc. Agron. Spec. Publ. 5, Madison, Wis.

Lag, J. 1951. Illustration of influence of topography on depth of A2-layer in Podzol profiles. Soil Sci. 71:125–27.

Losche, C. K. 1967. Soil genesis and forest growth on steeply sloping landscapes of the southern Appalachians. Ph.D. thesis, N.C. State Univ. (Order 67-11, 282) Univ. Microfilms. Ann Arbor, Mich. (Dissertation Abstr. 28B:765-B)

Norton, E. A., and R. S. Smith. 1930. The influence of topography on soil profile character. J. Am. Soc. Agron. 22:251–62.

Rode, A. A. 1959. Hydromorphism of soils. Soviet Soil Sci., 1959, no. 10, pp. 1131–40. (Transl. 1961 from Pochvovedeniye, 1959, 10:1–13, by Am. Inst. Biol. Sci., Washington.)

Ruhe, R. V. 1969. Quaternary landscapes in Iowa. Iowa State Univ. Press, Ames.

Russell, R. J. 1967. River plains and sea coasts. Univ. Calif. Press, Berkeley.

Soil Survey Staff. 1951. Soil survey manual. U.S. Dept. Agr. Handbook 18. U.S. Govt. Printing Office, Washington.

Terminology Committee. 1965. Glossary of soil science terms. Soil Sci. Soc. Am. Proc. 29:330–51.

Wilson, L. 1968. Slopes, pp. 1002–20. In R. W. Fairbridge (ed.), The encyclopedia of geomorphology. Reinhold, New York.

CHAPTER 10

Contributions of Climate to Total Soil Environment

W HEREAS the role of relief as a factor of soil formation can be satisfactorily studied within the confines of a region or local area, it is necessary to examine worldwide relationships to evaluate properly the role of climate in soil formation. Certain local variations in climate (microclimates) provide evidence of the influence of the climatic factor on soil formation within relatively small areas. However, the magnitude of climatic control on soil formation can best be seen by making comparisons that are global in scope. The two most commonly measured features of climate that have been correlated to soil properties are rainfall and temperature.

Climate not only exhibits its influence on soil formation by control of some of the chemical and physical reactions taking place in the soil but also by its control of the organic factor and to some extent of the factors of relief and time through erosion and deposition of soil materials. Also, climate is not unchanging. Climates shift with time and even though changes take place over long periods, some do occur well within the age of many soils. Important as average climatic conditions are to soil genesis, the extremes of weather occurring in a given climatic region may be more influential in the development of certain properties of the soil.

RAINFALL–SOIL
PROPERTY RELATIONSHIPS. Water is a necessary agent in the formation of soil. Water dissolves soluble material; it is necessary for the growth of plants and other organisms that contribute organic material to the soil; it transports material from one part of the soil to another; it physically ruptures material when it freezes; and it has several other important functions. The type and rate of reactions in which water is involved are certainly dependent to some extent on other factors such as temperature, pH, and redox potential. But, to obtain a general view of the role of water in soil formation, we may simply evaluate the amount of water in the soil profile that is available to carry out these functions.

Except for the role of water in the erosion or deposition of soil material at the soil-air interface, the important functions of water are within the soil profile. As was discussed earlier (Chapter 9), whether rainfall enters the soil or runs off is regulated to some extent by relief. But texture, porosity, and shrink-swell potential of the soil are also involved. For the purpose of discussion in this section, it will be assumed that the relief factor, and infiltration rates, are comparable in the areas compared.

Average annual precipitation has been measured and reported to vary between 1.27 mm (0.05 in.) at Iquique, Chile, South America, and 11,505 mm (460.20 in.) at Waialeale, Hawaii (Reed 1941). Little if any data exist from soils formed under these extreme conditions, but much work has been carried out in intermediate climates to show how soil properties relate to the total amount of rainfall. In summary of the work, Jenny (1941) reported the following relationships in the regions having 380 to 890 mm (15 to 35 in.) annual rainfall:

1. The hydrogen ion concentration in the soil increases (pH decreases) with increasing rainfall.

2. The depth to carbonates in the soil increases as rainfall increases.

3. The nitrogen content of the soil increases with increasing rainfall.

4. The content of clay in the soil solum increases as rainfall increases.

Although the above relationships have clearly been shown to exist, their extrapolation to the extremes of climatic variability present in the world would be hazardous without more data.

Only a portion of the total rainfall moves through the entire soil profile. Loss of water occurs from the soil mainly in the form of evaporation from the soil surface or via the plant (transpiration). These two forms of water loss from the soil are often considered collectively as evapotranspiration. From the standpoint of studies in soil genesis, it is desirable that this quantity be evaluated because it is necessary for the water to get into the soil before it can carry out many of the reactions of which it is capable. Several methods can be used for measuring evapotranspiration in any given area. However, not enough data are available from these direct measurements to use in studying relationship to soil genesis. Likewise, elaborate formulae are available to calculate potential evapotranspiration values from various climatic parameters (Palmer and Havens 1958). Although more recent methods are more accurate than ever before, the limitation on amount of data available hinders their use in studies relating soil development to climatic parameters. One early attempt at evaluating the effective rainfall, or that rainfall not lost to evapotranspiration, is Lang's rain factor (Lang 1920), which is mean annual precipitation (mm) divided by mean annual temperature (° C). This factor exhibits some degree of correlation with the distribution of soil. Comparison of soil maps with maps showing rainfall as altered by evapotranspiration data can be expected to show only general trends. These trends may even be misleading because of the dependence of the use of the climatic factor as a criterion in constructing the soil map and legend. Furthermore, alteration of climate by local features of relief often plays a major role in altering the direct effect of climate.

Mosaics of contrasting parent materials and of soils of widely differing ages may interfere with the development of climatic zonal patterns, as has been observed in Australia.

TEMPERATURE–SOIL
PROPERTY RELATIONSHIPS. Temperature as a component of the climate has been considered as an independent soil-forming factor (Jenny 1941). Temperature influences in many ways the reactions involved in the processes of forming soil. It is the principal component in the calculation of potential evapotranspiration and thus has a large measure of control on the amount of effective rainfall. When the soil system cools to the point where water solidifies, all the chemical reactions involving water cease, but physical breakdown of larger particles by frost action may occur. Temperature exerts a marked influence on the type and quantity of vegetation present in an area and thus on the amount and kind of organic matter produced. The rate of organic matter decomposition is almost directly controlled by temperature. The influence of temperature can readily be seen by considering Van't Hoff's temperature rule (Van't Hoff 1884) which can be stated: "For every 10° C rise in temperature the speed of a chemical reaction increases by a factor of two to three."

Except for additions of heat energy transmitted to the soil from subterranean sources, which may be of local importance in geyser basins, metabolic heat from organisms in the soil, and heat from exothermic chemical reactions in the soil, all of which are considered minor, the predominant source of energy is the sun. Thus, the energy transfer involved in heating and cooling the soil takes place at the upper boundary of the soil. Solar energy arriving at the soil surface varies widely, depending on several factors. Solar radiation increases with elevation. The rate of radiation increase is most rapid in the lower dust-filled layers of the air (Geiger 1957). Solar radiation reaching the soil is also reduced by cloud cover and changes markedly with the season of the year. Once the incoming solar radiation has reached the soil surface, it has to be absorbed before it is changed into heat. The absorption of the solar radiation at the soil surface is affected by many variables such as soil color, orientation of the surface with respect to the incoming solar radiation, and vegetative cover. In general, the darker the color, the more radiation is absorbed and the lower the albedo. The nearer the soil surface approaches an angle normal to the direct rays of the sun, the greater the absorption. The effect of vegetative cover on absorption varies with density, height, and color of the vegetation. In many forested areas, the vegetation almost completely shades the soil surface. In the case of deciduous trees, however, the soil surface may have little shade in the early spring and almost complete shade in the summer.

Heat absorbed during the day is lost during the night. This loss takes place by reradiation (radiation from the soil into space), a process that goes on all the time. In general, the rate of reradiation for soil surfaces can be calculated as if the soil were a black body. According to the Stefan-Boltzmann equation, the rate of reradiation in calories cm^{-2} min^{-1} (S) is:

TABLE 10.1 🌿 **Relation of surface temperature to reradiation**

Ground surface temperature (° C)	−10	0	10	20	30	40	50	60	70	80
Reradiation (cal cm⁻² min⁻¹)	0.395	0.459	0.530	0.609	0.696	0.792	0.899	1.015	1.143	1.200

SOURCE: Geiger 1957.

$S = \sigma (T)^4$ where σ has a constant value of 8.25×10^{-11} and T is temperature in absolute or Kelvin degrees, that is, in centigrade $+ 273$. From this formula we can compute the temperature-reradiation relations shown in Table 10.1.

The radiation received at the earth's surfaces varies considerably, depending on the factors previously discussed. Table 10.2 presents data to show the variation in incoming radiation at Potsdam with season, orientation of the surface, and amount of cloudiness.

Considering only the data in Tables 10.1 and 10.2, it can be seen that maximum surface temperatures of 60 and 70 C are to be expected on a horizontal surface on clear days in June. Higher temperatures are not possible because the reradiation rate at that temperature equals the incoming radiation rate. A maximum surface temperature of 10 to 20 C is all that can be expected on cloudy days, although it is not safe to use the same reasoning for cooler seasons because of other heat sources in the soil profile.

Heating and cooling of a soil take place for the most part from the soil surface. The magnitude of the temperature change in the soil depends not only on the features affecting radiation but also on the thermal properties of the soil and amount of turbulence created by wind. A schematic graph of the diurnal temperature change in a soil profile is given in Figure 10.1. It will be noted that the most drastic temperature changes take place near the surface of the soil and that diurnal temperature changes are absent below about 50 cm (20 in.).

Several soil properties have been shown to be temperature dependent (Jenny 1941). With increasing temperature, soil colors tend to become less gray and more reddish. Bases are most completely leached in warm areas, according to Jenny. Nitrogen and organic matter contents decrease as the temperature increases. Clay contents increase with increases in temperature.

CLIMATIC INDICES AND EVAPOTRANSPIRATION
IN FUNCTIONAL SOIL ANALYSIS. Several indices have been used to characterize climate. Most of these involve ratios of precipitation to temperature. The values obtained from these calculations can in turn be plotted on maps correlated with soil features. While it is not reasonable to assume that complete correlation will exist, because of effects of the other soil-forming factors, a study of some of these indices and their relation to soil properties provides an insight into the function of climate as a soil-forming factor. Meyer (1926) obtained an NS quotient from the following formula:

$$\frac{\text{Precipitation (mm)}}{\text{Absolute saturation deficit of air (mm Hg)}} = NS$$

TABLE 10.2 ❧ Incoming solar radiation in cal cm⁻² min⁻¹ at midday during the middle of the month at Potsdam (52.4° N Lat)

	Jan.	Feb.	Mar.	Apr.	May	June	July	Aug.	Sept.	Oct.	Nov.	Dec.
No Clouds												
Total radiation (surface normal to radiation)	1.014	1.120	1.121	1.275	1.278	1.274	1.196	1.169	1.220	1.167	0.987	0.967
Horizontal surface	0.287	0.471	0.649	0.937	1.065	1.113	1.027	0.918	0.796	0.569	0.324	0.240
Average Cloud Cover												
Total radiation	0.264	0.325	0.437	0.536	0.626	0.650	0.574	0.573	0.573	0.467	0.266	0.242
Horizontal surface	0.075	0.137	0.253	0.394	0.522	0.568	0.493	0.450	0.374	0.228	0.087	0.060

SOURCE: Geiger 1957.

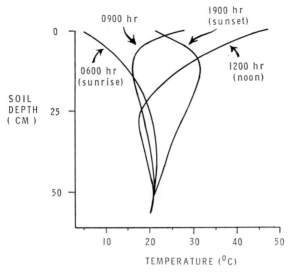

FIG. 10.1. Schematic graph of diurnal soil temperature variation with depth.

Jenny used the *NS* quotient to compare the soils found in the United States and in Europe (Table 10.3). Vilensky, Prescott, and Thornthwaite have prepared climatic systems of soil classification (Jenny 1941). Such comparisons are of limited value because they are correlated with soil names that were not adequately defined in terms of soil properties.

A more complete method of defining the moisture regime of the soil is by water balance calculations. Such calculations are based on the measurements of rainfall distribution, temperature, and a calculation of the potential evapotranspiration.

Thornthwaite (1948) developed an empirical formula to express the relationship between evapotranspiration and temperature. Palmer and Havens (1958) facilitated the use of the Thornthwaite formula by developing nomographs. The Soil Conservation Service (USDA 1960), seeing the usefulness of the Palmer-Havens nomographs, reproduced them on a larger scale.

TABLE 10.3 ❧ **Parallelism of soil-climate relationships in Europe and the United States [Temperate regions only, 4–12 C (39–54 F)]**

Climatic Soil Type	Moisture Factor (annual NS quotient)	
	Europe	United States
Gray Desert	100	30–110
Arid Brown	. . .	60–120
Chestnut	140–270	100–180
Chernozems	130–250 (350)	140–250 (320)
Arid-humid boundary	200	200–250
Degraded Chernozem	230–350	. . .
Prairie	. . .	260–350 (420)
Brown Forest and Gray-Brown Podzolic	320–460	280–400
Podzols	400–1,000	380–750

SOURCE: Jenny 1941.

The Thornthwaite technique, or a number of other techniques, can be used for arriving at a potential evapotranspiration value. This value calculated on a monthly basis in turn can be used in a water balance formula to calculate the moisture regime in the soils of an area if the following assumptions are made: (1) There are no additions or losses of water by runoff; (2) vegetation is growing on the site and actively transpiring; and (3) dew formation is negligible. With these assumptions, a simple water balance for a soil is: $P = e + s$, where P = precipitation = 125 cm (49 in.); e = evapotranspiration = 94 cm (37 in.); s = surplus water percolating below the lower boundary of the soil = 30 cm (12 in.).

The water-storage capacity of the soil can be taken into consideration as illustrated in Figure 10.2, and the following observations can be made:

1. The average rainfall shows a decrease in late summer.

2. Evapotranspiration is highest in the summer and lowest in the winter.

3. Precipitation exceeds evapotranspiration from November through May. (Note: P and e for each month are plotted at the end of the month.)

4. Evapotranspiration exceeds rainfall June through October.

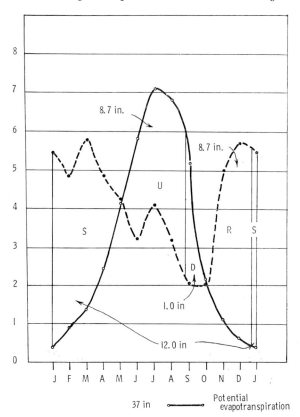

FIG. 10.2. Average yearly water balance at Tallulah, Louisiana, for a soil storing 22 cm (8.7 in.) of water.

37 in ⊶——⊶ Potential evapotranspiration

49 in •— — —• Precipitation

Thus, the average moisture regime in the soil can be interpreted as follows: The soil, after being depleted of moisture in the summer, begins to store water about the first of November. It stores water (R) until nearly the end of January at which time it has 22 cm (8.7 in.) of water, that is, its water-holding capacity. At that time the excess water (P) being greater than e, percolates as surplus water (S) below the depth of the plant root, that is, bottom of the soil, presumably reaching the groundwater. By the end of May, potential evapotranspiration (e) has, with the increasing summer temperatures, exceeded the precipitation (P) and the plants start to draw on the stored soil water. The plants continue to utilize (u) the stored soil water freely until mid-September when the soil is at or near the wilting point. Thus, from mid-September to the end of October the soil is near the permanent wilting point and the plants have to either die or be hardy enough to struggle along with only the rainfall during that period. The soil is seen to have a moisture deficit (D) for this period.

Another quantity, calculated actual evapotranspiration (Ea), can be determined from the water balance calculations. To calculate Ea, a water-holding capacity for the soil root zone needs to be determined or estimated. To again use Figure 10.2, where the water-holding capacity is 22 cm (8.7 in.), the calculated actual evapotranspiration is the sum of the potential evapotranspiration values (e) for the months where precipitation (P) exceeds evapotranspiration (e), that is, November through May; plus the potential evapotranspiration for that period where stored water is utilized, that is, June, July, and part of August; plus the precipitation (P) for these months where there is a moisture deficit (D), that is, the later part of August, September, and October. Arkley (1967) used water balance calculations to characterize several kinds of soil in the United States. He reasoned that the degree of leaching would be best characterized as the annual sum of monthly P-e values from those months when $P > e$ or the amount of precipitation in the wettest month, whichever was greatest. He called this value leaching index (Li). Arkley reasoned that a greater amount of actual evapotranspiration would be an index of several soil properties. In particular, he reasoned that more energy would be related to greater organic matter production and weathering. When climatic dates associated with great soil groups, as mapped on available small-scale maps, were plotted on charts with leaching index (Li) as one axis and calculated actual evapotranspiration (Ea) as the other axis, clear parameters could be defined for each great soil group. Considerable overlapping of great soil groups was observed but much of this could be expected because of differences in water-holding capacity (a figure standardized at 6 in. in Arkley's calculations), the general nature of the maps available, and the lack of a quantitative range of characteristics used in defining the great soil groups of the 1938 Soil Classification System (Baldwin, Kellogg, and Thorp 1938).

Water balance calculations have also been used to calculate the age of soils containing calcium carbonate (Arkley 1963). By determining the water-holding capacity of the various soil horizons, carbonate solubility, and carbonate content of the parent material, and by calculating the amount of water passing a given depth in the soil from water balance data, soil ages were obtained that closely correlated ages indicated from geological data.

MICROCLIMATE AND SOIL GENESIS. Microclimate can be thought of as that climate in the first few feet above the soil surface. It varies markedly with relief features, soil color, and vegetation.

In the Northern Hemisphere, south-facing slopes tend to be warmer and thus more droughty than north-facing slopes. In a comparison of north- and south-facing slopes in Michigan, detailed comparisons were made during the growing season in 1957 (Cooper 1960).

On the slopes where the data in Table 10.4 were obtained, the soils on the south-facing slope were found to have lighter brown-colored A horizons and redder B horizons than those on the north-facing slope. Soils on the south-facing slope contained more clay in the B horizon (avg 12.6%) than did the north-facing soils (avg 7.7%). Also, the north-facing soils had both deeper solums and A horizons, 91–103 cm and 65–66 cm, respectively, compared to south-facing soils, 64–85 cm and 40–64 cm, respectively.

Shul'gin (1957) reported that similar north-south relationships were found in Russia and also that soils on west-facing slopes were warmer than east-facing slopes. He attributed this to a greater expenditure of solar insolation for evaporation of dew on east-facing slopes than on west-facing slopes which were drier during the afternoon when they received more direct insolation. He reported that a field sloping 1° toward the south was as warm as a level field 100 km farther south. The magnitude of the temperature-slope aspect decreases in more equatorial areas.

TABLE 10.4 ❧ **Summary of microclimatic conditions on north and south slopes in Michigan during the 1957 growing season**

Measurement	Slope facing	Mar. 12– Apr. 14	Apr. 15– June 1	June 2– July 15	July 16– Aug. 23	Aug. 24– Sept. 16	Sept. 17– Oct. 16	Oct. 17– Nov. 9
Rel. light int.	North	34.2	21.1	2.5	1.8	1.8	2.4	20.5
(% in open)	South	42.4	30.7	7.0	3.5	3.8	5.5	36.8
Max. air temp.	North	48.5	72.9	79.6	85.8	75.8	70.4	61.3
(° F) 50 cm above soil	South	52.5	76.7	87.7	95.1	80.3	76.9	66.7
Min. air temp.	North	25.7	46.5	58.1	57.1	51.8	39.4	29.4
(° F) 50 cm above soil	South	27.1	47.7	58.2	57.5	52.2	39.8	29.6
Evaporation	North	N.R.*	99.4	75.1	84.8	61.4	59.8	N.R.*
cc/wk	South	N.R.*	144.6	118.1	119.3	95.8	91.3	N.R.*
Soil temp.	North	34.8	57.6	66.4	70.8	65.6	55.8	47.4
(° F) at 2 cm	South	41.1	61.1	69.5	76.1	68.2	60.3	51.0
Soil temp.	North	33.0	50.5	61.8	65.5	62.4	55.0	48.4
(° F) at 20 cm	South	38.8	55.6	64.9	70.6	65.6	59.6	50.7
Percent moisture	North	30.2	27.4	19.4	13.4	11.6	9.2	16.2
at 2 cm (% by wt.)	South	17.5	19.4	15.4	8.8	10.2	6.8	14.4

SOURCE: Cooper 1960.
* N.R. = not recorded.

Cold air ponding or cold air drainage into depressions in the landscape alters the climate of the local sites. Cold air, because of its greater density, tends to move down slopes during the night and collect in depressional areas. It is in these depressions that coldest air temperatures have been recorded. Frosts are more likely to occur in these areas.

The properties of the soil in some measure tend to control the climate in which they form. Light, or whitish-colored soil surfaces tend to reflect more radiation than darker-colored soils. When incoming solar radiation is reflected, there is less net radiation to be absorbed and heat the soil. Snow is especially effective in reflecting the incoming radiation.

The amount of water present in the soil at any given time greatly controls the temperature of the soil. Water has a high specific heat capacity (1 cal g^{-1} C), whereas dry soil has a specific heat capacity of about 0.2 cal g^{-1} C. However, once a wet soil is warmed, it takes longer to cool than a dry soil because of the same mechanism. Therefore, under the same climatic conditions, a sandy soil containing only about 15% water at field moisture capacity cools and heats more rapidly than a clay soil holding 30% water.

CLIMOSEQUENCES. The various soil properties discussed in relation to moisture and temperature regimes form a continuous pattern over the landscape.

Examples of climosequences are most often studied in mountainous areas where the temperatures decrease at a rate of about 6 C per 1,000 meters elevation (Petterssen 1941) and rainfall usually increases with elevation. Although the rate of change varies, the following observations have been reported (Fridland 1959; Martin and Fletcher 1943; Vologuev 1964; Whittaker et al. 1968): Organic matter content, nitrogen content, and C/N ratio increase with increased elevation, while pH value and Ca, Mg, and K contents decrease with increased elevation.

Although climosequences are dramatic when compressed into the space of a few miles on a mountainside, such settings are usually atypical of the soils present. Climosequences are best observed over longer transects where the changes in soil characteristics can be observed in several settings.

PALEOCLIMATES. Climate is not constant with time. Although meteorological measurements do not reach into history for any great length of time, there is much geologic and botanical evidence to show that climate has changed through time (Schwarzback 1963). Evidence is steadily accumulating to show that climatic change may be important within the history of many of the soils we presently observe on the earth's surface. Soil is changed in response to changes in its environment so that only irreversible characteristics are likely to remain (Nikiforoff 1953). Therefore, it is difficult to interpret paleoclimatic changes in soils; however, postulation of paleoclimatic effects are frequently used to explain more weathered clay mineral suites than are likely under ambient climates.

Shifts in climate are generally viewed as changes in the general circulation pattern. This means that for a given time the mean position of the major circulation patterns may be slightly shifted so as to modify the climate in one area by increasing the rainfall, decreasing rainfall in another, and probably not creating any appreciable change in many areas. An example could be made by postulating a southerly shift in the mean position of the subtropical high pressure cells over the Atlantic and Pacific oceans to their present winter position. This would probably mean that southern California, where most of the rain falls in the winter, would receive much more annual rainfall. New Mexico, where rainfall comes primarily in the summer when a northwesterly flow of air from the Gulf of Mexico brings moisture, would be drier. In Arizona, where rainfall is divided into a summer and winter rainy period, it is probable that little difference in annual rainfall would result.

CLIMATIC MEANS VS CLIMATIC EXTREMES. Climate can be considered as an average of weather events. We have all observed that one year is not exactly like any other year. We commonly speak of drought years and years of floods. Throughout the history of soil science, attempts have been made to establish a quantitative measure of the relationship of a certain soil property to various climatic elements. This work has led to a better understanding of soil-climate relationships discussed in this chapter. It should not be anticipated, however, that these relationships can ever be exact when only climatic averages are used. The event in the weather that produces a given soil property may be caused by certain infrequent climatic circumstances. For example, in an arid region 1 year in 10 may produce enough rainfall to leach the soluble salts from the solum into the groundwater table. If there is no rapid process by which the salt is replenished, the soil solum remains relatively salt-free. In similar fashion, a few hot days may be more effective in decomposing organic matter than weeks of "average" temperature.

CLIMATE AND SOIL CLIMATE. In addition to thinking of climate as a soil-forming factor, pedologists recently have come to consider the soil climate as a property of the soil. Thus, both mean temperature and annual temperature fluctuations are considered in characterizing soils for classification. In a similar fashion, the content of water in the rooting zone of the solum is used as criteria for soil classification. It is obvious that these parameters cannot be measured with any meaning by one visit to a soil site, and for this reason they have been severely criticized. Also, soil climatic properties are very closely related to ambient atmospheric climates (Smith et al. 1964). Often students have difficulty in seeing a clear separation. Some consider it wrong to use temperature and moisture as soil properties.

It is our conviction that while soil climate is closely related to atmospheric climate, the logic of using soil climate as a soil feature is just as sound as to use texture which in many cases is a direct result of initial material texture. Further, climatic properties of the soil have a direct influence on soil use and can be quantitatively measured and recorded.

Measurements are presently being done on a probability basis, that is, dry for more or less than 60 consecutive days in 7 out of 10 years. (See Chapter 14 for a further discussion on the use of soil climatic data in soil classification.)

LITERATURE CITED

Arkley, R. J. 1963. Calculation of carbonate and water movement in soil from climatic data. Soil Sci. 96:239–48.

——. 1967. Climates of some great soil groups of the western United States. Soil Sci. 103:389–400.

Baldwin, M., C. E. Kellogg, and J. Thorp. 1938. Soil classification, pp. 979–1001. In Soils and men, yearbook of agriculture. U.S. Dept. Agr. U.S. Govt. Printing Office, Washington.

Cooper, A. W. 1960. An example of the role of microclimate in soil genesis. Soil Sci. 90:109–20.

Fridland, V. M. 1959. Vertical soil zonality in North Vietnam. Soviet Soil Sci., 1959, no. 11, pp. 1263–71. (Transl. 1961 from Pochvovedeniye, 1959, 11:8–18, by Am. Inst. Biol. Sci., Washington.)

Geiger, R. 1957. The climate near the ground. (Transl. by M. N. Stewart et al.) Harvard Univ. Press, Cambridge, Mass.

Jenny, H. 1941. Factors of soil formation. McGraw-Hill, New York.

Lang, R. 1920. Vervitterung und Bodenbildung als ejnfuhrung in die Bodekunde. Schweizerbart'sche Verlag, Stuttgart.

Martin, W. P., and J. E. Fletcher. 1943. Vertical zonation of great soil groups on Mt. Graham, Arizona, as correlated with climate, vegetation and profile characteristics. Univ. Ariz. Agr. Exp. Sta. Tech. Bull. 99:89–153.

Meyer, A. 1926. Uber einige Zusammerhange Zwischew Klima und Boeden in Europa, Chemie der Erde 2:209–347.

Nikiforoff, C. C. 1953. Pedogenic criteria of climatic changes, pp. 189–200. In H. Shapley (ed.), Climatic change: Evidence, causes and effects. Harvard Univ. Press, Cambridge, Mass.

Palmer, W. C., and Havens, A. V. 1958. A graphical technique for determining evapotranspiration by the Thornthwaite method. Monthly Weather Rev. 86:123–28.

Petterssen, S. 1941. Introduction to meteorology. McGraw-Hill, New York.

Reed, W. W. 1941. The climates of the world, pp. 665–84. In Climate and man, yearbook of agriculture. U.S. Dept. Agr. U.S. Govt. Printing Office, Washington.

Schwarzbach, M. 1963. Climates of the past: An introduction to paleoclimatology. (Transl. and edited by R. O. Muir.) Van Nostrand, London.

Shul'gin, A. M. 1957. The temperature regime of soils. Gidromeleorologrcheskoe Izdatalel'stovo. (Transl. by A. Gourevich.) Israel Prog. for Sci. Trans. Cat. no. 1357, 1965.

Smith, G. D., F. Newhall, L. H. Robinson, and D. Swanson. 1964. Soil temperature regimes: Their characteristics and predictability. SCS-TP-144. Soil Conserv. Serv., U.S. Dept. Agr.

Thornthwaite, C. W. 1948. An approach toward a rational classification of climate. Geograph. Rev. 38:55–94.

U.S. Department of Agriculture. 1960. Palmer-Havens diagram for computing potential evapotranspiration by the Thornthwaite method. M-3047A, Soil Conserv. Serv., Portland, Oreg.

Van't Hoff, J. H. 1884. Études de dynamique chimique. (Studies of dynamic chemistry.) Frederik Muller & Co., Amsterdam, Holland.

Vologuev, V. R. 1964. Ecology of soils. (Transl. from Russian by A. Gourevich.) Israel Prog. for Sci. Trans., Jerusalem. Publ. in U.S. by Davey, New York.

Whittaker, R. H., S. W. Buol, W. A. Niering, and Y. H. Havens. 1968. A soil and vegetation pattern in the Santa Catalina Mountains, Arizona. Soil Sci. 105:440–50.

✢

Organisms: Biological Portion
of the Soil and Its Environment

THE ROLE of the multilayered biotic community that surmounts and permeates the horizonated soil body is significant in soil genesis. Gleason and Cronquist (1964) liken the lithosphere to an easel, the pedosphere to a canvas, and the biosphere to a painting on the canvas. Yet interactions occur among the three and each layer is changing in character, the surficial one the most rapidly.

Vysotskii (Vilenskii 1957) spoke of the soil as the arena of life and water as the life blood of the soil. "Life is the process of maintaining bodies of protein" in which the soil is intimately involved. Wilde (1954) points out that the biologic factor in soil formation is so influential and versatile that even soil reaction is not always a dependable criterion in pedologic theory, as shown by the existence of calcareous Podzol soils and very acid nonpodzolic mull soils. Polynov (1951) noted that the relative mobility from land to sea of the four elements Na > K > Ca > Mg has been changed by soil and biota to Ca > Na > Mg > K. Since life first appeared on this planet the biosphere has accumulated oxygen in the atmosphere, calcium in fossiliferous limestone, and carbon in fossil fuels. The evolutionary succession of conifers, angiosperms, and grasses has meant increasing efficiency in biocycling phosphorus and in protecting the soil against erosion.

THEORETICAL MODELS. The equation (Jenny 1941)

$$S = f \, o \, \text{(organisms)}_{cl, \, r, \, p, \, t \ldots} \tag{11.1}$$

means that the soil is a function of organisms acting together as an independent variable when the other factors, climate, relief, parent material, and time, are kept constant. We may consider the biotic factor, organisms, as one of a number of flux potentials (Px of Jenny 1961) introduced into the ecosystem from outside. The independent variable concept is useful in theoretical considerations despite the fact that in most actual situations dependent variable relationships exist.

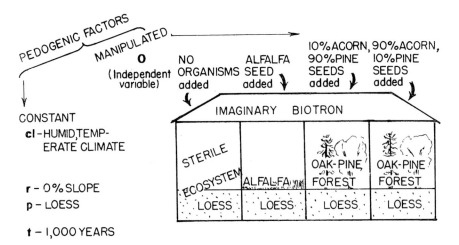

FIG. 11.1. Theoretical model illustrating the concept of the active factor of soil formation, organisms, as an independent variable. (Adapted from Jenny 1958.)

The four chambers in the imaginary biotron of Figure 11.1 (adapted from Jenny 1958) illustrate this approach. Each contains a layer of loess with a level surface maintained under an identical humid temperate climate for 1,000 years. The only pedogenic factor which is unique in each chamber is the organic one. No organisms are introduced in the first room, alfalfa seed is in the second, 10% acorns and 90% pine seeds are in the third, and the reverse proportion is in the fourth. The alfalfa is harvested annually to stimulate growth, and reseeding takes place as needed to maintain a uniform stand. Plant nutrients are supplied in amounts equal to those removed by harvest. Necessary symbiants are added in the last three chambers.

It is postulated that at the end of 1,000 years the profiles in the four rooms would be: (1) no soil, but a very slightly leached and eroded loess profile; (2) a young alfalfa *(Medicago sativa)* soil; (3) identical young oak-pine *(Quercus-Pinus)* soils under cover of mixed pine-oak forest, having about equal proportions of each species. The supposed convergence in makeup of the forest communities in two of the chambers demonstrates the relationship between the independent variable (o) and the two dependent ones (o', o''). The first is one mix of seed in chamber 3 and another mix in chamber 4. The second two are the actual above-ground (o') and below-ground (o'') portions of the biotic communities that developed dependently on initial material, levelness of terrain, climate, and span of time. Because these last four factors were constant, the two forests gradually assumed identical characteristics.

The direct impact of organisms (seeds, spores, invading fauna) from the outside onto fresh material in the hypothetical experiment can occur naturally, as on fresh mudflow, new volcanic ash layer, or sea-bottom exposed above tide by uplift. With establishment of biota on the new terrain, organic agencies external (*o*) to the ecotessera (Jenny 1965) (Fig. 11.2) are screened from contact with initial material (*p*) by layers of the ecosystem of chambers 3 and 4 (Fig. 11.1). Two subspecies of the oak-pine soil would be present because the tessera (a sampled column, including a tree plus soil solum) would be different at a pine tree from that at an oak tree. One would expect to find intergrade soil profiles between trees of different species. Further pedologic complexity would develop because of stem flow of rainwater and other variations in microclimatic factors (*cl′* and *cl″*). Yet most variations of soil properties in chambers 3 and 4 would be within the range of a single soil series, type, and slope phase.

The independent variables may now be listed in an expanded equation:

$$o', \; cl', \; t + o'', \; cl'', \; r'' \;\; p'', \; t, \; s_1, \; s_2, \; s_3, \; \ldots = f \; o_{(cl, \, r, \, p, \, t)} \tag{11.2}$$

Here the dependent assemblages of organic and microclimatic factors (*o′*, *cl′*) and the soil-enclosed organisms, climate, and particles of initial material (*o′*, *cl′*, *p′*) and various soil properties (s_1 . . .) in the rooting zone are functions through time of the external organic factor (*o*) which is varied while external climate, initial material, and relief are held constant.

NATURAL SYSTEMS. The influence of the organisms (*o′*) on soil formation may be illustrated by observations in contrasting biotic communities, and under certain components of them, such as individual trees and colonies of insects.

Pedologists have long been interested in the contrasts between soils formed in fairly close proximity under prairie, forest, and prairie-forest tran-

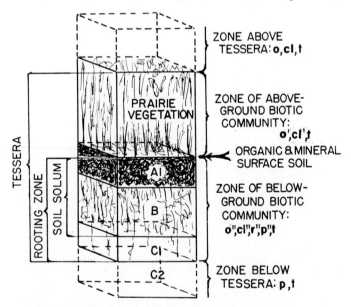

FIG. 11.2.   Diagram of an arbitrarily delimited column in an ecosystem (a tessera) with zones in which the active factor of soil formation, organisms, acts independently of other factors, and dependently on them.

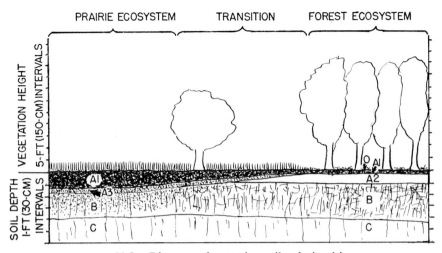

FIG. 11.3. Diagram of organism-soil relationships at a prairie-forest border.

sition (Fig. 11.3) (Jones and Beavers 1964; White and Riecken 1955). Data in Table 11.1 show how soil properties (s_1, s_2 . . .) vary from one of these ecosystems to another on uniform initial materials in the north central region of the United States.

Within a distance of 40 kilometers in Menominee County in northeastern Wisconsin, a forested area that escaped clear-cutting in logging days

TABLE 11.1 ❧ Some properties of soils developed under prairie, prairie-border, and nearby deciduous forest cover in the north central region of the United States

Soil Property	Soil Formed under Prairie	Soil Formed under Prairie-border Conditions	Soil Formed under Deciduous Forest
Clay content (%)			
A horizon	28	23	21
B horizon	34	36	36
Ratio, B:A	1.21	1.57	1.71
Total nitrogen (%)			
A11 (0–7 cm)	0.35	0.35	0.35
A12 or A2 (20–30 cm)	0.21	0.11	0.08
B2 (55–65 cm)	0.09	0.06	0.05
Soil reaction (pH)			
A11 (0–7 cm)	5.1	6.1	6.5
B2 (55–65 cm)	5.2	5.0	5.2
C (115–130 cm)	6.2	6.0	5.2
Base saturation (%)			
A11 (0–7 cm)	90	68	80
A12 or A2 (20–30 cm)	68	80	42
B2 (55–65 cm)	78	87	78
Silt-size opal phytoliths (mT/Ha) in solum (largely in A)	20	10	5

SOURCE: Based on representative data from Daniels, Brasfield, and Riecken (1962); Jones and Beavers (1964); Riecken (1965); and White and Riecken (1955).

by virtue of being an Indian reservation, three major forest communities occur with correspondingly distinct soil profiles: (1) hemlock forest [Podzol (Spodosol) soil], (2) hemlock-northern hardwood forest [double profile soil: weak Podzol (Spodosol) over weak Gray-Brown Podzolic (Alfisol) soil], and (3) northern hardwood forest [weakly developed Gray-Brown Podzolic (Alfisol) soil] (Milfred, Olson, and Hole 1967). Disturbances by fire and windstorm without fire alter the hemlock forest ecosystem most easily and the hardwood forest the least, with corresponding changes in soil profiles (Fig. 11.4). This biosequence differs from the one of Figure 11.3 particularly in being dependent on a lithosequence (sand soils-sandy loam soils-loam soils), and to that extent is more divergent from the theoretical model of Figure 11.1.

Studies of soils in ancient coastal redwood (*Sequoia sempervirens,* Don, Endl.) and subalpine bristlecone pine *(Pinus aristata* Engelm.) forests of California (LaMarche 1968; Zinke and Crocker 1962) reveal highly weathered "moderately" developed soils (Dystrochrepts, Argixerolls, Hapludults) under the redwoods and an unusual very gravelly Ustochrept under the pine. The Ustochrept on Reed Dolomite appears to be a carbonate analogue of a Humod with O-A2-Bh-Ccam-Cca horizon sequence 60 cm (24 in.) deep (Hole 1968). The Bh horizon has an organic matter content of about 15%, dry weight.

Leachates from bark and other organic debris under the Kauri tree *(Agathis australis)* in New Zealand have an unusual capacity for podzolization of the underlying soil profile. The A2 horizon may be as much as 183 cm (5 ft) thick under very old trees. Because of its basinlike form (Fig. 11.5), the body of Podzol (Spodosol) soil under each tree has been called "basket"-Podzol or "egg-cup" Podzol (Spodosol). Properties of one particular "basket"-Podzol having an unusually thick A1 are characterized by data in Table 11.2 (Swindale 1955).

Colonies of the eastern mound-building ant, *Formica exsectoides,* build mounds as much as a meter high and two meters in diameter in openings in deciduous forests of northeastern United States (Salem and Hole 1968). The soil profile at a mound (Fig. 11.6) is very different from between-mound profiles (Table 11.3) and may bear evidence of the ant activity centuries after this has ceased.

SOIL PATTERN AS A COMPLEX MOSAIC. It is evident that an ecosystem typically consists of a complex mosaic of overlapping tesseras that reflects within the limitations of time and materials not only present but also past patterns of distribution of biota. A soil profile adjacent to the bole of a beech tree *(Fagus grandifolia)* may differ from the less watered profile 2 meters away from the tree trunk (Gersper and Holowaychuk 1970) as much as a Brown-Forest-Gray-Brown Podzolic intergrade soil differs from a well-developed Gray-Brown Podzolic soil.

Complexity also exists on microscopic and submicroscopic scales (Fig. 11.7) (Gray 1967; Jackson 1965; Jenny and Grossenbacher 1963; McLaren and Skujins 1968). The generalization is made that a gram of soil may

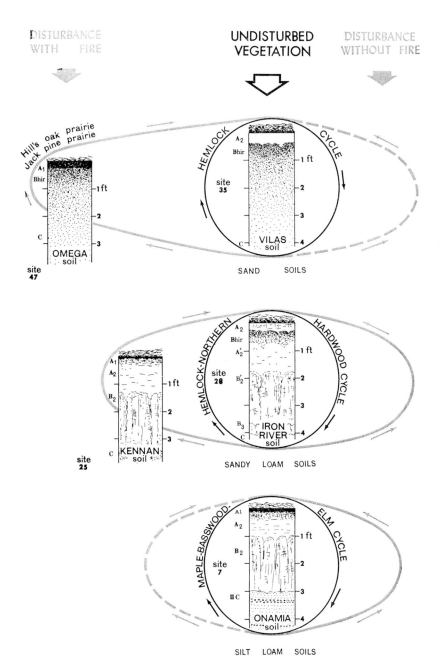

DISTURBANCE
WITH   FIRE

UNDISTURBED
VEGETATION

DISTURBANCE
WITHOUT FIRE

FIG. 11.4.   Diagram illustrating five important kinds of
soil-plant communities on a lithosequence in forest
lands of Menominee County in northern Wisconsin.
(Courtesy of Geological and Natural History Survey, Univ.
Extension, University of Wisconsin.)

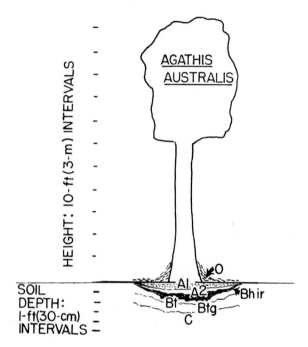

FIG. 11.5.   Diagram of a cross section through a "basket-Podzol" soil formed under a Kauri tree in New Zealand. (After Swindale 1955.)

contain a million bacteria, a kilometer of fungal hyphae, and thousands of algal and protozoan cells. Yet there is a considerable proportion of scantily populated area within the same gram of soil, partly as a result of insufficiency of certain growth requirements and partly the effect of diffusion of growth-inhibiting substances.

GENERAL BIOTIC PROCESSES
OF SOIL FORMATION. Probably the four most important pedogenic processes going on in a tessera are capture of energy and substance through photosynthesis, the reverse of it which is decomposition of plant residues, cation exchange, and formation of organomineral complexes. The materials which the biota of an ecosystem mobilize and circulate tend to be deposited within the tessera and recycled many times in it before eventual escape, as on the occasion of a forest fire that suddenly releases stores of nutrients. Local nitrogen and calcium cycles are just two examples of the many biocycles within a biota-sola combination. Many organisms produce substances that are toxic to other organisms and thereby set limits, at least temporarily, to populations and distributions in a landscape. Leaf litter accumulates on the ground surface. Coatings of organic matter form on ped surfaces along with clays and chelated iron oxides. Most of the organic material in these concentrations is subject to rapid breakdown. An exception to this may be proteinaceous materials of C/N ratio as low as four that are complexed with clay in the subsoil. Probably much more of the energy captured by photosynthesis of terrestrial plants is liberated by organisms

TABLE 11.2 ❧ Some properties of a "basket"-Podzol soil profile formed under a Kauri tree (*Agathis australis*) in New Zealand

Soil Horizon (depth in cm)	Moist Color	Particle Size Distribution			Texture	Consistence	Structure	Free Fe	Quartz/ Feldspar	Clay Mineralogy	
		Clay	Silt	Sand						Kao- linite	Gibb- site
		(%)	(%)	(%)				(%)		(%)	(%)
A1 (0–<27)	10YR 4/1 dark gray	6	47	47	sandy loam	friable	massive	0.01	3.5	3	...
A2 (0<27–40)	2.5Y 7/2 light gray	8	40	52	sandy loam	cemented	massive	0.02	2.7	10	1
Bh (40–48)	5YR 2/1 black	10	36	54	sandy loam	...	massive	0.53	1.8	5	3
Bir (48–49)	2.5YR 3/4 dark reddish brown	...	...	...	...	indurated	massive	...	...	19	33
Bt (49–60)	7.5YR 5/6 strong brown	18	44	38	loam	firm	massive	5.35	1.3	56	7
Btg (60–90)	2.5Y 7/4 pale yellow, mottled	35	53	12	silty clay	...	...	0.78	0.6	...	...
C (90–120)	...	9	60	41	silt loam	...	...	0.73	0.3	26	...
C2	Greenish brown greasy strongly weathered clay										
R	Putahi Hill rhyolite										

SOURCE: Swindale 1955.
NOTE: "Basket"-Podzol refers to Parahakai sandy loam developed under an annual rainfall of 62 in. and mean annual temperature of 58 F, Bay of Islands County, Omapers S.D., New Zealand.

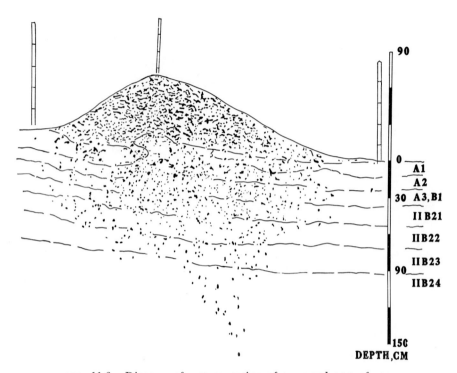

FIG. 11.6. Diagram of a cross section of a mound nest of a colony of the eastern mound-building ant *(Formica exsectoides)*. (Courtesy of Salem and Hole 1968.) (Horizontal and vertical scales are the same.)

in the soil profile than through food chains above it. Mineral components released through decomposition of organic matter are largely reabsorbed by biota. Respiration of soil fauna and flora during the decomposition process can evolve five to ten liters of CO_2 per square meter of land surface per day. Rearrangement of soil materials by plants and animals involves roughening of the soil surface, formation of channels, and their partial filling. Daily, tidal, seasonal, and annual rhythms and cycles of biotic activity and transformations take place on and in the soil. Cycles of stability and instability of a landscape surface result from climatic changes or interruptions of plant successions (Fig. 11.4), with corresponding effects on soil profiles.

SOME BIOGENIC SOIL FEATURES. In the infinite array of soil features formed by plants and animals, several stand out as important and easily observable. These include biotically shaped peds and voids, concentrations of both organic and mineral matter, and certain irregularities in the soil surface.

BIOTICALLY SHAPED PEDS AND VOIDS. Much of the organic material of the organic horizons has acquired a granular or fine blocky struc-

TABLE 11.3 ❧ Properties of soil profiles at (AM) and near a mound (NM) of the eastern mound-building ant (*Formica exsectoides*)

Horizon (AM)	(depth in cm) (NM)	Bulk Density AM	Bulk Density NM	Vol. of Ant Channels AM (%)	Vol. of Ant Channels NM (%)	pH (paste) AM	pH (paste) NM	Clay AM	Clay NM	Silt AM	Silt NM	Sand AM	Sand NM	Free Iron Oxides AM (%)	Free Iron Oxides NM (%)	Organic Matter AM (%)	Organic Matter NM (%)
O1 and Ar* (0–0.3)	...	0.8	...	...	...	7.0	...	49	...	46	...	5	...	5	...	4	...
C (0.3–40)	...	0.8	...	21	...	7.2	...	50	...	47	...	3	...	5	...	3	...
A21 and A3b (40–81)	O1 (2.5–0) and A1, A2 (0–35)	1.0	1.0	8	0	6.3	6.2	20	20	77	75	3	5	4	2	3	3
B1b (81–91)	B1 (31–45)	1.2	1.4	6	0	5.7	5.6	27	27	71	70	2	3	2	2	1	1
IIB21b (91–100)	IIB21 (45–55)	1.2	1.5	4	0	5.7	5.4	27	30	69	66	4	4	2	2	0.8	0.3
IIB22b (100–120)	IIB22 (55–73)	1.2	1.5	2	0	5.5	4.5	43	47	55	51	2	2	3	3	0.6	0.3
IIB23b (120–135)	IIB23 (73–93)	1.5	1.5	0.4	0	4.8	4.3	74	83	25	15	1	2	7	6	0.3	0.3

SOURCE: Salem and Hole 1968.
* Ar refers to a concentration of chert, coarse sand, and fine gravel on the mound surface.

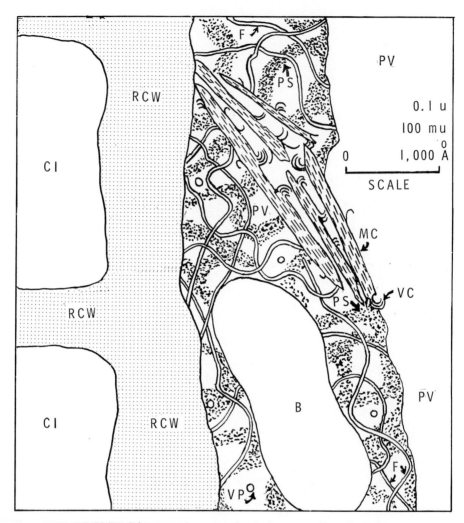

FIG. 11.7.   Diagram of contact between root cell wall *(RCW)*
and organic mucigel (made up of cellulose microfibrils, *F;*
and pectin substances, *PS*) and clay *(MC* is mica clay;
VC is vermiculite clay). *B* is bacterium; *VP,* virus particle;
PV, pore (void);   *CI,* cell interior.

ture from passage through alimentary tracts of insect larvae, worms, and
associated soil fauna.   The granular structure of dark surface soils (A1
horizon) is largely attributed to the work of earthworms, one species of
which pulls into its burrows virtually every leaf that falls to the forest
floor (Nielsen and Hole 1964).   Burrowing cicada nymphs create a special
"cylindrical blocky" structure in subsoils of well-drained silty soils of Brown
and Sierozem soil groups (Hugie and Passey 1963).   Pressure of growing roots
helps shape small soil peds.

Forms of voids as well as of solids can be attributed to biota. Insects smaller than the thickness of a leaf make tunnels through foliage even before it falls to the soil surface. Standing trees and fallen logs are riddled with insect galleries and hollows made by rodents and birds. Burrows of animals ranging in size from earthworms to rabbits and badgers interlace soil profiles, along with their fillings (krotovinas), usually of dark A1 horizon material. Channels made by roots and rhizomes influence the flow of water and air, thereby setting the pattern for mottling and iron oxide precipitates in many soils. Roots have considerable powers of penetration and expansion. A quack grass *(Agropyron repens)* rhizome grew straight through a carrot in the garden of one of the authors and hence could obviously have pierced a firm soil ped.

BIOLOGIC CONCENTRATIONS OF ORGANIC MATTER. Four major kinds of organic horizons in mineral soils have been widely recognized: mor, mull, sward, and Orterde.

The term *mor,* first used by a Danish scientist, P. E. Müller (Wilde 1958), refers to the surficial forest soil horizon of acid litter and humus. This horizon is best developed under coniferous trees and heath. Fungi are more important than bacteria in the decomposition process and produce significant amounts of antibiotic compounds. Available nutrient levels are low. Microfauna, such as collembola and orbitids, are abundant and account for a respiration rate per unit of biomass more than twice that for mull. An accumulation of feces of these little animals is a powdery mass called moder. The ratio of volume of living organisms to volume of organic material is about 1:30,000 in the litter layer (O1) horizon and 1:15,000 in the humus layer (Eaton and Chandler 1942).

The term *mull* designates a forest soil horizon (A1) of intimately mixed mineral matter and amorphous humus. It is slightly acid and is best developed under base-rich litter such as beneath maple trees *(Acer).* Bacterial activity exceeds that of fungi. Available nutrient levels are high. Earthworms and millepeds are abundant. The number of animals per unit volume of mull is about a third more than in mor, and oxygen consumption is three times greater.

Sward is a term coined by Wilde (1958) for the dominantly rhizogenous A1 horizon in grasslands as contrasted with the principally zoogenous mull horizon of forest soils. About 90% of the total organic matter of a prairie tessera is below ground, and the thick dark sward horizon gives evidence of this. The Mollic epipedon of the new USDA soil classification is a more general unit that includes this horizon as well as many Ap horizons formed by cultivation of forest soils.

The B horizon of a humus Podzol (Humod), called Orterde by German pedologists, is a soft, weakly granular layer that is dark reddish brown to black, and with an organic matter content ranging as high as 15%, dry weight.

Another form of dark organic matter in soils is the humic film ("organ" of Brewer 1964) on the surface of peds of B horizons, as in some Brunizems

(Argiudolls), Solodized Solonetz soils (Natrustolls), and Gray-Brown Pod-zolic soils (Hapludalfs).

Light-colored A horizons may contain considerable accumulations of light-colored organic matter as reported by Wilde (1950) for a forest soil in Indiana with as much as 6% organic matter content in the first 15 cm.

Irregularities in the lower boundaries of soil horizons may be deter-mined by biotic activity. To some extent the tonguing of a Podzol B hori-zon may have been initiated by tree roots. Pencil-size extensions of the A1 horizon into the A2 in Gray-Brown Podzolic soils (Hapludalfs) are probably krotovinas in worm channels.

SURFICIAL DEPOSITS AND CHANNELS OF BIOLOGIC ORIGIN. Deposition of soil on the land surface is accomplished by trees, animals, and men.

Baxter (in Milfred, Olson, and Hole 1967) notes that large trees with sound trunks and shallow rooting systems have blown down in violent windstorms in northern Wisconsin to produce "cradle knolls" with a relief of about a half meter and a width from mound crest to cradle center of about 1.2 meters. Most of the approximately 350 cradle knolls per hectare were made during storms of more than a century ago, or at least long enough ago to allow for the complete decomposition of the fallen trees. Rate of production of these features has been about 1% of the land surface every 500 to 1,000 years.

Work of mound-building ants has already been alluded to. Lyford's study (1963) of ants that make small (1 cm high and 7 cm wide) surficial deposits of soil revealed that the upper 35 cm of the Acid Brown Forest soils (Dystochrepts) of the Harvard Forest have been completely churned by these insects. On the Great Plains, graveled bare areas as much as 6 meters in diameter are created by ants (Thorp 1949). Pathways made by ants in tropical rain forests are extensive. In any case, surface clearings and accumulations of soil by ants probably affect 1 to 4% by area of a land-scape. This may include wetlands where the insect nests are elevated.

The termite is a primitive insect that works in a humid atmosphere at all times and is particularly active in intertropical regions. Termataria as large as 3 meters in height and 15 meters in diameter and as numerous as 75 per hectare are reported in the literature (see Fig. 23.2). In a pasture in the USSR, bald spots made by these nests occupied 20% of the area. In semiarid regions, termites tunnel as deep as 8 meters to maintain a humid environment in the nest. Accumulations in termataria include or-ganic matter, salts, and carbonates as well as moisture. It is possible that over significant areas soil profiles of intertropical regions have been thor-oughly reworked, and in that sense rejuvenated by termites for hundreds of thousands of years.

Channels are excavated in the soil to depths of usually between 1 and 30 centimeters by grasshoppers, wild bees, wasps, and spiders.[1] Burrows of

1. S. E. Reichert of the University of Wisconsin Zoology Department (Madison), a specialist in prairie spiders, is of the opinion (1970) that spiders effect more pedoturbation per unit area in most Wisconsin prairies than do ants.

crustacea, reptiles, amphibia, moles, voles, shrews, rodents, foxes, badgers, and birds are locally important. Thorp (1949) estimated that prairie dogs and badgers have moved 90 metric tons per hectare of gravelly subsoil to the surface. Surface soil texture over a third of the area of the experimental plots at Akron Station in Colorado had been changed from silt loam to loam by these animals. Crabs and crayfish may move several tons of soil to the surface of a hectare each season. In New Zealand a petrel *(Pachyptela turtur)* nests in extensive burrows that it makes in the soil. The beaver covers soil with dams and also with impounded water. When dams break and are abandoned, sedge prairies and Mollisols may develop in the midst of forest on the old pond sites. Crows roosting in pine trees in the University of Wisconsin arboretum cough up fine gravel with bud scales from their craws at a rate that over a period of a thousand years could change the texture of the silt loam beneath them to a gravelly silt loam.

The work of man (Bidwell and Hole 1965) in furrowing fields, removing hills, filling in low places, reducing (or increasing) nitrogen content of soil, and accelerating erosion is familiar to us. In rice paddies, farmers develop new kinds of soil profiles in place of the natural ones. In modern cities, engineers prevent soil formation by sealing the land surface with pavement and buildings.

CONCLUDING CONSIDERATIONS. Detailed as Table 11.4 may appear, it gives only a grossly simplified indication of the variety, numbers, and total weight of organisms in the upper 15 cm of soil. The estimated dry weights are about a fourth to a fifth of live weights (biomass), and percentages by dry weight of soil are probably less than a third of the proportionate volumes occupied by biota. The table makes no attempt to indicate the wide ranges of fluctuations of populations within a matter of hours for the smallest organisms, to seasons and years for large animals and plants and man himself. Animal biomass in a natural ecosystem is typically less than 1% of the plant biomass from which primary production comes.

Net primary production of organic matter in an ecosystem is about half of gross primary production because of the large amount of material consumed in respiration by the plants. Ovington (1962) reports representative figures for net primary production in multiples of 10^3 kg/ha ranging from 40 in tropical forests to 14 in a Danish beech forest, 9 for a maize field, 3 for a Minnesota grassland, and 1 for an alpine meadow. Total biomass of fauna and flora during a favorable season is probably 2 to 30 times this amount, the larger factor being applicable to forest ecosystems where large tonnages of dead heartwood are usually counted as part of the standing live crop.

Attempts to equate different animals in terms of biomass or of a particular function (Mellanby 1960) have been less than successful. Thirteen million slugs and four sheep per hectare may have about the same biomass but do not perform equivalent work in terms of soil information. With the help of modern machinery and fossil fuel representing work of organisms that stored energy millions of years ago, a farmer weighing 100 kg

TABLE 11.4 &⅃ Estimates of amount of organic matter and proportions, dry weight, and number of living organisms in a hectare of soil to a depth of 15 cm in a humid temperate region

Item	Dry Weight		Estimated Number of Individuals
	(%)	(kg/ha)	
Organic matter, live and dead	6	120,000	...
Dead organic matter	5.28	105,400	...
Living organisms	0.72	14,600	...
Bacteria	0.10	2,600	2×10^{18}
Protozoa	0.005	100	7×10^{16}
Fungi	0.10	2,000	8×10^{16}
Actinomycetes	0.01	220	6×10^{17}
Algae	0.0005	10	3×10^{14}
Higher plants, roots	0.5	10,000	...
Animals			
Nematodes	0.001	20	2.5×10^{9}
Gastropods (snails, slugs)	0.001	20	2×10^{3}
Annelidae (earthworms, potworms)	0.005	100	7×10^{3}
Arthropods			
Crustacea (isopods, crayfish)	0.0005	10	4×10^{17}
Arachnidae			
Acarine (mites)	0.0001	2	4×10^{5}
Araneae (spiders)	0.0001	2	5×10^{5}
Myriapoda (millepedes, centipedes)	0.001	20	1×10^{3}
Opiliones (harvestmen)	0.00005	1	2.5×10^{4}
Diplopoda, Chilopoda, Symphyla	0.0011	25	3.8×10^{7}
Hexapoda			
Collembola (springtails)	0.0001	2	4×10^{5}
Hymenoptera (ants)	0.0002	5	5×10^{6}
Diptera, Coleoptera, Lepidoptera	0.0015	35	5×10^{7}
Vertebrates			
Mice, voles, moles	0.0005	10	4×10^{5}
Rabbits, squirrels, gophers	0.0006	12	10
Foxes, badgers, bear, deer	0.0005	10	<1
Birds	0.0005	10	100

can in less than four days in a year do all the necessary operations on a hectare to plant, cultivate, manage, and harvest a crop of maize, all with effect on the soil. Each kind of organism and symbiotic unit of two or more probably has unique pedologic significance.

Not only does the tessera bear the brunt of impact of rain, snow, and hail, but it is a consumer of water on a large scale. Three hundred to a thousand pounds of water are moved through the plant in the process of production of each new pound (dry weight) of plant material. The role of the vegetation in water disposal becomes evident in particularly sensitive landscapes. For example, Thorp (1957) reports that in a landscape of western Australia the annual precipitation of 76 cm (30 in.) is just adequate to leach cyclic salts from the solum as fast as they are brought in by air. When European settlers ringed and killed trees of the open eucalyptus forest to increase pasturage, water no longer consumed by the trees on hillsides flushed salts from a deeper soil zone to the surface on footslopes, where new Solonchak (Natrustolls) soils and some bare salt spots developed. After about a century, the excess sodium was removed from these areas and the readjusted ecosystem was again able to cope with the annual quota of salts without interruption of vegetative cover.

LITERATURE CITED

Bidwell, O. W., and F. D. Hole. 1965. Man as a factor of soil formation. Soil Sci. 99:65–72.

Brewer, R. 1964. Fabric and mineral analysis of soils. John Wiley & Sons, New York.

Daniels, R. B., J. F. Brasfield, and F. F. Riecken. 1962. Distribution of sodium hydrosulfite extractable manganese in some Iowa soil profiles. Soil Sci. Soc. Am. Proc. 26:75–78.

Eaton, T. H., Jr., and R. F. Chandler, Jr. 1942. The fauna of forest-humus layers in New York. Cornell Agr. Exp. Sta. Memoir 247.

Gersper, P. L., and N. Holowaychuk. 1970. Effects of stemflow water on the Miami soil under a beech tree. I. Morphological and physical properties. Soil Sci. Soc. Am. Proc. 34:779–86.

Gleason, H. A., and A. Cronquist. 1964. The natural geography of plants. Columbia Univ. Press, New York.

Gray, T. R. G. 1967. Stereoscan electron microscopy of soil organisms. Science 155:1668–70.

Hole, Francis D. 1968. What is a well-developed soil? Soil Survey Horizons 9:12–13.

Hugie, V. K., and H. B. Passey. 1963. Cicadas and their effect upon soil genesis in certain soils in southern Idaho, northern Utah, and northeastern Nevada. Soil Sci. Soc. Am. Proc. 27:78–82.

Jackson, M. L. 1965. Clay transformations in soil genesis during the Quaternary. Soil Sci. 99:15–22.

Jenny, H. 1941. Factors of soil formation. McGraw-Hill, News York.

———. 1958. Role of the plant factor in the pedogenic functions. Ecology 39:5–16.

———. 1961. Derivation of state factor equations of soils and ecosystems. Soil Sci. Soc. Am. Proc. 25:385–88.

———, 1965. Tessera and pedon. Soil Survey Horizons 6:8–9.

Jenny, H., and K. Grossenbacher. 1963. Root-soil boundary zone as seen in the electron miscroscope. Soil Sci. Soc. Am. Proc. 27:273–77.

Jones, R. L., and A. H. Beavers. 1964. Variation of opal phytolith content among some great soil groups in Illinois. Soil Sci. Soc. Am. Proc. 28:711–12.

LaMarche, V. C., Jr. 1968. Rates of slope degradation as determined by botanical evidence, White Mountains, Calif. U.S. Geol. Surv. Professional Papers 352:341–77.

Lyford, W. H. 1963. Importance of ants to Brown Podzolic soil genesis in New England. Harvard Forum Paper 7. Petersham, Mass.

McLaren, A. D., and J. Skujins. 1968. The physical environment of microorganisms in soil, pp. 3–24. In T. R. G. Gray and D. Parkinson (eds.), The ecology of soil bacteria: An international symposium. Univ. Toronto Press, Toronto.

Mellanby, K. 1960. The biological importance of the invertebrate soil fauna. Soils Fertilizers 23:8–9.

Milfred, C. J., G. W. Olson, and F. D. Hole. 1967. Soil resources and forest ecology of Menominee County, Wisconsin. Nat. Hist. Survey, Univ. Extension Bull. 85, Soil Series 60. Madison, Wis.

Nielsen, G. A., and F. D. Hole. 1964. Earthworms and the development of coprogenous A1 horizons in forest soils of Wisconsin. Soil Sci. Soc. Am. Proc. 28:426–30.

Ovington, J. D. 1962. Quantitative ecology and the woodland ecosystem concept. Adv. Ecol. Res. 1:103–92.

Polynov, B. B. 1951. Modern ideas of soil formation and development. (Transl. from Russian. Pochvovedeniye, 1948, pp. 3–13.) Soils Fertilizers 14:95–101.

Riecken, F. F. 1965. Present soil-forming factors and processes in temperate regions. Soil Sci. 99:58–64.

Salem, M. Z., and F. D. Hole. 1968. Ant *(Formica exscetoides)* pedoturbation in a forest soil. Soil Sci. Soc. Am. Proc. 32:563–67.

Swindale, L. D. 1955. Mineralogy and genesis of some rhyolite-derived soils of New Zealand. Ph.D. thesis, Univ. Wis., Madison.

Thorp, James. 1949. Effects of certain animals that live in soils. Sci. Monthly 68: 180–91.

———. 1957. Report on a field study of soils of Australia. Sci. Bull. 1, Earlham College, Richmond, Ind.

Vilenskii, D. G. 1957. Soil science, 3rd ed. (Transl. by A. Birrow and Z. S. Cole.) Israel Prog. for Sci. Trans., Jerusalem, 1960. Available U.S. Dept. Commerce, Washington.

White, E. M., and F. F. Riecken. 1955. Brunizem-Gray-Brown Podzolic soil biosequences. Soil Sci. Soc. Am. Proc. 19:504–9.

Wilde, S. A. 1950. Crypto-mull humus: Its properties and growth effects (a contribution to the classification of forest humus). Soil Sci. Soc. Am. Proc. 15:360–62.

———. 1954. Reaction of soils: Facts and fallacies. Ecology 35:89–92.

———. 1958. Forest soils: Their properties and relation to silviculture. Ronald Press, New York.

Zinke, P. J., and R. L. Crocker. 1962. The influence of giant sequoia on soil properties. Forest Sci. 8:2–11.

Time as a Factor of Soil Formation ❧

W E MAY REGARD time, like space, as continuous, yet recognize a "time zero" for a given soil. Time zero is the point in time at which a pedologically catastrophic event is completed, initiating a new cycle of soil development. The catastrophe may be a sudden change in topography of land surface or of water table, as caused by geologic uplift or tilting of a mass of the lithosphere; rapid shift of a retreating slope due to geologic erosion (Ruhe 1960; the "topographic unconformity" of Thwaites 1946); or man-initiated accelerated erosion or land forming. Sudden change in vegetation may be caused by the felling of a forest in lumbering operations and introduction of agricultural crops, or by burning of a forest and its replacement by prairie. Sudden climatic change may be of macroclimate (Bryson and Wendland 1967) or microclimate. Microclimate varies with condition of vegetative cover. A change in initial material may be by a significant sedimentary deposit (loess, volcanic ash, sand, colluvium), lava flow, or addition of fertilizers or salts from salty water or wind-blown dust.

Theoretically, one can keep other state factors constant (climate, organic disseminules, topography, parent material) and allow time to be the only variable of soil development.

The relation of soils to time may be discussed with respect to (1) relative stage of development, (2) absolute dating of horizons and profiles, (3) rate of formation, (4) relation to age of slope and land form and associated weathering complex, and (5) inspection of man-made experiments out-of-doors and in the laboratory. Furthermore, we may consider the relevance to soil genesis of the geomorphic theories of Davis (1899), Penck (1924), and Hack (1960).

RELATIVE STAGE OF SOIL DEVELOPMENT. The Davisian terms— youth, maturity, and old age (senile), originally developed for description and identification of landscapes—have been applied to soils. Azonal (A-C) soils, including Entisols and some Inceptisols, may be considered *young*, or youthful. Immature soils include the intrazonal ones, the development of which has been dominated by excessive water, salts, or carbonates. *Mature soils* are thought of as zonal ("normal") soils, in equilibrium with the environment. *Senile soils* are pedogenic accumulations of inert material—sesquioxides and heavy minerals. Possible fates of a soil are (1) to continue indefinitely as a zonal soil whose profile sinks into initial material as fast as erosion removes soil, (2) to become a senile soil, (3) to become the parent material of another soil because of a great and relatively sudden change in climate and/or biota, (4) to be buried, or (5) to disappear altogether as a victim of an erosional event. In the first case, time ceases to be important and the soil never becomes senile. In the second case, time is not important after the stage of senility has been reached. In the third case, a succession of soils may reflect plant succession. It is important to note that some soil profiles undergo a decline in permeability with time. Pseudogleys in Europe and Planosols in central United States are examples. In the older terminology, therefore, an azonal soil may become a zonal one and, in turn, be transformed into an intrazonal one.

ABSOLUTE DATING OF SOIL HORIZONS
AND PROFILES. Dating of soils has been by tree-ring counts and by the ^{14}C method. Calcareous dunes in England were dated by tree-ring counts on *Salix repens,* and time required for decalcification of the dunes was determined to be about 300 years. The development of a stable content of organic matter (8%) took 240 years (Salisbury 1925). Dutch polder clays took about the same length of time to be decalcified, but at a slower initial rate (Hissink 1938). Chandler (1937) studied forest soils on glacial moraines in Alaska and found a litter layer well developed in 15 years, a brownish A horizon in silt loam by 250 years, and a Spodosol (Podzol) profile 10 inches thick in 1,000 years. Soil development from bare rock to muskeg took about 2,000 years at one site in Alaska (Crocker and Major 1955). Soil on Weichsel glacial drift (72,000 years old) in western Europe is classified as Braunerde. Soil on Saale drift (112,000 years old) is a pseudo-gley soil (Stremme 1960). A Spodosol (Podzol) buried under peat in Northern Ireland (Proudfoot 1958) was dated by artifacts and paleobotany and shown to have formed in about 1,000 years during the period 3,000–2,000 B.C.

Dickson and Crocker (1954) used tree rings to establish ages of five hornblende/andesitic tuff-breccia sandy loam mudflows in northern California at 27, 60, 205, 566, and 1,200 years, all with a simple slope of about 5% and with Ponderosa pine vegetative succession, in a wet-dry climate with the dry period in the summer. All soil samples were taken under pine trees.

As soon as the vegetation and N-fixing bacteria became established on a body of this mudflow material, soil genesis began. A weathering gradient (involving temperature and water) and organic matter gradient were established. During the 60 to 205-year interval, competition between plants was marked, and organic matter was rapidly added to the soil as the plant population diminished. There was slight, if any, formation of secondary clays in the 1,200 years, and no textural profile was developed. The soil reaction profile (pH) varied with an unknown factor, possibly quality of organic matter or of clay minerals. Base saturation percentage declined with time until, after 205 years, base cycling by trees began to influence the soil profile. The decrease in bulk density throughout the profile was apparent but seemed not to be due to weathering of minerals but rather to a swelling of the entire soil as organic matter was introduced. Root growth and wetting and drying were very likely important in this swelling process. Original porosity of the parent material was 30%, and this reached 60% in the A1 horizon. It is not known why iron was not released until after 205 years. Apparently the seat of accumulation of nitrogen was in the soil. At age 205 years, the forest floor weighed about 76.8 tons/acre, on a dry basis. At age 1,200 years the weight was half this, or about 38.4 tons/acre (Dickson and Crocker 1953a, b).

Of unknown ages are chronosequences of soils on alluvial fans in California, and on Mississippi River delta deposits in Louisiana and Mississippi (Muckenhirn et al. 1949). Robinson (1950) used depths of leaching of carbonates to estimate ages of loess-derived soils in southwestern Wisconsin.

On the Bermuda Islands, Terra Rossa soils (Rhodustalfs) about 2 feet thick (with tongues extending into eolinanite and limestone as deep as 4 meters) and with a half-inch-thick porcelainlike cemented layer (containing crandallite) just below the B horizon have been estimated to be 200,000 years old by Sayle, assuming that the soils are upland residual soils. Ruhe considers them young, depressional, cumulative soils (Ruhe, Cady, and Gomez 1961).

Crowther (1935) suggested a logarithmic pT scale:

$$pT\ 1 = 10^1 \ \text{years (very recent agricultured cultivation)}$$
$$2 = 10^2 \ \text{years}$$
$$3 = 10^3 \ \text{years (old European agricultured cultivation)}$$
$$4 = 10^4 \ \text{years}$$
$$5 = 10^5 \ \text{years}$$
$$6 = 10^6 \ \text{years}$$
$$7 = 10^7 \ \text{years}$$
$$8 = 10^8 \ \text{years (Cretaceous)}$$
$$9 = 10^9 \ \text{years (Precambrian)}$$
$$10 = 10^{10} \ \text{years (Pre-Solar system)}$$

Table 12.1 presents a time scale based on work of Hough (1958), Kulp (1961), and Jackson (1963). Table 12.2 is adapted from Wright and Frey (1965).

Radiocarbon dating of buried wood overlying an ancient Entisol at the Forest Bed in Wisconsin between the Cary and so-called Valders tills has given a date of about 11,000 years (Hole 1967). Dating of some modern soils, as reported by Broecker, Kulp, and Tucek (1956) (also see Ruhe 1969a), is listed in Table 12.3.

Buol (1965) reported ^{14}C dating of carbonate carbon in caliche layers in a Red Desert (Argid) soil. At a depth of 100 cm, the carbon was 2,300

TABLE 12.1 ❧ **The geologic time scale from Recent to Precambrian**

Era	Period	Epoch	Beginning of Interval		
			Kulp (1961)	Hough (1958)	Hammond (1970)
			(millions of years)		
Cenozoic	Quaternary	Recent			
		Pleistocene	1	1	0–2
	Tertiary	Pliocene	13	12	2
		Miocene	25	28	11
		Oligocene	36	40	25
		Eocene	58	60	40
		Paleocene	63	60	60
Mesozoic	Cretaceous		135	130	70
	Jurassic		181	155	135
	Triassic		(230)	185	180
Paleozoic	Permian		280	210	225
	Pennsylvanian	Carboniferous	...	235	
	Mississippian				
	Devonian		405	320	
	Silurian		(425)	360	
	Ordovician		500	440	
	Cambrian	Upper	530	...	
		Lower	600	520	600

years old; at 150 cm, 9,800 years old; and at 213 cm, 32,000 years old. The conclusion is drawn that the uppermost caliche is related to the modern soil and that the buried caliches are related to former soil regimes.

RATE OF SOIL FORMATION. The question, How long does it take to form an inch of soil? is asked frequently in connection with soil losses by erosion. The pedologist does not think in terms of inches or centimeters, but rather in terms of horizons, sola, and profiles. The difference in rate of soil formation on solid rock and on unconsolidated deposits is evident. Table 12.4 presents some estimates of ages and rates of formation (years per centimeter) of some soil horizons and profiles, but the significance of these is questionable. Radiocarbon dates of Clarion and Webster soils in glacial drift in central Iowa indicate a rate of formation of A horizon of about 7.3 cm per century (Simonson 1959).

Judging by material in solution in rivers, Clarke (1924) concludes that the land surface of the earth is being lowered one foot every 30,000 years. A great deal of material must be leaving mature soils, as well as young and senile ones.

Factors affecting rate of pedogenesis include intensity of weathering, retrogression, and pedologic inertia. Intensity of weathering per unit volume of soil is greater in a shallow profile developing relatively slowly from a fine-textured igneous rock than in a deeper profile of the same age and climate forming from a coarse-textured gabbro. Retrogression is the reversal of a process, as in the case of seasonal resalinization of a soil which has been leached gradually. Solonetzic soils of southern Illinois began as moderately well-drained soils undergoing progressive leaching, but arrived at a condition of sodium accumulation because of a pedogenically related decline in permeability of the underlying glacial till (Wilding et al. 1963). Pedologic inertia of soil refers to its resistance to change in response to changed environmental conditions. Some clayey calcareous and ferruginous soils are particularly resistant to change. Relict or incongruous Oxisols are examples. They are red, iron-rich, highly weathered soils now in dry lands and must have been formed in a more humid climate.

Simonson (1941, 1954) compared a paleo-prairie-Planosol (now classed as an Alboll) with a present-day Putnam soil, the first being a soil formed mainly in the Yarmouth period and perhaps 150,000 years old, while the Putnam is 17,000 to 30,000 years old. The younger soil has 58% clay, and the older, buried one has 63% clay in the maximum Bt horizon. The Bt of the younger soil is maximal at 50 cm, while the older soil has a clay maximum at 63 cm. This worker also compared an Alfisol (Gray-Brown Podzolic) from the Midwest buried in the Yarmouth period to a present-day Weller soil in the same area. The differences in depth and percent of clay are 61% at 33 cm in the ancient soil and 50% clay at 68 cm in the Weller.

TABLE 12.2 ❧ **Paleoclimatic sequence in central United States**

Years before Present	Climatic Trend or Event
75	Warmer, drier climate. Annual precipitation decreased at Omaha, Madison, and LaCrosse in about 1885. Mountain glaciers retreated.
250	Cool period. Mountain glaciers advanced.
350	Warm period. Mountain glaciers retreated.
500	Cool, moist period.
700	Great drought. Prairie expanded at expense of forests.
900	Warm period. Norsemen occupied Greenland.
1,200	Cool, moist period. Permafrost was raised.
3,000	Wet period.
4,500–6,500	Great drought. Prairie expanded at expense of forest.
6,500–8,500	Deciduous forest replaced coniferous forest.
8,500–9,500	Coniferous forest. Climate cool, moist. Forests advanced. Sloth and mammoth disappeared.
9,500–11,000	Valderan glaciation. End of the Wisconsinan glacial stage.
11,000–12,500	Two Creekan interstadial.
12,500–22,000	Woodfordian glaciation (Cary glaciation was 14,000 to 16,000 years BP).
22,000–28,000	Farmdalian deglaciation.
28,000–40,000	Rockian glaciation of the Altonian substage.
40,000–70,000	Other Altonian glaciation. Beginning of the Wisconsinan stage.
70,000–120,000 ?	Sangamonian interglacial stage (at this time sea level was relatively high).
110,000 ?	Illinoian stage of glaciation.
200,000 ?	Yarmouthian interglacial stage (at this time sea level was relatively high according to Broecker 1966).
260,000 ?	Kansan stage of glaciation.
350,000 ?	Aftonian interglacial stage.
1,000,000 ?	Nebraskan stage of glaciation.

SOURCE: Ruhe and Scholtes 1956; Wright and Frey 1965.

TABLE 12.3 ❧ **Some radiocarbon dates for selected soil horizons**

State	Soil	Horizon	Age (by ^{14}C)
South Dakota	Barnes loan (Chernozem)	0–4″	350 yr (± 120)
Iowa	Clarion (Brunizem)	A11 0–6″	440 yr (± 120)
	Webster (Humic Gley)	A11 0–6″	270 yr (± 120)
	Cresco-Kenyon	A11 0–4″	210 yr (± 120)
		A12 4–8″	Less than 100 yr
	Edina (Planosol)	A1 0–6″	410 yr (± 110)
Indiana	Humic Gley	A2 8–12″	840 yr (± 200)
Dates for lake bottom sediments in Wisconsin are:			
Wisconsin	Lake Mendota bottom sediments, S. Wis.	0–27″	5,250 yr (± 1,600)
	Trout Lake bottom sediments, N. Wis.	0–27″	1,800 yr (± 1,600)

SOURCE: Broecker, Kulp, and Tucek 1956.

TABLE 12.4 ❧ **Some estimates of rate of soil formation**

Soil Horizon or Profile	Age upon Completion of Formation (yr)	Depth of Soil (cm)	Rate of Formation (yr/cm)	Literature Citation
Azonal soil (Entisol) on volcanic ash	45	35	1.3	Mohr and van Baren 1954
Hardening of a tropical clay surface soil to laterite, following deforestation	35	15	2.3	Aubert and Maignien 1949
A1 (mull) horizon in a Gray-Brown Podzolic (Hapludalf) soil formed from weathered loess in Wisconsin	265	7	3.8	Nielsen and Hole 1964 Van Rooyen, 1973
A1 horizons of a Brunizem (Hapludoll) soil formed from weathered loess in Iowa	400	33	12.0	Simonson 1959 Arnold and Riecken 1964
Two-meter-thick organic (Histosol) soil in a bog in Wisconsin	3,000	200	15.0	. . .
Formation of a Podzol (Spodosol) soil in sandy glacial drift with 10 cm LFH and 10 cm A2 horizons	1,200	57	21.0	Tamm and Östlund 1960
Solum, including a textural B horizon, of a Gray-Brown Podzolic (Hapludalf) soil formed from weathered loess in Iowa	4,000	100	40.0	Arnold and Riecken 1964
Decalcified loess in southern Wisconsin	8,000	100	80.0	Robinson 1950
A1–A2 horizon sequence in a Gray-Brown Podzolic (Hapludalf) soil formed from weathered loess in Iowa	2,500	30	83.0	Parsons, Scholtes, and Riecken 1962
Solum of a Red-Yellow Podzolic (Ultisol) soil in Australia	29,000	300	97.0	Butler 1958
One-meter-thick solum of a tropical soil (Oxisol) in Africa	75,000	100	750.0	Aubert 1960

These observations suggest that an Alfisol (Gray-Brown Podzolic) Bt horizon may build up from below, while the Alboll (Planosol) B forms first above, and the zone of maximum clay accumulation gradually moves down. Simonson found buried, well-drained Alfisols (Gray-Brown Podzolic soils) under 244 cm of loess, on the east side of the Mississippi River valley in the state of Mississippi.

Ruhe, Daniels, and Cady (1967) report evidence that surfaces of glacial till of Kansan age in southern Iowa have been progressively lowered by a process of pediment formation. Each lower pediment, as these gently sloping erosional-depositional surfaces are called, is occupied by a younger soil. On the till lies a gritty "till-like" mass-wasted material called "pedisediment," which is believed to have streamed slowly across the pediment from the actively retreating break. On the highest divide (Fig. 12.1) in Adair County, Iowa, the Sharpsburg silty clay loam has formed in a loess deposit that is 15,000 to 27,000 years old, overlying a buried 220,000(?)-year-old soil, the Clarinda silty clay loam, a paleo-Alboll (Humic Gley-Planosol), which outcrops locally. On a pediment lying about 16 meters (50 ft) lower is an Adair silt loam, formed where loess of the same age as that on the divide thins to 50 cm (20 in.) and rests on a buried paleo-Alfisol (Gray-Brown Podzolic soil) that is possibly 70,000 years old. On a still lower pediment surface is the Shelby clay loam, formed on a <14,000-year-old surface on Kansan glacial till. At the foot of these slopes are Fluvents (alluvial soils), with the youngest layer at the surface.

Three successive soil profiles, two Spodosols (Podzol soils) and one Cryaquept or Cryochrept Tundra soil, were found by Bryson, Irving, and Larsen (1965) in southern Canada, in the vicinity of Ennadai Lake (latitude 60° N) and Dubawnt Lake (more than 62° N). Table 12.5 presents information concerning this chronosequence.

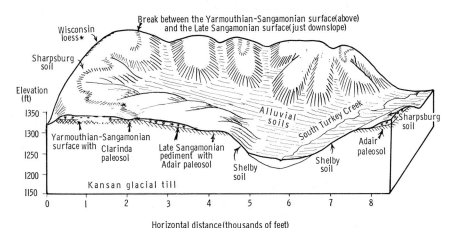

FIG. 12.1. Block diagram of part of the South Turkey Creek area in Adair County, Iowa, showing landscape positions of soils on several geomorphic surfaces. (Generalized from maps and figures by Ruhe, Daniels, and Cady 1967.)

Walker (1966) estimated erosion of glacial drift upland soils in the vicinity of the Colo bog in central Iowa at less than one ton per acre per year during the last 13,000 years. This contrasts with a rate of one to eight tons for geologic erosion of loess-derived soils in southwestern Iowa (Ruhe and Daniels 1965) and the post settlement rate of three to ten tons in the same part of Iowa.

Thorp (1965) concluded that some Ultisols (Red and Yellow Podzolic soils) in southeastern United States may predate the Illinoian glaciation (about 110,000 years BP). Some of these soils are buried under a succession of loess deposits on the eastern escarpment of the Mississippi River valley deep into south central United States. Blocks of Aftonian soils (see Table 12.2) have been found incorporated in Kansan till, of Yarmouthian soils in Illinoian till, and of Sangamonian soils in Wisconsinan till. On interfluves near Lincoln, Nebraska, a truncated Yarmouthian soil formed in Kansan drifts is overlain, in succession, by a Sangamonian soil and a modern (Sharpsburg) soil developed in loess. On the eroded valley walls, from which the loess has been stripped, younger Burchard and Steinauer soils

TABLE 12.5 ✁ **Some information on a chronosequence of soils in southern Canada**

Date	Event	Evidence
5500 BP	Drainage of extensive glacier-dammed lakes, exposing glacial drift.	Stratigraphy in adjacent areas.
3500 BP (1500 BC)	Burning of forest which had spread as far north as latitude 63° N and under which a Podzol soil had formed. "Buffalo culture" of the Great Plains had extended throughout the forest. The forest did not regenerate again north of latitude 62° N, where tundra and the "Arctic culture" have been present since the burn.	A buried Spodosol (Podzol), under a surficial Podzol (see below), contained charcoal dated at 1604 BC near Ennadai Lake and at 1450 BC farther north at Debawnt Lake.
1000 BP (AD 900)	Burning of forests which had spread north to about latitude 62°. Thereafter "Caribou-Eskimo culture" prevailed.	Charcoal in a forest soil at south end of Ennadai Lake was dated at AD 1080 ± 100 years. Charcoal in a surficial Spodosol (Podzol) in a double Podzol sequence near the southern end of Ennadai Lake was dated at AD 860. A Spodosol is buried under sand in which a Cryochrept (tundra) soil now exists, at the north end of Ennadai Lake.
Present	The present forest-tundra (forest-"barrens"; Spodozol-Cryochrept; Podzol-Tundra) soil boundary is about 12 miles (20 km) wide and meets the shores of the southern part of Ennadai Lake at about latitude 60° N, which is the present climatic frontal position.	Current observations.

Source: Bryson, Irving, and Larsen 1965.

have developed in Kansan till. On the high plains of Texas, a Yarmouth surface has been deflated in places, exposing the caliche below the Yarmouth paleosol, and in other places has been covered with recent dunes. Near Melbourne, Australia, a Red-Brown earth, possibly 70,000 years old, occurs on a high terrace above the Maribyrnon River, and a Chernozem soil, possibly 8,000 years old, occurs on a lower terrace from which a human skull was excavated and accompanying charcoal dated at 8,500 years BP. A precursor of the Ultisol (Red-Yellow Podzolic) Tifton soil in southeastern Georgia is preserved buried under early Pleistocene sediments and does not have the development of iron concretions in the A and B horizons that the normal Tifton soil has. This latter soil may well be Pliocene in age, or something over a million years old.

Rate of formation of a soil horizon is related to pedogenesis days per year, somewhat comparable to growing days per year for plants. Subsoils in permafrost regions may be immobilized for centuries or millenia. Upon exposure by erosion to freeze-thaw cycling, soil formation may resume, possibly under the influence of plants, such as lupine, that develop from germinating 10,000-year-old seeds buried in the newly thawed soil horizon.

TIME FACTOR AS ELUCIDATED BY
MAN-MADE EXPERIMENTS. Laboratory experiments have been useful in elucidating processes of soil formation, such as eluviation, formation of cutans, weathering of primary mica to clay minerals, and transformation of one clay mineral into another. One may conclude that processes which operate in the laboratory are relatively rapid. This does not tell us when and how they may operate in the soil under natural conditions. By manipulating biota and freshly exposed parent material out-of-doors, information on rate of formation or degradation of A1 horizons, and of platy, granular, and blocky structure has been obtained.

Young soil developed a darkened A1 horizon 15 cm thick over a period of 50 years in a borrow pit in North Dakota (Simonson 1959). A humus-calcareous soil formed in a century in an abandoned limestone quarry in the USSR (Smirnov 1960).

CONCEPT OF THE "MATURE" SOIL
AND THE TIME FACTOR. Marbut (1928) conceived of a mature soil as one on somewhat sloping land and hence subject to geologic erosion. To him an Alboll (Planosol) on a level upland was an intrazonal soil that must inevitably become a zonal soil, as dissection progresses and drainage is improved. Others have thought of the Planosol as the more mature soil of the two. Nikiforoff (1942, 1949) described the mature soil as "a steady stage of its parent material adjusted to the environment. . . . The time factor has no significance after the soil reaches maturity." Palmquist (1964) concluded from a study of recent soil survey reports and other data (Leopold, Wolman, and Miller 1964; Ruhe, 1960) that at about 20,000 years of age an equilibrium has been approached in parts of Iowa with respect to drainage density, thickness of soil solum, thickness of B horizon of the soil, and clay content

in the B horizon. However, Butler (1958) concluded that changes in these characteristics have continued to occur in Australia. He made a study of soils of different ages formed from alluvium that indicated that soil development continues throughout the history of a soil. In that sense maturity is not a steady state but rather is a changing one. He defined the "soil cycle" as the sequence: (1) exposure of a new surface on initial material; (2) soil catenal development; and (3) burial of the catena or its removal by erosion. He referred to the soil cycle as the *"k-cycle,"* which begins with the *ku* phase, the unstable phase of erosion and deposition, and concludes with a longer *ks* phase, the stable phase during which soil formation takes place both on the newly exposed surface of initial material and on related fresh deposits. In New South Wales, *ku* phases, he reports, have occurred during dry climatic periods characterized by severe erosion by rainstorms. *Ks* phases have coincided with humid periods. Buried soils in some basins record three *k*-cycles: k_3 produced "red and yellow Podzolic" soils 29,000 years ago; k_2 yielded "Gray-Brown soils" 3,740 years ago; and k_1 has resulted in a minimal "prairie soil" 390 years old. Locally, a k_0 cycle 1 to 120 years old has begun to replace the k_1 soil surface.

RECONCILIATION OF THE SCHOOLS OF
DAVIS (1899), PENCK (1924), AND HACK (1960)
REGARDING SLOPE DEVELOPMENT. Davis (1899) imagined a land mass which was initially uplifted rapidly and then degraded to a plain. In his model, degradation was at first principally by stream erosion, which produced deep steep-walled valleys. As the limit of base level decelerated this process, talus slopes grew upward and buried cliffs of youthful valley walls. Weathering and other processes of soil formation yielded deep soils which were subject to creep and some sheet erosion. Graded slopes, like graded streams, are products of a balance between processes of weathering and processes of removal. Davis did not explain how youthful cliffs retreat or how soils developed on the uplands between the youthful valleys. He thought in terms of a closed system with respect to energy inherited from diastrophism. Some of his concepts were later elaborated by Penck.

Penck (1924) and later investigators with similar approaches and concepts, King (1962) and Ruhe (1960), emphasized parallel retreat of slope as a landscape development process. King suggests that regardless of climate there are four elements of slopes: (1) crest (waning slope where soil forms and creep occurs); (2) scarp (zone of backwasting by rill erosion and slippage); (3) debris slope (zone of weathering of debris from the scarp); and (4) footslope (zone of movement of soil material by sheet and linear flow). In regions of high relief where scarps have been initiated tectonically (geologic uplift) or by rapid stream incision in strata of resistant and weak members, parallel retreat of slopes occurs. This is particularly evident in arid regions. In regions of weak rocks and low relief, only crest and pediment members are present. Ruhe added another element, (5) the alluvial toeslope. He explained stone lines in soil profiles (Ruhe 1956b, 1959, 1960; Ruhe and Daniels 1958) as features of the base of pedisediment. Noncumulative soils on the crest form in a relatively closed system, with virtually no erosion. Soils on the scarp (pediment backslope) and debris slope (pediment footslope) are in equilibrium between weathering and erosion in the first instance, and between weathering plus deposition and removal in the second instance.

Hack (1960) conceived of landforms and processes as being in equilibrium, and hence time independent. Therefore, geologic patterns, not evolutionary processes, are the important phenomena. Removal of quartzite requires more energy than does removal of mica schist. Therefore, steeper and longer slopes develop on quartzite than on schist. Most soils are viewed as time independent, extending themselves into the initial materials at the same rate as erosion lowers the hills, both tops and sides. After a brief (20,000 years?) period of disequilibrium, a steady state of both soil formation and downwasting of the landscape is reached.

Strahler (1950), Scheidegger (1960, 1961), and others suggest that parallel retreat of slopes is active only where (1) streams are close to the footslope and help in rapid removal of debris, as in youthful stages of degradation; and (2) bedrock consists of a resistant cap over a weak understratum. In arid regions erosion is great, and hence scarps and pediments are well developed. Frye (1959) noted that in southeastern Ohio slopes are convexoconcave, in south central Kansas scarps are prominent but pediments small, and in central west Texas crest and pediment are dominant. The sequence is from humid to drier regions, and to regions of less drainage density and less vegetation. Davis, Penck, and Hack explain different aspects of geomorphology and pedology.

A new and very descriptive set of landscape terms has been developed by Ruhe (1969a) with reference to hillslopes: summit, shoulder, backslope, footslope, and toeslope. He states, "Hillslopes probably comprise the most important part of the earth's surface as a setting for soils" (Ruhe 1969b). Studies by him and his associates clearly show that a general characteristic of many landscapes is a set or "series of stepped levels like a staircase" (Ruhe 1969b). Study areas and environments where these principles have been evolved and tested include Congo—humid tropical (Ruhe 1956a); Iowa—temperate subhumid (Ruhe, Daniels, and Cady 1967; Ruhe 1969a); southern New Mexico—arid (Gile and Hawley 1966; Ruhe 1969b); North Carolina—warm temperate humid (Cady and Daniels 1968; Daniels, Gamble, and Nettleton 1966; Daniels et al. 1966). Time zero of soil formation, and hence soil age, differs among the soils under the surfaces of these stair steps; and hence their properties differ though they may have formed from similar initial material. Soils under the surfaces of a simple, one-step hillslope may also differ in age, if the principle of ascendancy (Ruhe 1969a) applies; that is, "a hillslope is younger than the higher surface to which it ascends." The converse principle of descendancy is that "a hillslope is of the same age as the alluvial valley fill to which it descends." In this case, soils on the descending hillslope are in the same time frame as the soils on the associated alluvial valley fill. In the case of the New Mexico study area (Ruhe 1969b), key radiocarbon dates permitted establishment of fairly precise dates (before present) of the time zero of soil formation on the stepped surfaces.

In North Carolina studies by Craig (1963) and by Daniels et al. (1966) of coastal plain soils from similar parent material but on landforms and under surfaces of greatly different age allows a good comparison of soil differences due to age (ruling out significant climatic-vegetational shifts affecting soil properties). One soil has formed on the stable interstream divide under a depositional surface which may date to late Pliocene or early Pleistocene (Daniels, Gamble, and Nettleton 1966); whereas its nearby associate, formed in the same sediment, is in a sideslope position under a surface which may be no older than late Pleistocene (Cady and Daniels 1968; Daniels et al. 1966). That is, their age difference may be of the order of a million years. The soil of the older landform has a solum nearly three meters thick, has plinthite in the lower B horizon, has a thick A2 horizon resulting from profound clay translocation and destruction, and has a clay mineralogy suite of A1-interlayered vermiculite (2:1–2:2 intergrade), kaolinite, and gibbsite. But the soil on the sideslope has a solum less than a meter thick, no plinthite, a thin A2 horizon, no gibbsite, and contains appreciable illite as well. These differences are illustrated in Figure 12.2.

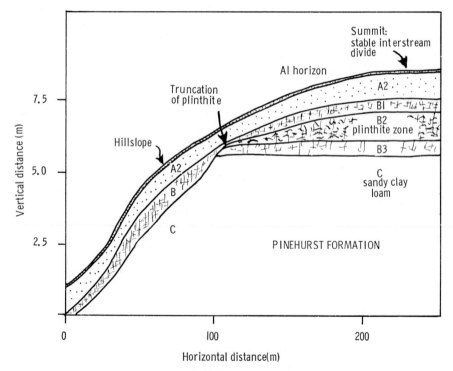

FIG. 12.2. Comparison of soil development on "old" stable interstream divide and "young" hillslope in identical initial materials, coastal plain of North Carolina. (After Gamble, Daniels, and McCracken 1970.)

We must not underestimate the importance of an understanding of landscape evolution and its strong effect on the time factor in soil formation. Although a given landscape may appear quite uniform and simple from a casual glance, chances are that it has a complex geomorphic history and that some soils differ on this landscape because of differences in their time zero of soil formation related to the landform they occupy.

LITERATURE CITED

Arnold, R. W., and F. F. Riecken. 1964. Grainy gray ped coatings in Brunizem soils. Proc. Iowa Acad. Sci. 71:350–60.

Aubert, G. 1960. Influences de la vegetation sur le sol en zone tropicale humide et semi-humide. Rapports du Sol et de la Vegetation. Colloq. Soc. Nat. Fr., 1959, pp. 11–22.

Aubert, G., and R. Maignien. 1949. L'Erosion eolienne dans le nord du Senegal et du Senegal et du Soudan Francais. Bull. Agr. Congr. Belg. 40:1309–16.

Broecker, W. S. 1966. Absolute dating and the astronomical theory of glaciation. Science 151:299–304.

Broecker, W. S., J. L. Kulp, and C. S. Tucek. 1956. Lamont natural radiocarbon measurements. III. Science 124:154–65.

Bryson, R. A., and W. M. Wendland. 1967. Radiocarbon isochrones of the retreat of the Laurentide ice sheet. Tech. Rept. 35. Dept. Meteorol., Univ. Wis., Madison.

Bryson, R. A., W. N. Irving, and J. A. Larsen. 1965. Radiocarbon and soil evidence of a former forest in the southern Canadian tundra. Science 147:46–48.

Buol, S. W. 1965. Present soil-forming factors and processes in arid and semiarid regions. Soil Sci. 99:45–49.

Butler, B. E. 1958. Depositional systems of the riverine plain of south-eastern Australia in relation to soils. Commonwealth Sci. and Ind. Res. Organ. Soil Publ. 10.

Cady, J. G., and R. B. Daniels. 1968. Genesis of some very old soils—the Paleudults. Trans. 9th Intern. Congr. Soil Sci. (Adelaide, Australia) 4:103–12.

Chandler, R. F., Jr. 1937. A study of certain calcium relationships and base exchange properties of forest soils. J. Forestry 35:27–32.

Clarke, F. W. 1924. Data of geochemistry, 5th ed. U.S. Geol. Survey Bull. 770. Dept. Interior, U.S. Govt. Printing Office, Washington.

Craig, R. M. 1963. Morphology and genesis of soils formed in various parent materials on degradational sideslope landforms of the upper coastal plain of North Carolina. M.S. thesis, N.C. State Univ., Raleigh.

Crocker, R. L., and J. Major. 1955. Soil development in relation to vegetation and surface age at Glacier Bay, Alaska. J. Ecol. 43:427–48.

Crowther, F. 1935. Some inductive methods in pedology. Trans. 3rd Intern. Congr. Soil Sci. (Oxford, England) 1:339–43.

Daniels, R. B., E. E. Gamble, and W. D. Nettleton. 1966. The Surry scarp from Fountain to Potters Hill, North Carolina. Southeastern Geol. 7:41–50.

Daniels, R. B., E. E. Gamble, W. H. Wheeler, and W. D. Nettleton. 1966. Coastal plain stratigraphy and geomorphology near Benson, North Carolina. Southeastern Geol. 7:159–82.

Davis, W. M. 1899. The geographical cycle. Geograph. J. 14:481–504.

Dickson, B. A., and R. L. Crocker. 1953a. A chronosequence of soils and vegetation near Mt. Shasta, California. I. Definition of the ecosystem investigated and features of the plant succession. J. Soil Sci. 4:123–41.

———. 1953b. A chronosequence of soils and vegetation near Mt. Shasta, California. II. The development of the forest floors and the carbon and nitrogen profiles of the soils. J. Soil Sci. 4:142–54.

Dickson, B. A., and R. L. Crocker. 1954. A chronosequence of soil and vegetation near Mt. Shasta, California. III. Some properties of the mineral soils. J. Soil Sci. 5:173–91.

Frye, J. C. 1959. Climate and Lester King's "uniformitarian nature of hillslopes." J. Geol. 67:111–13.

Gamble, E. E., R. B. Daniels, and R. J. McCracken. 1970. A2 horizons of coastal plain soils: pedogenic or geologic origin. Southeastern Geol. 11:137–52.

Gile, L. H., and J. W. Hawley. 1966. Periodic sedimentation and soil formation on an alluvial-fan piedmont in southern New Mexico. Soil Sci. Soc. Am. Proc. 30:261–68.

Hack, J. T. 1960. Interpretation of erosional topography in humid temperate regions. Am. J. Sci. 258A:80–97.

Hammond, A. L. 1970. Deep sea drilling. Science 170:520–21.

Hissink, D. J. 1938. The reclamation of the Dutch saline soils (Solonchak) and their further weathering under the humid climatic conditions of Holland. Soil Sci. 45:83–94.

Hole, F. D. 1967. An ancient young soil. Soil Survey Horizons 8:16–19.

Hough, J. L. 1958. Geology of the Great Lakes. Univ. Ill. Press, Urbana.

Jackson, M. L. 1963. Aluminum bonding in soils: A unifying principle in soil science. Soil Sci. Soc. Am. Proc. 27:1–10.

King, L. 1962. Morphology of the earth: A study and synthesis of world scenery. Hafner, New York.

Kulp, J. L. 1961. Geologic time scale. Science 133:1105–14.

Leopold, L. B., M. G. Wolman, and J. P. Miller. 1964. Fluvial processes in geomorphology. W. H. Freeman, San Francisco.

Marbut, C. F. 1928. Soils: Their genesis and classification. Pub. 1951 by Soil Sci. Soc. Am., Madison, Wis.

Mohr, E. C. J., and F. A. van Baren. 1954. Tropical soils: A critical study of soil genesis as related to climate, rock, and vegetation. The Royal Tropical Inst., Amsterdam. Interscience, New York.

Muckenhirn, R. J., E. P. Whiteside, E. H. Templin, R. F. Chandler, Jr., and L. T. Alexander. 1949. Soil classification and the genetic factors of soil formation. Soil Sci. 67:93–105.

Nielsen, G. A., and F. D. Hole. 1964. Earthworms and the development of coprogenous Al horizons in forest soils of Wisconsin. Soil Sci. Soc. Am. Proc. 28:426–30.

Nikiforoff, C. C. 1942. Fundamental formula of soil formation. Am. J. Sci. 240:847–66.

———. 1949. Weathering and soil formation. Soil Sci. 67:219–30.

Palmquist, R. D. 1964. The relationship between topography and pedogenesis. Unpublished Lit. Rev. Dept. Soil Sci., Univ. Wis., Madison.

Parsons, R. B., W. H. Scholtes, and F. F. Riecken. 1962. Soils of Indian mounds in northeastern Iowa as benchmarks for studies of soil genesis. Soil Sci. Soc. Am. Proc. 26:491–96.

Penck, W. 1924. Die Morphologische Analyse. (Morphological analysis of land forms.) Stuttgart. (Transl. by Hella Czech Katherine Cumming Boswell, 1953.) Macmillan, London.

Proudfoot, V. B. 1958. Problems of soil history. Podzol development at Goodland and Torr Townlands, County Antrim, Northern Ireland. J. Soil Sci. 9:186–98.

Robinson, G. H. 1950. Soil carbonate and clay contents as criteria of rate and stage of soil genesis. Ph.D. thesis, Univ. Wis., Madison.

Ruhe, R. V. 1956a. Landscape evolution in the High Ituri, Belgian Congo. INEAC Series Sci. 66. (Scientific reports from former Belgian Congo agronomic research institute.)

———. 1956b. Geomorphic surfaces and the nature of soils. Soil Sci. 82:441–55.

———. 1959. Stone lines in soils. Soil Sci. 87:223–31.

———. 1960. Elements of the soil landscape. Trans. 7th Intern. Congr. Soil Sci. (Madison, Wis.) 4:165–69.

———. 1969a. Quaternary landscapes in Iowa. Iowa State Univ. Press, Ames.

———. 1969b. Principles for dating pedogenic events in the Quaternary. Soil Sci. 107:398–402.

Ruhe, R. V., and R. B. Daniels. 1958. Soils, paleosols, and soil-horizon nomenclature. Soil Sci. Soc. Am. Proc. 22:66–69.

———. 1965. Landscape erosion—geologic and historic. J. Soil Water Conserv. 20:52–57.

Ruhe, R. V., and W. H. Scholtes. 1956. Ages and development of soil landscapes in relation to climatic and vegetational changes in Iowa. Soil Sci. Soc. Am. Proc. 20:264–73.

Ruhe, R. V., J. G. Cady, and R. S. Gomez. 1961. Paleosols of Bermuda. Geol. Soc. Am. Bull. 72:1121–41.

Ruhe, R. V., R. B. Daniels, and J. G. Cady. 1967. Landscape evolution and soil formation in southwestern Iowa. U.S. Dept. Agr. Tech. Bull. 1349.

Salisbury, E. J. 1925. Note on the edaphic succession in some dune soils with special reference to the time factor. J. Ecol. 13:322–28.

Scheidegger, A. E. 1960. Analytical theory of slope development by undercutting. J. Alberta Soc. Petrol. Geol. 8:202–6.

———. 1961. Mathematical models of slope development. Geol. Soc. Am. Bull. 72:37–50.

Simonson, R. W. 1941. Studies of buried soils formed from till in Iowa. Soil Sci. Soc. Am. Proc. 6:373–81.

———. 1954. Identification and interpretation of buried soils. Am. J. Sci. 252:705–32.

———. 1959. Outline of a generalized theory of soil genesis. Soil Sci. Soc. Am. Proc. 23:152–56.

Smirnov, M. P. 1960. The rate of soil formation in old quarries of calcareous rocks and secondary changes in buried Light-Gray Forest soils. Soviet Soil Sci., 1960, no. 4, pp. 417–27. (Transl. 1961 from Pochvovedeniye, 1960, 4:91–102.)

Strahler, A. N. 1950. Equilibrium theory of erosional slopes approached by frequency distribution analysis. Am. J. Sci. 248:673–96.

Stremme, H. 1960. Bodenbildung auf Geschiebelehman Verschiedenen alters in Schleswigholstein. Zeit. Deutch. Geol. Gesell. 112:299–308.

Tamm, C. O., and H. G. Östlund. 1960. Radiocarbon dating of soil humus. Nature 185:706–7.

Thorp, J. 1965. The nature of the pedological record in the Quaternary. Soil Sci. 99:1–8.

Thwaites, F. T. 1946. Outline of glacial geology. Edwards Brothers, Ann Arbor, Mich.

Van Rooyen, D. J., 1973. I. Organic carbon and nitrogen states in two Hapludalfs under prairie and deciduous forest, as related to moisture regime, some morphological features and response to manipulation of cover. II. Estimate of the influence of the agricultural and industrial revolutions on moisture regimes of two representative Wisconsin soils. Ph.D. Thesis, University of Wisconsin, Madison.

Walker, P. H. 1966. Postglacial erosion and environmental changes in central Iowa. J. Soil Water Conserv. 21:21–23.

Wilding, L. P., R. T. Odell, J. B. Fehrenbacher, and A. H. Beavers. 1963. Source and distribution of sodium in solonetzic soils in Illinois. Soil Sci. Soc. Am. Proc. 27:432–38.

Wright, H. E., Jr., and D. G. Frey. 1965. The Quaternary of the United States. Princeton Univ. Press, Princeton, N.J.

❧

Principles and Historical Development of Soil Classification ✣

$\mathbf{M}$AN SEEMS to have a natural tendency and urge to sort out and classify the natural objects of his environment. Soils are no exception, being objects of common experience and observation— undergirding agricultural production and supporting buildings and highways. Early systems of soil classification were quite simple and highly practical. But with increasing sophistication of agriculture, greater knowledge about soils as a collection of independent natural bodies, and greater complexity and diversity of soil uses, the classification of soils has become more scientific and organized. And the classification system used in a particular field or discipline tends to reflect the "state of the art" and the current thinking in that field. Kubiëna (1948) states, "Show me your [classification] system and I will tell you how far you have come in the perceptions of your research problems." The renowned physicist, Ampere, is reported to have said, "Perfect scientific classification is first possible when one knows everything concerning the classified natural objects" (Kubiëna 1948).

PURPOSES OF CLASSIFICATION. Before proceeding to a detailed discussion of the principles of soil classification and its evolution, let us review the reasons man classifies and the uses he makes of a formal classification process. Mill (1925) discussed the logical bases and principles of the classification process. We draw upon his writings for some of the following discussion.

We classify natural phenomena in order to

1. Organize knowledge (thereby contributing to economy of thought).

2. Bring out and understand relationships among individuals and classes of the population being classified.

3. Remember properties of the objects classified.

4. Learn new relationships and principles in the population we are classifying.

5. Establish groups or subdivisions (classes) of the objects under study in a manner useful for practical, applied purposes in:

a. predicting their behavior.
b. identifying their best uses.
c. estimating their productivity.
d. providing objects or units for research and for extending and extrapolating research results or our observations.

In general and simple classification systems, we arrange or structure the system so that it does some, if not all, of the above five functions. Such classification for a specific, applied, practical purpose is a *technical grouping* (Cline 1949). However, in scientific activities and in other cases where we are organizing our knowledge without reference to a specific, applied objective, we wish a classification system set up in such a way that each group has as many unique, natural properties as possible and its name and its properties relate it to, yet separate it from, all others. Such a system may be called a scientific or natural classification. Stated another way, a natural classification is one in which the purpose is, insofar as possible, to bring out relationships of the most important properties of the population being classified, without reference to any single specified and applied objective (Cline 1949). In a natural classification, all the attributes of a population are considered and those which have the greatest number of covariant or associated characteristics are selected as the ones to define and separate the various classes (Mill 1925). In our present soil classification, we try to approach a natural classification system as an ideal, though we tend to give weight to properties of higher agricultural relevance.

DEFINITIONS. Certain terms are used in taxonomy (classification) with narrow and specific meanings:

Class—a group of individuals or other units similar in selected properties and distinguished from all other classes of the same population by differences in these properties.

Taxon—a class at any taxonomic level of generalization.

Category—a series or array of taxa produced by differentiation within the population at a given level of abstraction or generalization, and is composed of all the classes at one level of generalization. (Note that in everyday speech, category and class are used as synonyms, but in classification usage they have different and specific meanings—a category being composed of a group of classes.)

Differentiating characteristic—a property chosen as the basis for grouping individuals (or classes of a lower level of generalization) into groups. Individuals similar with respect to that characteristic are placed in the same group (Cline 1949).

Multiple category system—a hierarchical system of categories designed to classify a large and complex population of units. If a particular population of natural objects is so complex that a single grouping process (using a single differentiating characteristic to produce classes within a single category) does not bring out all the relationships and produce a useful and satisfactory classification, those classes are further subdivided according to one or more other differentiating characteristics to produce a new category at a lower level of generalization or abstraction (Cline 1949). In the case of such a widely varying and complex population as soils, it is necessary to establish several categories of classification. Such a procedure is called a multiple category system of classification. The highest categories have small numbers of classes defined in broad general terms by means of a few differentiating characteristics. We say these are at a high level of generalization or abstraction. In the lower categories, such as the soil series, there are large numbers of classes of narrow range defined in quite specific terms by a large number of differentiating characteristics. We indicate these as being at a low level of generalization or abstraction.

PRINCIPLES. Now we need to consider some principles as background for our discussion of soil classification systems. Some aspects of these are unique to pedology, but mainly they are general principles of any natural classification or approximation to such a classification of a population of natural objects. (We have drawn in part on a paper by Cline 1949 for this discussion. He, in turn, used expressions coined by C. C. Nikiforoff.)

Genetic thread principle—The theories of soil genesis provide a framework for aiding in determining the significance and relevance of soil properties for use as differentiating characteristics. This principle is followed in the use of theories of evolution as a framework for taxonomy in the plant and animal kingdoms.

Principle of accumulating differentia—In a multiple category classification system, differentiating characteristics accumulate or pyramid from the higher levels of generalization to the lower levels. As a result, classes at the lower levels are defined and differentiated not only by the differentiating characteristic(s) used at a given categorical level but also by those which have been used as differentia at the higher levels. In the lowest category, a large number of differentiating characteristics have been accumulated such that the classes are quite narrowly and completely defined.

Principle of wholeness of taxonomic categories—All individuals of the population must be classified in each category, according to the characteristics selected as differentiating at that level. Some of our earlier soil classification systems violated this principle by omission of certain kinds of soils from the classification at one or more of the categorical levels. Another way of stating this is that any differentiating characteristic should classify all the individuals of a given population.

Ceiling of independence principles—A property or characteristic used as a differentiating characteristic in a category must not separate similar individuals in a lower category. Every characteristic is limited in its use by a "ceiling" categorical level above which it cannot be applied without injecting confusing and inappropriate cleavages in lower categories.

SUCCESSION OF CLASSIFICATIONS. A classification undergoes revision as the body of knowledge on which it is based expands. In this sense, classification succession is a phenomenon common to all disciplines.

We should always keep in mind that classification systems are developed by human beings to organize ideas and properties in the most useful manner (Cline 1963). Classification systems are abstracts of "knowledge" and of concepts derived from knowledge, based on the past experiences (collection and study of data from experiments and field observations) and present biases of man (Cline 1961). The data which we consider to be "fact" are facts only within the context and perspective of the operations by which these data were obtained (Bridgman 1927). An example in the field of soil science is the previously held concept, and hence "fact," that exchangeable hydrogen was the main source of soil acidity in acid mineral soils (Coleman, Weed, and McCracken 1959; Jenny 1961). A comparison of the general basis and class definitions for a soil classification system published in the 1938 USDA *Yearbook of Agriculture* (Baldwin, Kellogg, and Thorp 1938) with those in *Soil Classification—A Comprehensive System, 7th Approximation* (Soil Survey Staff 1960) is a more complete and complex example of this point. In the development of the total field of science, we can see many examples of the uncovering of new facts requiring rather complete reorganization of the theories and laws making up the body of understanding in a particular field and thereby requiring extensive changes in the classifications based on this understanding (Bridgman 1927; Cline 1961). The point here, then, is that we must accept the provisional, ephemeral, changing state of current knowledge and consequently of classifications based on it, particularly in a relatively new and rapidly developing field such as soil science. We must be prepared to accept additional changes in soil classification; indeed, we should help make them.

AVOIDING RIGOR MORTIS
IN CLASSIFICATION. Another point should be considered before we open our discussion of the development and evolution of soil classification systems. This is that it is easy for a classification system to prejudice the future and for us to become prisoners of our own taxonomy (Cline 1961). This has been a particular problem with soil classification systems, in which conventional wisdom has at times and in places frozen soft, tentative hypotheses into hard dogma, preventing acceptance of new ideas and concepts; or, worse, restricting research patterns so as to preclude acquiring of new facts in certain subject matter areas. Therefore, a classification system, particularly in a field such as soil science, must have a self-destruct mechanism—a procedure for continuing reevaluation of the body of theories making up the genetic thread in our taxonomy. Also, we must avoid selection of soil genesis theories and hypotheses as basic differentiating characteristics, although we may use them, with caution, as guides to relevant properties of the soils themselves which may be used as differentia.

HISTORIC PERSPECTIVE OF
SOIL CLASSIFICATION. The evolution of soil classification can be subdivided into five general periods: (1) an early technical era, (2) the period of the founding of pedology by the Russian group of soil geneticists, (3) the

early American period, (4) the middle period of general development of soil classification and soil surveys in the world, and especially in the United States, which we refer to as the Marbut period, and (5) the present modern period of quantitative pedology.

EARLY TECHNICAL PERIOD. The early technical period in classification of soils had its inception and flowering in western Europe in the middle and later parts of the nineteenth century. For example, Thaer (1853) published a classification which combined textural (particle-size distribution) properties as a primary breakdown or higher category, with agricultural suitability and productivity as classes of a lower category. He established six kinds of soil: clay, loam, sandy loam, loamy sand, sand, and humus. As an example of classes within these six kinds of soil, he recognized four classes within the group of clay soils: black klei-heavy wheat soils, strong wheat soil, weak wheat soil, thin wheat soil. Fallou (1862) devised a soil classification largely based on geologic origin and lithologic composition of what we now call parent material:

Class 1—Residual soils: Class 2—Alluvial soils:
   Soils of limestone rocks   Gravel soils
   Soils of feldspathic rocks   Marl soils
   Soils of clay rocks   Loam soils
   Soils of quartz-bearing rocks   Moor soils

Richthofen (1886) worked out a system of soil classification with a strong geologic basis and nomenclature and thus similar to Fallou's system:
A. Residual soil types
   1. Disintegrated rock
   2. Deeply weathered rock
   3. Eluvial soils of plateaus
   4. Colluvial loam
   5. Laterite
   6. Organic soils: humus, moor, peat
   7. Undissolved residues
B. Accumulated soil types
   8. Coarse sediments of continental waters
   9. Fine-grained sediments of continental waters
  10. Chemical deposits in fresh waters
  11. Marine soils
  12. Glacial deposits
  13. Volcanic ash
  14. Eolian accumulations

From these examples, it can be seen that these were technical systems of classification prepared for a specific purpose or objective and using factors or characteristics not properties of the soils themselves as differentiating. We have presented them here for comparison with later, more comprehensive systems, as a historical record and as indication of the first stirrings of a new field—soil science with its subfield of pedology.

PERIOD OF FOUNDING OF PEDOLOGY. During the latter part of the early technical period, soil studies were underway on the Moscow Plains in Russia which were to have tremendous impact in soil science and which were to be the foundations of modern pedology. On the great Central Russian Upland of the present-day Soviet Union, rather uniform loesslike parent materials extend for hundreds of miles. An increasing temperature gradient is imposed on this area from north to south, and an increasing annual rainfall and moisture gradient extends from east to west. Associated with these are important vegetational pattern differences, especially the major shift from forest to steppe (prairie). These factors have left their imprint on the relatively uniform parent material, producing distinct soil differences. These differences were noted by the founder of modern pedology, Dokuchaev, who first understood the full significance of these soil differences and thereby established the concept of soil as an independent natural body.

V. V. Dokuchaev (1846–1903) was born in a middle-class family in Smolensk where he graduated from a seminary. Upon entering Petersburg University, he received training as a geologist. He did his first field work in the Smolensk area, based on which he prepared his first published work, *Origin of River Valleys*. About this time he met by chance the great Russian statistician and natural scientist, Chevlovsky, who had prepared the first soil map of Russia, using folk definitions and names—"poor" and "rich" soils. As a result, Dokuchaev became particularly interested in the Chernozem, the "richest" of Russian soils. He received a grant from the Free Economic Society in St. Petersburg for studies of these soils (apparently the first grant for scientific research on soils). In 1883 he published his classic monograph, *Russian Chernozem* (Dokuchaev 1883). This was the first published work on soil as a natural body formed by action of a set of soil-forming factors producing genetic layers in the parent material. He followed this monumental monograph with a series of publications on soil genesis and classification, including the first publication with a classification of soils based on the properties and soil-forming factors of the soils themselves (Dokuchaev 1886). Dokuchaev was concerned about and interested in not only the technical and scientific aspects of soil classification but also in practical applications of soil classification. For example, he interpreted his classification of soils of the Nizhnii-Novgorod (now Gorki) region in terms of tax value assessment (Dokuchaev 1886). Also, he became concerned about very serious drought in parts of the steppes of Russia and established a series of experimental plots on different soils. From the results obtained, he recommended forest belts and other procedures for wind protection and water conservation. He has been erroneously regarded and described by certain latter-day soil scientists as a theoretical scientist who related soil formation and classification solely to climates; for example, a "climatic soil scientist." This misconception apparently stems in part from a statement made by Glinka (1931) in a widely translated book in which he listed five main soil types as proposed by Dokuchaev, linked to climate as the sole or main soil former. Also, much of Dokuchaev's writing emphasized the importance of climate. However, we should point out that a sentence translated from Dokuchaev's classic work, *Russian Cherno-zem*, states that "soils must be classified and studied according to their profiles." And we have previously pointed out Dokuchaev's interest and activities in soil interpretation and experimentation to aid in the solution of applied, practical problems of soil and water conservation.

Dokuchaev is reported as being a very sociable man who liked people. His writings and dynamic personality attracted numbers of capable students. Those of his students most closely involved in soil classification were N. M. Sibirtsev and K. D. Glinka. Their contributions also stand as significant monuments in the Russian pedological founding period.

Sibirtsev apparently was the most talented and brilliant of Dokuchaev's students, but unfortunately he met an untimely death of tuberculosis at the age of 39. However, Sibirtsev was able to complete his classic text, *First Course in Soil Science,* the first text on soils and soil classification. He developed the concept of soil zones (Sibirtsev 1901), a powerful idea that certain kinds of soils are associated with certain climate-vegetational or ecological regions. This concept is a basic part of many soil classification systems used today.

Glinka (1867–1929) was the most influential and prolific writer of Dokuchaev's pupils, and the best known in the Western world because of widespread translations of his works: *The Types of Soil Formation, Their Classification and Geographical Distribution* (Glinka 1914), *The Great Soil Groups of the World* (Glinka 1927), and his classic *Treatise on Soil Science* (Glinka 1931). His first book introduced the new Russian concepts of soils, soil classification, and the major soil type names of Chernozem, Podzol, and Solonetz to the Western world. Glinka emphasized soil geography, soil formation, and weathering processes. He was a brilliant lecturer and organizer, being responsible for the organization of soil science in Russia.

A number of other outstanding Russian soil scientists were active and prominent in the latter part of the Russian pedologic founding period, but we shall discuss only the pioneers of this particular period.

EARLY AMERICAN PERIOD. We now turn to the early American period, approximately 1899–1922. In this period the emphasis and bias were more on technical or single-factor classifications in operational programs of the United States soil survey, with emphasis and bias toward geologic techniques and nomenclature, though there were some notable exceptions. Earlier, Ruffin (1832) pointed out the need for a soil classification program in the United States. Hilgard (1833–1906) pioneered early soil classifications and mapping in the United States. He was a geologist for the state of Mississippi. He published a classic pioneer work on soils of Mississippi and later was responsible for establishment of work on soil science in California, especially with respect to sodic and saline soils. He apparently was first in America to conceive of soil as a natural body and pointed out correlations between soil properties on the one hand and vegetation and climate as causal factors. It has been suggested that Dokuchaev was a "follower of Hilgard." Although Dokuchaev's work came a little later, there is no indication known to us that the two were in contact or knew of each other's work. Hilgard's concepts and ideas about soil and soil genesis were not applied in operational soil surveys in America, and introduction of similar ideas was not to come until more than 50 years later.

Milton Whitney developed the first American soil classification system related to soil survey and used as a basis for soil-mapping operations. This system was published in 1909 (Whitney 1909), but actual soil surveys started in the United States about 1899 (U.S. Dept. Agr. 1899). This was a broad classification, mainly according to physiographic regions or provinces and the texture (particle size distribution) of the soil. Whitney and associates established as the highest taxonomic category the soil province, composed of soils within the same physiographic regions, such as coastal plains or piedmont. Soils within a province formed from similar geologic materials (such as marine sediments or glacial till) were defined as a *series* (a term still in use today in the United States for the taxa of the lowest taxonomic category). Soils were subdivided within the series according to texture to form the lowest taxonomic category and the mapping unit—the type. We should point out that Whitney and his associates held a broader meaning of texture than is now given the term in the United States. They meant it to include not only particle size distribution but also soil consistence, organic matter content, aggregation, and related properties. This system was widely used as a basis for soil surveys in many parts of the United States.

Coffey (1912) was apparently the first in the United States to propose soils as independent natural bodies that should be classified on the basis of their own properties and that differences in these properties were due to climatic and associated vegetational differences from place to place. He proposed five great soil groups (apparently the first use of this term which is now widely used): (1) arid, (2) dark-colored prairie, (3) light-colored timbered, (4) black swamp, and (5) organic.

However, his concepts and proposals apparently were not generally accepted and were not made the basis for any operational soil survey program. But his ideas did serve as a forerunner or signal of changes to come in soil classification in America.

MIDDLE AMERICAN PERIOD. C. F. Marbut was the central figure in what we call the middle or Marbut period in the evolution of soil taxonomy. Born and reared on a Missouri farm, he was trained in geology, especially geomorphology. He undertook graduate study in that field at Harvard University under the well-known American geomorphologist William Morris Davis. He became attracted to and interested in soils, entering on duty with the U.S. Department of Agriculture Bureau of Soils in 1910. He introduced the concepts of Dokuchaev and his students Glinka and Sibirtsev to America after translating into English a German edition of Glinka's work on types of soil formation and soil groups of the world. This caused him to introduce the soil-forming factors of climate and vegetation and to reduce the emphasis on geologic nature and origin of the soil materials as developed by Whitney. He evolved his ideas on classification in successive steps (Marbut 1922, 1927), which culminated in his master work on soil classification published in the *Atlas of American Agriculture* (Marbut 1935). A summary outline of this 1935 classification is presented in Table 13.1.

This dedicated man must be considered the founder of American pedology, based on his many contributions in addition to his worldwide influence. Some of his many contributions were:

1. Establishment of the soil profile as the fundamental unit of study. He focused attention on properties of soils themselves rather than their geologic relationships or broad soil-forming factors.

2. Preparation of the first truly multicategorical system of soil taxonomy.

3. Establishment of the criteria for soil series which are still in use today.

Additional information and further developments and improvements in soil genesis have brought out some difficulties and problems in Marbut's classification that are now being rectified in our present efforts in soil classification. Some of these are:

1. His multiple category system was not truly comprehensive. He omitted the classification of "immature and abnormal" soils in one or more categories. This was because of his emphasis on "normal" soils on "normal landscape," meaning the well-drained soils of the hillslopes. That is, there was some violation of the principle of wholeness of taxonomic categories.

TABLE 13.1   ❧   Soil classification by Marbut 1935

Category 6	Pedalfers	Pedocals
Category 5	Soils from mechanically comminuted materials Soils from siallitic decomposition products Soils from allitic decomposition products	Soils from mechanically comminuted materials
Category 4	Tundra Podzols Gray-Brown Podzolic Red soils Yellow soils Lateritic soils Laterite soils	Chernozems Dark Brown soils Brown soils Gray soils Pedocalic soils of Arctic and tropical regions
Category 3	Groups of mature but related soil series Swamp soils Glei soils Rendzina Alluvial soils Immature soils on slopes Salty soils Alkali soils Peat soils	Groups of mature but related soil series Swamp soils Glei soils Rendzina Alluvial soils Immature soils on slopes Salty soils Alkali soils Peat soils
Category 2	Soil series	Soil series
Category 1	Soil units or types	Soil units or types

2. Certain of his criteria for differentiation (differentiating characteristics) based on assumed genesis or genetic inferences have been shown by later study to be incomplete or incorrect. For example, his assumption that zonal ("normal") soils could be divided into two broad classes—one in which calcium carbonate accumulates (Pedocals) and another in which aluminum and iron accumulate (Pedalfers)—has been shown to be inadequate and not satisfactory. As a differentiating characteristic, this assumed difference does not produce mutually exclusive classes. A soil which is a normal Pedocal in one region was considered an Intrazonal Pedocal in another region due to slight differences—parent material or landscape position. Some soils accumulate both $CaCO_3$ and Al-Fe compounds as well. For these and related reasons, this differentiating characteristic and these particular classes have had to be abandoned.

3. His "normal soil on a normal landscape" concept as a basic frame of reference for soil classification has not been found appropriate because of its tendency to ignore more poorly drained soils and because of the complexity in landform and consequent differences in soil age, as well as differences in climate over time, which make it difficult, if not impossible, to establish which is the normal soil of reference in many landscapes.

4. It also appears that more emphasis is needed on the three-dimensional aspects of soils rather than the extreme emphasis given the two-dimensional soil profile by Marbut.

These points are not intended in any sense to discount or minimize the many important contributions of Curtis Fletcher Marbut, a pioneer and leader in his time.

Efforts have been made to revise the Marbut system as new information has been obtained and evolution of concepts has taken place. A comprehensive effort along these lines was made by Baldwin, Kellogg, and Thorp (1938) in their classification of all known United States soils published in the 1938 USDA *Yearbook of Agriculture*. An outline of the orders and suborders proposed by them is presented in Table 13.2. This classification marked the start of truly comprehensive and quantitative soil classification.

A return to the zonality concept of Sibirtsev was made in this system through recognition of zonal, intrazonal, and azonal classes at the highest categorical level. The Pedalfer-Pedocal concept was deemphasized. More emphasis was given to soil as a three-dimensional body. However, a few problems have come to light which are now being met in this modern period of soil classification. For example, their two highest categorical levels are defined in genetic terms, not on the basis of properties of the soils themselves. The order classes are not mutually exclusive since a soil can be zonal in one region and intrazonal in another. Not all soils were placed in families, and families were not clearly defined as taxa at this intermediate level. Great soil groups were not clearly defined, their definitions were comparative and qualitative, and it was difficult to obtain agreement among different persons on the appropriate great soil group in which certain soil should be placed. And nomenclature problems were great, especially for great soil groups, with the hybridization of folk and coined names from many different sources.

TABLE 13.2 ❧ **Soil classification in 1938 *USDA Yearbook of Agriculture* (highest two categories only)**

Category 6 Order		Category 5 Suborder
Zonal soils	Pedocals	Soils of the cold zone 1. Light-colored soils of arid regions
		2. Dark-colored soils of the semiarid, arid, subhumid, and humid grasslands
	Pedalfers	3. Soils of forest-grassland transition
		4. Light-colored podzolized soils of the timbered regions
		5. Lateritic soils of forested warm-temperate and tropical regions
Intrazonal soils		1. Halomorphic (saline and alkali) soils of imperfectly drained arid regions and littoral deposits
		2. Hydromorphic soils of marshes, swamps, seep areas, and flats
		3. Calomorphic
Azonal soils		No suborders

MODERN QUANTITATIVE PERIOD. Revisions of the 1938 USDA *Yearbook* classification by Thorp and Smith (1949) and by Riecken and Smith (1949) signaled the start of the modern period of classification. In these revisions new great soil groups were added and definitions were revised and sharpened. We set the actual start of this period as 1951, as this was the year in which a decision was made in the United States to develop a new system of classification (Smith 1968). This date approximately corresponds with the time periods in which many other countries started efforts to improve and/or further develop their soil classification systems.

Reasons for undertaking development of a new system of classification in the United States have been summarized in the following points (drawn in part from Kellogg 1963; Simonson 1952a, b; Smith 1968). However, it should be pointed out that the 1938 system has been very useful and has played a major role in the recent growth and development of the whole field of soil science.

1. The highest category of the 1938 system, based on zonality, did not provide mutually exclusive taxa; it was not possible to define clearly the differences between Zonal and Intrazonal soils.

2. Classification at higher levels of the 1938 system, as well as those of other existing systems, was based on external environmental factors and assumed genesis as differentiating characteristics, not properties of soils themselves. Thus there was risk of "prejudicing the future" and the difficulty of not being able to classify certain soils with good agreement because of uncertainties or disagreement concerning their genesis.

3. Some definitions of taxa were based on virgin soil profiles under their native vegetation, without allowing for modifications due to tillage and/or erosion.

4. Too much emphasis had been placed on soil *color* as a differentiating characteristic, without consideration of its relevance or number of accessory characteristics associated with it.

5. Taxa of the lower categorical levels especially were defined in terms of comparative and subjective definitions, and/or not enough differentiating characteristics were used. Quantitative, objective differentiating characteristics were needed for interpersonal agreement on classification and for the taxa to be used as to behavior and use; that is, to make them suitable objects for research.

6. No suitable classes were provided for some soils at some of the categorical levels. Provision had to be made for all known soils to be classed at all categorical levels and for flexibility to accommodate classification of newly discovered and defined soils in developing areas of the world.

7. Soil families had not been clearly defined as a category and as taxa within this category.

8. The nomenclature was a collection from several sources, both folk names in several languages and coined names, and was therefore awkward; the same term held different meaning for different people, translation was difficult, and the naming of soil intergrades was difficult if not impossible.

As a consequence, development of a completely new system, above the levels of the soil series, was started within the USDA under the leadership of G. D. Smith, with the cooperation of soil scientists in the United States universities and of certain overseas scientists. The development of this new comprehensive system was by a series of *approximations* which were circulated for criticism and comment. The *7th Approximation* was published in 1960 (Soil Survey Staff 1960) to insure wider circulation and hence a broader spectrum of comment and criticism. Comments received and further studies were used as a basis for supplements published in 1964 and 1967 (Soil Survey Staff 1964, 1967). It is these publications, plus supplements and a few additional later revisions, which we use as a basis for presentation and discussion of modern soil classification in the United States in the following chapter.

LITERATURE CITED

Baldwin, M., C. E. Kellogg, and J. Thorp. 1938. Soil classification, pp. 979–1001. In Soils and men, yearbook of agriculture. U.S. Dept. Agr. U.S. Govt. Printing Office, Washington.

Bridgman, P. W. 1927. The logic of modern physics. Macmillan, New York.

Cline, M. G. 1949. Basic principles of soil classification. Soil Sci. 67:81–91.

———. 1961. The changing model of soil. Soil Sci. Soc. Am. Proc. 25:442–46.

———. 1963. Logic of the new system of soil classification. Soil Sci. 96:17–22.

Coffey, G. N. 1912. A study of the soils of the United States. U.S. Dept. Agr. Bur. Soil Bull. 85. U.S. Govt. Printing Office, Washington.

Coleman, N. T., S. B. Weed, and R. J. McCracken. 1959. Cation-exchange capacity and exchangeable cations in piedmont soils of North Carolina. Soil Sci. Soc. Am. Proc. 23:146–49.

Dokuchaev, V. V. 1883. Russian Chernozem (Russkii Chernozem). In Collected writings (Sochineniya), vol. 3. (Transl. from Russian by N. Kaner.) Israel Prog. for Sci. Trans., Jerusalem, 1967. Available U.S. Dept. Commerce, Springfield, Va.

———. 1886. Report to the provincial zenstvo (local authority) of Nizhnii-Norgorod (now Gorkii), no. 1. Main phases in the history of land assessment in European Russia, with classification of Russian soils. In Collected writings (Sochineniya), vol. 4. Acad. Sci., USSR, Moscow, 1950.

Fallou, F. A. 1862. Pedologie oder allgemeine und besondere Bodenkunde. Dresden, Germany.

Glinka, K. D. 1914. Die typen der bodenbildung, ihre klassifikation und geographische verbreitung. Gebrüder Borntraeger, Berlin.

———. 1927. The great soil groups of the world. (Transl. from German by C. F. Marbut.) Edwards Brothers, Ann Arbor, Mich.

———. 1931. Treatise on soil science (Pochvovedeniye). 4th ed. (Transl. from Russian by A. Gourevich.) Israel Prog. for Sci. Trans., Jerusalem, 1963. Available from U.S. Dept. Commerce, Washington.

Jenny, H. 1961. Reflections on the soil acidity merry-go-round. Soil Sci. Soc. Am. Proc. 25:428–32.

Kellogg, C. E. 1963. Why a new system of soil classification? Soil Sci. 96:1–5.

Kubiëna, W. L. 1948. Entwicklungslehre des Bodens. Springer-Verlag, Wien.

Marbut, C. F. 1922. Soil classification. Am. Assoc. Soil Survey Workers, 2nd Annual Rept., Bull. 3:24–32.

———. 1927. A scheme for soil classification. Proc. and Papers, 1st Intern. Congr. Soil Sci. (Washington) 4:1–31.

———. 1935. Soils of the United States. In USDA Atlas of American agriculture, part 3. Advance sheets no. 8.

Mill, J. S. 1925. A system of logic, 8th ed. Longmans, Green, and Co., London.

Richthofen, F. F. von. 1886. Führer für Forschungsreisende. Berlin.

Riecken, F. F., and G. D. Smith. 1949. Lower categories of soil classification: Family, series, type, and phase. Soil Sci. 67:107–15.

Ruffin, E. 1832. An essay on calcareous manures. Edited by J. Carlyle Sitterson. Belknap Press of Harvard Univ. Press, Cambridge, 1961.

Sibirtsev, N. M. 1901. Soil science (Pochvovedeniye). In Selected works (Izbrannye Sochineniya), vol. 1. (Transl. from Russian by N. Kaner.) Israel Prog. for Sci. Trans., Jerusalem, 1966. Available from U.S. Dept. Commerce, Springfield, Va.

Simonson, R. W. 1952a. Lessons from the first half-century of soil survey. I. Classification of soils. Soil Sci. 74:249–57.

———. 1952b. Lessons from the first half-century of soil survey. II. Mapping of soils. Soil Sci. 74:323–30.

Smith, G. D. 1968. Soil classification in the United States, pp. 6–24. In World soil resources report 32. FAO, Rome, Italy.

Soil Survey Staff. 1960. Soil classification, a comprehensive system—7th approximation. U.S. Dept. Agr. U.S. Govt. Printing Office, Washington.

———. 1964. Supplement to Soil classification system—7th approximation. U.S. Dept. Agr. U.S. Govt. Printing Office, Washington.

———. 1967. Supplement to Soil classification system—7th approximation. U.S. Dept. Agr. U.S. Govt. Printing Office, Washington.

Thaer, A. D. 1853. Grundsätze der rationellen Landwirtschaft, 5th ed. (Originally published as vols. 1–4, 1809, –10, –12.) Not seen but cited by J. S. Joffe, 1949, Pedology, 2nd ed. Pedology Publ., New Brunswick, N.J.

Thorp, J., and G. D. Smith. 1949. Higher categories of soil classification: Order, suborder, and great soil groups. Soil Sci. 67:117–26.

U.S. Department of Agriculture. 1899. Field operations of the division of soils. Rept. 64. U.S. Govt. Printing Office, Washington.

Whitney, M. 1909. Soils of the United States. U.S. Dept. Agr. Bur. Soils Bull. 55. U.S. Govt. Printing Office, Washington.

Modern Soil Classification Systems &

In THIS CHAPTER we have selected examples of classifications of soil that illustrate different approaches and concepts, with emphasis on the United States Comprehensive Soil Classification System because it is used as a basis for our later chapters, we have worked with it extensively, and because it has been and is being used as a basis for classifying very significant portions of the earth. We will also review soil classification systems being used in the USSR, France, Belgium, the United Kingdom, Australia, Canada, and Brazil. A discussion of the systems recently proposed by Kubiena (1948, 1953) in a western European context is also presented.

We have also included a discussion of the FAO/UNESCO Soil Map of the World, which, though it is not a formal soil taxonomic system, does present a systematized, partial classification system in connection with the description and definition of the map units. It is being used to inventory and describe the world's resources on a comparable basis. Several countries in the developing parts of the world are using this map legend and its set of defined soil units to inventory the soil resources of their lands.

In the latter part of the chapter we will present a discussion of numerical taxonomy, a relatively new technique in soil classification.

MODERN USSR SOIL CLASSIFICATION. Modern soil classification in the Soviet Union continues the general approach first outlined by Dokuchaev and Sibirtsev. This includes a strong genetic emphasis, evaluating soil properties and the pedogenic processes in the solum in relation to soil-forming factors. The study of soil profiles with respect to their classification has three main components in the analyses used by most USSR pedologists today: the soil properties, the soil-forming or pedogenic processes, and the agents or factors of soil formation (Rozov and Ivanova 1968). This approach has been designated as "ecological-genetic" by some of the Soviet soil scientists. A point of emphasis in their systems is the action of soil-forming factors to produce soil properties in kinds of profiles called "soil types," or genetical soil types (Gerasimov and Ivanova 1959).

The category of "soil type" in the Soviet system more nearly corresponds to the level of generalization in the order and suborder categories of the more recent United States systems, whereas the designation "soil type" was given to the lowest categorical level in those United States systems before the present one. The soil type (as previously used and designated in the United States) as the lowest category has been eliminated from the formal United States taxonomic system in the new Comprehensive Classification and relegated to use as a phase for use and management interpretations (Soil Survey Staff 1960). The Russian use of the term soil type at a high level of classification (for *broad* groups of soil) has precedence in prior usage in time over the American usage of the term for narrowly defined lowest-level taxa; unfortunately, the American usage was introduced by Whitney (1909) before contact was made with the Russian school of pedology and became well established in America before the conflicting nature of the usage was discovered.

The soil type in the Russian sense is defined by Prasolov as having "unity of origin, substance transformation, migration, and accumulation; that is, they are generalizations of many actual soils which have a common origin and similar pedogenic processes" (Prasolov 1931; Rozanov and Ivanova 1968). Definition and diagnosis of soil types in the present Soviet classification used by the Dokuchaev Institute and the Ministry of Agriculture (Rozov and Ivanova 1968) are based on the soil profile morphology, mineral and chemical composition, organic matter composition, liquid and gaseous phases of the soil, physicochemical properties, and soil moisture, gas, and moisture regimes in the soil.

In the Soviet Union, pedology is ordinarily divided into two aspects: *soil classification,* which deals with the categorical levels of soil types and higher categories and is concerned mainly with soil genesis on a broad scale; and *soil systematics,* which is concerned with soil taxonomy at levels lower than the soil type, and deals mainly with mapping problems.

Soil classification in the Soviet Union at present is based on schemes such as the one published by Gerasimov, Zavalishin, and Ivanova (1939), with more recent modifications (Gerasimov 1968; Kovda, Rozanov, and Samoylova 1969; Kowalinski 1966; Liverosky 1969; Rozov and Ivanova 1968).

Soil systematics is rather well worked out and is used in rather detailed farm mapping and approved by the Soil Science Society Classification Committee; on the other hand, classification above the soil type level has not been as completely worked out (Rozov and Ivanova 1968).

A summary of the main features of each of the categories and their nomenclature is presented in the following discussion, based on summarization from several Soviet sources (mainly Rozov and Ivanova 1968) and discussions with Soviet soil scientists (the order of presentation commences with the highest categorical level and proceeds downward through lower levels).

Class: Taxa are defined according to broad temperature belts and designed as a global classification.

Subclass: Automorphic (approximately equal to "zonal"), hydromorphic, semihydromorphic, and alluvial are the subclasses more commonly recognized. This category is also designed for global scale classifications.

Type: This is the level most commonly used for broad regional comparisons and generalizations, as previously mentioned. Approximately 110 soil types are now recognized in the USSR. Each type has developed in a single set of bioclimatic and hydrologic conditions and is characterized by a clear manifestation of basic soil morphology, including similarities in type of organic matter accumulation, type of decomposition and synthesis of minerals, translocations of soil materials, and structure. Similar measures are needed for maintenance and increase of soil productivity. As to nomenclature, generally the name for the color of the A horizon is combined with "zem" (meaning land) to produce terms such as Chernozem and Krasnozem. Some names indicating dominating soil features are also used, such as Solonetz and Solonchak. A few descriptive folk terms are used, such as Podzol.

Subtype: This category is composed of taxa within the types differing qualitatively in expression of one of the soil-forming processes and/or intensity with which they reflect the main pedogenic process of the type. The nomenclature reflects location and temperature differences from north to south within the USSR with use of the adjectives northern, southern, warm, or cold, or some transitional modification as expressed in soil properties with adjectives such as leached and sod podzolic.

Genera: Taxa are defined according to properties of the parent material as reflected in texture and composition, or according to special dominating effects of chemical composition of the groundwater, or according to some relict or fossil features. These effects are expressed in major modifications of one or more of the key horizons, a shift in horizon sequence, or in horizons additional to those normal for the type. As examples, some (not all) of the genera within the Chernozemic subtypes are listed: typical, noncalcic, calcic, solonetzic, solodized, vertisolic, iron-illuvial, and humus-illuvial.

Species: Taxa are defined within the genera according to degrees of development or expression of the main pedogenic processes within the type (for example, the Podzols are subdivided as weak, medium/strong, and very strong), or according to depth of effect of the pedogenic process (such as podzolization), or according to humus content of the A1 horizon in Chernozems. Ordinarily, one or more of three types of soil properties are used as differentiating characteristics: (1) amount or supply of a certain material in the solum, expressed as kg/m^2 or metric tons/ha, (2) thickness of a certain horizon in cm, (3) content of a substance in a given horizon expressed as percent.

Varieties: This category differentiates according to texture (at a lower level of generalization than used at the level of the genera).

A summary of the classes of higher levels in a scheme of classification published earlier (Ivanova 1956) in the USSR is presented in Table 14.1. Indications of a few modifications since the publishing of this system can be found in Rozanov and Ivanova (1968).

SOIL CLASSIFICATION IN

WESTERN EUROPE. Pedology in this part of the world has been strongly influenced by the Russian pioneers in pedology and by certain European soil scientists who emphasized or extensively used chemical methods of differentiating among soils (Ramann 1911, 1918; Sigmond 1938). In general, there has been great emphasis on genesis and on zonality concepts, with emphasis also on laboratory-derived parameters such as silica:sesquioxide and silica:alumina ratios for use as differentiating characteristics. We note that western European soil scientists generally place less emphasis on clay translocation and argillic horizons than do United States soil scientists, probably in part because of the extensive occurrence of soils without such features (former Sols Bruns Acides, Braunerde and Rendzina, Ranker Lithosols) on European landscapes due to a combination of parent material, climate, and youthfulness factors. It should be remembered that many western European soils have formed in glacial deposits under very young, Late Pleistocene surfaces. This resultant low degree of soil development has been reflected in the Western European approach to soil classification. Another important point is the high degree of significance attached to wet (hydromorphic) soils in most western European classifications—more so than in the United States. We need also to keep in mind that western European soils surrounding villages have been intensively cultivated and manured since medieval times, and thus their soils reflect the imprint of man's activities more than in most other parts of the world. And a final reminder as we open our discussion of western European soil classification, many areas of western Europe are ecologically different from most of the United States in that they either have or approach a Mediterranean climate, with cool, moist winters and hot, dry summers.

NATURAL SOIL SYSTEM OF KUBIENA. This system is generally within the context of the soils of Europe; however, it is designed as a system of worldwide application. A summary as prepared by Gerasimov and Ivanova (1959), based largely on Kubiena's 1948 classification and with some reference to his 1953 classification, is presented in Table 14.2. It carries a thread or theme of theoretical soil developmental paths. Stages of evolutionary development of soils within certain ecological conditions are prescribed. For example, in the "terrestrial" soils, one can see stages from soils with weak A horizons only to soils with well-developed profiles with a full ABC horizon sequence, moving from the top to the bottom of the table. Considerable emphasis is given to hydromorphic soils, as they are accorded equal status at the highest categorical level with the well-drained soils. A good deal of emphasis is laid on chemical and mineralogical properties as differentiating characteristics.

TABLE 14.1 ❧ Scheme for a general soil classification according to USSR system

Class of Soil	Kind of Soil Formation (Subclass)	Automorphic	Automorphic hydromorphic	Hydromorphic
I. Arctic-tundra soils	1. Tundra-Arctic 2. Subarctic sod 3. Bog tundra 4. Arctic Solonchak	Arctic soils Tundra soils Sod soils	 Arctic Solonchak soils	 Bog tundra soils
II. Boreal frozen taiga soils	1. Frozen taiga 2. Frozen bog	Taiga ferruginous Straw-colored taiga soils	 Straw-colored taiga gley soils	 Frozen bog soils
	3. Frozen solonetz	Frozen soluds	Frozen gleyey soluds	
III. Boreal taiga and forest soils	1. Taiga forest	Podzolic soils Gray forest soils	Podzolic bog soils Gray forest gley soils	
	2. Sod taiga 3. Bog	Sod taiga soils	Sod gley soils	 Bog soils
IV. Subboreal humid forest and meadow soils	1. Burozem	Moist forest acid unpodzolized soils Brown forest soils	Humid forest acid unpodzolized gley soils Brown forest gley soils	
	2. Meadow-Burozem		Meadow-Burozem soils	
	3. Sod moist forest	Humus-carbonate soils	Humus-carbonate gley soils	
	4. Bog			Bog soils
V. Subboreal steppe soils	1. Steppe	Chernozems Chestnut	Meadow Chernozems	
	2. Meadow 3. Meadow-bog			Meadow soils Meadow-bog soils
	4. Solonetz	Steppe solonetz	Meadow solonetz Solods	
	5. Solonchaks			Steppe Solonchaks
VI. Subboreal desert soils	1. Desert	Brown semidesert soils Gray-brown desert soils	Brown meadow-desert soils	
	2. Takyr 3. Solonetz desert	 Desert solonetz	Takyr soils Meadow-desert solonetz	
	4. Solonchak desert			Desert Solonchaks
VII. Subtropical humid forest soils	1. Zeltozem 2. Krasnozem 3. Bog subtropical	Zeltozems Krasnozems	Gley Zeltozems Gley Krasnozems	 Subtropical bog soils

TABLE 14.1   *   (continued)

Class of Soil	Kind of Soil Formation (Subclass)	Soil Type		
		Automorphic	Automorphic hydromorphic	Hydromorphic
VIII. Subtropical dry forest savanna and steppe soils	1. Cinnamon earth	Cinnamon soils	Cinnamon meadow-forest soils	
	2. Sod subtropical		Smolnitsas(?)	
	3. Meadow subtropical			Meadow subtropical soils
	4. Subtropical solonetz		Subtropical meadow-solonetz soils	
IX. Subtropical desert soils	1. Serozem-desert	Serozems	Meadow Serozem soils	
		Subtropical desert soils		
X. Tropical humid savanna and forest soils		(types not worked out)		
XI. Tropical dry forest and savanna soils		(types not worked out)		
XII. Tropical desert soils		(types not worked out)		

SOURCE: Ivanova 1956.

PEDOLOGY IN FRANCE. Significant themes and concepts in French pedology seem to be:

1. Classification is based on degree of evolution of the profile [AC→ A(B)C→ABC]. The (B) indicates a weak or color B horizon, without clay accumulation.

2. Characteristics of the entire profile are considered, with emphasis on alterations, humus type, "adsorption complex," structure, and wetness.

3. Properties due to hydromorphic conditions are stressed, being recognized at the highest categorical level.

4. The degree of lessivage (physical translocation of clay particles) is used for differentiating classes in lower categories.

The main elements of the pedologic system used by ORSTOM (the French agency responsible for soil classification) are presented in Table 14.3. This can be summarized: "Raw mineral soils, AC soils, vertisols and para-vertisols, calci- and magnesi-morphic soils, steppe soils, soils with mull, podzols, soils rich in sesquioxides, halomorphic and hydromorphic soils are subclassified according to climate, development, lithology, and hydromorphism" (Aubert 1965, 1968).

Comparisons between the French and American systems of soil classification are summarized:

1. In American classification, greater emphasis is given to B horizons

TABLE 14.2 ❧ Natural System of Kubiena 1953

A. Subaqueous or Underwater Soils
   AA. Subaqueous soils not forming peat
      1. Protopedon (subaqueous raw soil)
      2. Dy
      3. Gyttja
      4. Sapropel
   AB. Peat-forming subaqueous soils
      5. Fen

B. Division of Semiterrestrial or Flooding and Groundwater Soils
   BA. Semiterrestrial raw soils
      6. Rambla (raw warp soil)
      7. Rutmark (Arctic and snow basin)
      8. Raw gley soil
   BB. Anmoorlike soils
      9. Anmoor
      10. Marsh
   BC. Semiterrestrial peat soils
      11. Carr (Transition Wood Moor)
      12. High Moor
   BD. Salt soils
      13. Solonchak
      14. Solonetz
      15. Solod
   BE. Gley soils with land humus formation
      16. Gley
      17. Gray warp soils
      18. Rendzinalike warp soils
      19. Smonitza or Chernozemlike warp soils
      20. Vega

C. Terrestrial or Land Soils
   CA. Terrestrial raw soils
      21. Raw soils of cold deserts
      22. Dry desert raw soils
      23. Syrozem (raw soils of temperate zones)
   CB. Rankerlike soils
      24. Ranker
   CC. Rendzinalike soils
      25. Eurendzina
      26. Pararendzina
   CD. Steppe soils
      27. Serozem
      28. Burozem (Brown Desert steppe soils)
      29. Kastanozem (Chestnut-colored soil)
      30. Chernozem
      31. Para-chernozem
      32. Para-serozem
   CE. Terrae calxis
      33. Terra (includes Terra Fusca, Terra Rossa, etc.)
   CF. Boluslike silicate soils
      34. Braunlehm (Brown loams)
      35. Rotlehm (Red loams)
   CG. Latosols
      36. (Lateritic) Roterde
   CH. Brown earths
      37. Braunerde
   CI. Pseudogley
      38. Pseudogley
   CJ. Podzol
      39. Semipodzol
      40. Podzol

TABLE 14.3 ♣ **Soil classification system of the Pedology Section, ORSTOM, France—1964**

Class I.	Little weathered, skeletal soils 1.1 Due to climate: Groups—cold regions, deserts 1.2 Nonclimatic: Groups—eroded or skeletal recently deposited
Class II.	Poorly and minimally developed soils 2.1 Due to climate: Groups—Tundra, Rankers, Subdesert 2.2 Nonclimatic: Groups—Regosolic, Andosols, recent materials
Class III.	Calcomagnesimorphic soils 3.1 Rendzina: Groups—thin rendzina, rendzina with horizons, calcimorphic alluvial
Class IV.	Vertisols and Paravertisols 4.1 Topomorphic (depressional): Groups—Vertisols and Paravertisols 4.2 Lithomorphic: Groups—Vertisols and Paravertisols

Class V.	Isohumic soils
	5.1 Isohumic soils with partially saturated complex:
	    Groups—Bimodal: Brunizems, Brunizem w/textural Bl, Pseudogley;
	    Brunizem, Vertic Brunizem, Alkalic Brunizem
	5.2 Isohumic soils with saturated complex: Groups—Chernozem, Chestnut, Brown soils

Class VI.	Mull soils 6.1 Mull soils of temperate regions: Groups—Lessive, Brown soils 6.2 Mull soils of tropics
Class VII.	Podzols and Podzolic soils 7.1 Soils with mor and R_2O_3-enriched horizon 7.2 Soils with mor, R_2O_3-enriched horizon and gleyed
Class VIII.	Sesquioxide soils with rapidly mineralized organic matter 8.1 Red and Brown Mediterranean soils 8.2 Ferruginous Tropical soils 8.3 Ferrallitic soils
Class IX.	Halimorphic soils 9.1 Halomorphic without degraded structure 9.2 Halomorphic with degraded structure
Class X.	Hydromorphic soils 10.1 Organic hydromorphic 10.2 Medium organic matter hydromorphic 10.3 Low organic matter hydromorphic

SOURCE: Aubert 1968.

of clay accumulation (argillic horizons), though the French system does attach importance to distinguishing between sol lessives (those with argillic horizons) and podzols (those with spodic horizons).

    2. Greater emphasis is attached to soil properties associated with wetness in the French system. The basis is that wetness induces a peculiar evolution of profile that masks or prevents any other pedogenic process (Aubert 1965).

PEDOLOGY IN BELGIUM. Here we have an example of a relatively small country with a very intensive agriculture. Consequently, soil mapping has been done on a detailed, intensive basis, and hence their soil classification reflects this. *Principal series* are the main map and taxonomic units, defined by soil texture, drainage class, and profile development (Tavernier and Marechal 1962). Soil texture differentia are: seven textural (particle size) classes, organic material, and loess sediments.

Drainage classes are based on interpretations of morphological characteristics associated with wetness mottling, grayish base color, and depth at which these occur.

Profile development is recognized in several classes according to expression of horizon sequences: AC, A(B)C, ABC principally where (B) refers to a weakly expressed B horizon (cambic horizon).

PEDOLOGY IN THE UNITED KINGDOM. Soil groups approximately corresponding to the great soil group level of generalization in the 1938 USDA *Yearbook* classification (Baldwin, Kellogg, and Thorp) have been rather qualitatively defined in terms of similar kind of profile and similar genesis. Major soil groups of Wales (Ball 1964) are listed in comparison with great soil groups of the 1938 United States classification:

Welsh Major Soil Groups	*USA—1938 plus 1949 Revision*
Brown calcareous	(Brown forest)
Calcareous gley	Weisenboden or Humic gley
Brown earth	Sol Brun Acide
Brown earth with gley	Gray-Brown Podzolic
Noncalcareous gley	Humic gley and/or Low Humic gley
Brown Podzolic	(same)
Peaty Podzol	Podzol
Peaty gley	Half-bog
Organic soils	Bog
Alluvial gley	Alluvial
Skeletal and immature soils	Lithosols and Regosols

The British system of classification described by Avery (1956) has at its highest level a binary classification into automorphic (terrestrial) and hydromorphic (semiterrestrial) classes of soil. Ten major soil groups are recognized with differentiating characteristics based on soil moisture regimes and form or type of humus. The third category contains about forty subgroups; differentiating characteristics are based on general kind of profile (kind and arrangement of horizons). Some of the subgroups seem comparable to great soil groups of the 1938 United States classification, but others more nearly resemble those of Kubiena (1953) and other continental European classifications. Although the system is described as "morphogenetic," the classification appears to have a strong genetic bias. A summary of the higher levels of the classification is presented in Table 14.4.

SOIL CLASSIFICATION IN CANADA. Canadian soil classification (Leahey 1961, 1963, 1968; Stobbe 1962) is structured around the great group level of generalization, with nomenclature drawn both from abroad and locally coined names. This system has six categories in all: the order, great group, subgroup, family, series, and type. A summary of the orders and great groups is presented in Table 14.5.

TABLE 14.4 ❧ **British system of soil classification**

A. Automorphic (terrestrial) soils:
  Major soil groups
   1. Raw soils: physically weathered soils with weak A horizons
   2. Montane humus soils: physically weathered soils mainly of grasslands and higher mountains with moderlike (nonpeaty) humus
   3. Calcareous soils: neutral to alkaline forest, grassland, and cultivated soils with rendzina-moderlike humus, on calcareous materials
   4. Leached mull soils: neutral to moderately acid forest, grassland, and cultivated soils with mull humus
   5. Podzolized (Mor) soils: strongly acid heath and forest soils with raw humus and B horizons enriched in iron and/or humus

B. Hydromorphic (semiterrestrial) soils
   6. Warp (alluvial) soils: natural or man-made soils on recent alluvium with little or no gleying in the upper 40 cm
   7. Gray hydromorphic (gley) soils: periodically waterlogged forest, grassland, and cultivated soils with mull humus and gleyed subsoil horizons
   8. Gley-Podzolic soils: periodically waterlogged, strongly acid heath, moorland, and forest soils with raw humus or thin ($<$40 cm), peaty, humus formations and bleached A2 horizons, more or less masked by humus
   9. Peaty soils: alkaline to moderately acid dark-colored soils without B horizons, rich in largely decomposed organic matter, formed under waterlogged conditions with anmoor, peat-mull, or peat moder humus
  10. Peat (bog) soils: acid soils composed largely of partially decomposed plant materials accumulated under waterlogged conditions (including partially decomposed drained forms)

SOURCE: Avery 1956.

In this Canadian system, nature of the differentiating characteristics for each of the categories is summarized, drawing chiefly on publications of Leahey (1961, 1963, 1968).

Order: Major kinds of profiles, reflecting dominating pedogenic process.

Great group: Major pedogenic units based on presence or absence of major diagnostic horizons.

Subgroup: Based on genetic characteristics that define more specifically the soil profile as a whole, including central concept, intergrades, and other taxa with specific additional features of development.

Soil family: Based on characteristics that "have important bearing on soil properties," including texture, consistence, permeability, reaction, mineralogy, and horizon thickness; homogenous with respect to soil moisture regime and nutrient-supplying power of the soil.

TABLE 14.5 ❧ **Orders and great soil groups in the soil classification system of Canada**

1. Chernozemic order: Brown, Dark Brown, Black, and Dark Gray great groups
2. Solonetzic order: Solonetz and Solod great groups
3. Podzolic order: Gray-Brown Podzolic, Dark Gray Wooded, Gray Wooded, Humic Podzol, and Podzol great groups
4. Brunizolic order: Brown Forest, Brown Wooded, Acid Brown Wooded, Acid Brown Forest, Concretionary Brown, and Alpine Brown great groups
5. Regosolic order: Regosol and Podzo-Regosol great groups
6. Gleysolic order: Humic Gleysol, Gleysol, and Eluviated Gleysol
7. Organic Soil order: Fibrisol, Mesisol, and Humisol

Soil series: Uniform in kind and arrangement of horizons.

Soil types: Subdivisions based on "small differences in texture."

In reviewing this Canadian system, we should keep in mind the setting of a generally cool region with young depositional ground surfaces as a result of Late Pleistocene glaciation, but with a range in vegetation of hardwoods, conifers, prairie, boreal forest, and muskeg. In Canada there has been an emphasis on the landscape unit, composed of a catenal soil association, because of the close and intricate association of hydromorphic and better drained soils on the generally young and undissected landscapes with regionally high water tables. In the earlier days of Canadian pedology, "soil classification and soil mapping were considered to be synonymous. . . . It became evident that soil mapping and classification are two separate functions, both of which must be considered in soil survey operations" (Stobbe 1962). As a consequence, development of a Canadian national system was started in 1945 and accepted for use in 1960. At the higher levels, it is in greater part based on arrangement and presence or absence of diagnostic horizons which are quantitatively defined. The system is simpler than the United States Comprehensive Soil Classification System, and uses nomenclature and concepts tailored for Canadian conditions and needs.

PEDOLOGY IN AUSTRALIA. In reviewing classification systems and approaches in Australia, we must keep in mind the wide range of climate on this continent, from humid tropical in the north to extremely dry desert in the interior to warm subhumid temperate climates in western Australia. About one-third of Australia is arid and another third semiarid (Leeper 1950). There is the great abundance of paleosols (such as fossil laterite), reflecting the imprint of vastly different climates in the past, making for a complex landscape on a local scale. Ground surfaces (geomorphic surfaces or landscapes) have a wide range in age. Therefore, it should not be surprising to find that climatic gradients, vegetational differences, and paleosols are given some emphasis in Australian soil classification systems.

Among the first attempts at soil classification in Australia was that of Jensen (1914), who presented a system based on geology of the initial soil material. Prescott (1931) prepared a general soil map of Australia and a classification system based on soil properties. He strongly emphasized the role of climate and vegetation, and introduced the work and ideas of Dokuchaev. He also attached considerable importance to paleosols, mainly the laterite relict from previous humid tropical climates. His system had as its base "Major Soil Groups," based on soil color and kind of profile, related to present or past climate. However, he tended to use soil group and soil zone interchangeably. He established eighteen generalized soil environment groups, with a central concept soil profile for each. With additional studies, it became apparent that not all soils in a soil zone even approximately conformed to the generalized profile for that zone, hence this zonality concept was abandoned.

A morphogenetic soil classification was developed by Stephens, revising Prescott's approach through more emphasis on soil features. This culminated in his classification of 1962 (Stephens 1962), with 47 great soil groups defined qualitatively, primarily on the basis of morphology of central concept profiles (Table 14.6). Note the primary breakdown (highest level) is on the basis of "solum undifferentiated" (low degree of development) and "solum differentiated" (generally soils with appreciable horizon development). A secondary breakdown of the "solum differentiated" class into pedocals and pedalfers reflects the Australian conditions of large areas of arid and semiarid soils (two-thirds of the country, as we previously indicated), and hence their interest in and concern about calcium carbonate accumulation in soils, the differentiating feature of pedocals.

Northcote (1960) proposed a Natural Objective Classification, based on a bifurcating system (two classes per category), with specific values and limits for the properties of soils in each class. These soil properties were chosen to cover the degree and broad kind of soil profile differentiation, deemphasizing genetic connotations.

PEDOLOGY IN BRAZIL. Based on recent pedologic studies in Brazil, it appears that the well-drained "tropical soils" of that country can be placed in two broad classes: those with Latosolic B horizons (approximately equivalent to oxic horizon of the United States Comprehensive Soil Classification System), and those with textural B horizons (approximately equivalent to argillic horizons); soils with cambic horizons seem much less extensive and occur mainly in the higher altitudes (Costa de Lemos 1968).

The higher level classes recently recognized in Brazil are listed in Table 14.7, based on the presentation by Bennema and Camargo (1964). Soils of Brazil are now undergoing intensive additional investigation to obtain information for more refined and complete classification of soils.

SOIL UNITS FOR FAO/UNESCO SOIL

MAP OF THE WORLD. This joint project of FAO and UNESCO was started in 1961 with the aims of preparing a universal worldwide correlation of soil units, obtaining an inventory of the world soil resources through a set of soil maps with a common legend, and providing for ready transfer of land use and management knowledge. In its preparatory phases, it was under the coordination of D. L. Bramao of FAO; a special Project Advisory Panel was composed of pedologists from many countries. It has been necessary to develop a partial classification system to define the map units of this project. A bicategorical system has been established (Dudal, 1968a, b), with highest or upper class being approximately but not completely equivalent to the "great group" level of the United States and some other systems and to the soil type of the USSR system. The lower category is composed of intergrades or soils with special horizons or features of note. Phases have been proposed (Dudal 1968a) to subdivide the secondary classes according to differences in characteristics or qualities important in use and management of the soils: textural classes (3 classes), stoniness, presence of indurated (pan) layers, presence of hard rock, slope, and presence of salinity.

TABLE 14.6 ❧ Great soil groups arranged under solum classes, soil orders, and sub-orders—Australia

I. Solum Undifferentiated

1. Alluvial soils: showing only sedimentary horizons (amorphic)
2. Skeletal soils: shallow stony soils with no significant profile development (amorphic)
3. Calcareous coastal sands: with no significant profile development (amorphic)

II. Solum Differentiated

A. Pedalfers	B. Pedocals

a. Solum dominated by acid peat or peaty eluvial horizon
   4. Moor peats (phytomorphic)
   5. Alpine humus soils (mesomorphic)
   6. Moor podzol peats (polymorphic)
   7. Acid swamp soils (polymorphic)
b. Solum acid, and with organic, sesquioxide, and sometimes clay illuvial horizons
   8. Podzols (mesomorphic)
   9. Groundwater podzols (hydromorphic)
c. Solum acid, and with clay and sesquioxide illuvial horizons
   10. Lateritic podzolic soils (polymorphic)
   11. Gray-brown podzolic soils (mesomorphic)
   12. Brown podzolic soils (mesomorphic)
   13. Red podzolic soils (mesomorphic)
   14. Yellow podzolic soils (mesomorphic)
   15. Meadow podzolic soils (hydromorphic)
   16. Noncalcic brown soils (polymorphic)
d. Solum acid to neutral and lacking pronounced eluviation of clay
   17. Yellow earths (haemomorphic)
   18. Krasnozems (haemomorphic)
   19. Lateritic krasnozems (polymorphic)
   20. Lateritic red earths (polymorphic)
   21. Terra rossa (calcimorphic)
   22. Prairie soils (mesomorphic)

e. Solum dark colored and slightly acid to neutral in eluvial horizons. Calcareous illuvial horizons
   23. Black earths (mesomorphic)
   24. Wiesenboden (hydromorphic)
   25. Brown forest soils (calcimorphic)
   26. Rendzinas (calcimorphic)
   27. Groundwater rendzinas (polymorphic)
   28. Fen soils (polymorphic)
f. Solum saline or showing postsaline structure in the illuvial horizon
   29. Solonchaks (halomorphic)
   30. Solonetz (halomorphic)
   31. Solodized solonetz (halomorphic)
   32. Soloths (halomorphic)
   33. Solonized brown soils (polymorphic)
g. Solum with slightly acid to neutral eluvial horizons and calcareous illuvial horizons
   34. Red-brown earths (mesomorphic)
   35. Brown earths (mesomorphic)
   36. Brown soils of light texture (polymorphic)
   37. Arid red earths (polymorphic)
   38. Gray calcareous soils (calcimorphic)
h. Solum with neutral to alkaline weakly developed eluvial horizons and calcareous and/or gypseous illuvial horizons
   39. Gray soils of heavy texture (hydromorphic)
   40. Brown soils of heavy texture (mesomorphic)
i. Solum with deflated, slightly acid to alkaline, eluvial horizons, and calcareous and/or gypseous illuvial horizons
   41. Desert loams (mesomorphic)
   42. Gray-brown and red calcareous desert soils (calcimorphic)
   43. Red and brown hardpan soils (hydromorphic)
   44. Desert sand plain soils (polymorphic)
   45. Calcareous lateritic soils (polymorphic)
   46. Stony desert tableland soils (polymorphic)
   47. Desert sandhills (mesomorphic)

SOURCE: Stephens 1962.

TABLE 14.7 ♳ **Higher level soil classes recognized in Brazil**

1. Soil with Latosolic B horizon (includes Latosol Roxa; most other normal Latosols with CEC of clay <6.5 meq/100 g)
2. Soils with textural B horizon with plinthite
3. Soils with textural B horizon without plinthite (includes Rubrozems, Terra Roxa Estruturada, among others)
4. Soils with natric B horizon
5. Soils with incipient B horizon
6. Soils with hardpan below A horizon
7. Grumusols
8. Lithosols
9. Regosols
10. Sands
11. Podzols (including hydromorphic Podzols)
12. Other hydromorphic soils (including organic)

SOURCE: Bennema and Camargo 1964.

To properly define the classes, it has been necessary to agree upon a set of diagnostic horizons (Dudal 1968a), some of which are drawn from criteria for diagnostic horizons in the United States Comprehensive Soil Classification System and in part from other classification systems. The nomenclature for these diagnostic horizons and for the classes themselves has been drawn in part from "classic" soil names, mainly stemming from the original Russian soil type terminology, as well as some of the recently coined soil names used in western Europe and Canada, plus a few names developed especially for this purpose (for example, the Luvisols and Acrisols).

The chart which follows is a comparison of the diagnostic horizon terminology for the FAO/UNESCO system as proposed in 1968, with the nearest counterpart in the United States Comprehensive System.

FAO/UNESCO	U.S. Comprehensive
melanic A horizon	mollic epipedon
distinct A horizon (sombric)	umbric epipedon
weak A horizon (pallid)	ochric epipedon
peaty A horizon	histic epipedon
natric B horizon	natric horizon
argilluric B horizon	argillic horizon
planic B horizon (abrupt textural change)	
glossic B horizon (tongued upper boundary)	
plinthic horizon	plinthite
cambic B horizon (except gley horizons not included)	cambic horizon
gleyic horizon	(unique to FAO/UNESCO, for horizons with indications of pronounced wetness within 50 cm from the surface)
spodic B horizon	spodic horizon
oxic B horizon	oxic horizon
calcic horizon	calcic horizon
gypsic horizon	gypsic horizon
salic horizon	salic horizon
gibbsitic horizon	(unique to FAO/UNESCO, a layer with 30% or more gibbsite or 20% or greater by volume of gravel-size aggregates containing 30% or more gibbsite)

TABLE 14.8 ❧ Soil map units for the FAO/UNESCO soil map of the world

List of Soil Units

FLUVISOLS (water-deposited soils with little alteration)
Dystric
Eutric
Calcaric
Gleyic
RHEGOSOLS (thin soil over unconsolidated material)
Dystric
Eutric
Calcaric
ARENOSOLS (soils formed from sand)
Dystric
Eutric
GLEYSOLS (mottled or reduced horizons due to wetness)
Haplic
Humic
Calcic
Thionic
Plinthic
Histic
Tundric
RENDZINAS (shallow soil over limestone)
RANKERS (thin soil over siliceous material)
ANDOSOLS (volcanic ash with dark surfaces)
Haplic
Vitric
Gleyic
VERTISOLS (self-mulching, inverting soils, rich in montmorillonitic clay)
YERMOSOLS (desert soils)
Haplic
Calcic
Gypsic
Luvic
XEROSOLS (dry soils of semiarid regions)
Haplic
Calcic
Gypsic
Luvic
SOLONCHAKS (soluble salt accumulation)
Haplic
Humic
Takyric
Gleyic
SOLONETZ (high sodium content)
Haplic
Humic
Gleyic
PLANOSOLS (abrupt A-B horizon contact)
Haplic
Humic
Solodic
CASTANOZEMS (chestnut surface color, steppe vegetation)
Haplic
Calcic
Luvic

CHERNOZEMS (black surface, high humus under prairie vegetation)
Haplic
Calcic
Luvic
Gleyic
PHAEOZEMS (dark surface, more leached than Castanozem or Chernozem)
Haplic
Calcaric
Luvic
Gleyic
CAMBISOLS (light color, structure or consistence change due to weathering)
Haplic
Eutric
Calcaric
Vertic
Humic
Andic
LUVISOLS (medium to high base status soils with argillic horizons)
Haplic
Chromic
Ferric
Albic
Plinthic
Gleyic
PODZOLUVISOLS (leached horizons tonguing into argillic B horizons)
Haplic
Gleyic
PODZOLS (light-colored alluvial horizon and subsoil accumulation of iron, aluminum, and humus)
Humo-Ferric
Ferric
Humic
Ochric
Placic
Gleyic
ACRISOLS (highly weathered soils with argillic horizons)
Haplic
Humic
Plinthic
Gleyic
NITOSOLS (low CEC clay in argillic horizons)
Dystric
Eutric
FERRALSOLS (sesquioxide-rich clay)
Haplic
Ochric
Rhodic
Humic
Plinthic
HISTOSOLS (organic soils)
Dystric
Eutric
LITHOSOLS (shallow soils over hard rock)
Dystric
Eutric

SOURCE: Dudal 1969.

In Table 14.8 we present the list of soil units for this FAO/UNESCO World Soil Map as of 1969 (Dudal 1969). A qualitative description of each of the major soil units has been added in parentheses after each one.

U.S. COMPREHENSIVE SOIL CLASSIFICATION SYSTEM.

A classification system completely new in design and nomenclature above the categorial level of the soil series is now in use in the United States (Soil Survey Staff 1960, 1964, 1967). Background for this new approach, reasons for undertaking it, and methods of its preparation were described in Chapter 13 (see also Kellogg 1963). The diagnostic horizons for this new system were discussed in Chapter 2.

DEFINITION OF THE CLASSES.

One of the main differences between this system and others lies in the definition of the taxa. Differentiating characteristics selected are properties of the soils themselves, including soil temperature and moisture (moisture status throughout the year); genesis is not employed except as a guide to relevance and weighting of soil properties (Smith 1968). Definitions are precise and quantitative rather than comparative, and are written in "operational" terms (Smith 1968).

NEW NOMENCLATURE.

A completely new set of names for taxa above the level of the soil series was considered essential because of the problems discussed in Chapter 13. These were different concepts and interpretations held of the same name, different names being given to the same soil, the various sources of names from folk terms in different languages as well as coined terms many of which carry no connotation, the awkwardness in naming intergrades, and the difficulty in translating in different languages. And because many of the old, "classic" soil groups needed to be extensively redefined and their ranges of properties changed, new names were required for them.

Thus, a new nomenclature has been devised, using mainly classic Greek and Latin sources (Heller 1963). The names are connotative as far as feasible, those for the higher categories can be readily translated into western European languages, and they are relatively short (Smith 1968; Soil Survey Staff 1960). A formative element from each of the higher categories is successively carried down to and including the family category (soil series retain their geographic place names, following the procedure established by Whitney in the early 1900s) such that with a little experience, one can make several statements about soil properties simply from analyzing the name of the soil. These names may seem awkward and strange at first (indeed, one critical author ascribed an Alice-in-Wonderland aspect to them), but with a little study and experience one can appreciate and use the advantages of this nomenclature.

STRUCTURE OF THE SYSTEM.

The system contains six categories. From highest to lowest levels of generalization these are order, suborder, great group, subgroup, family, and series. Note the addition of the subgroup category in comparison with the 1938–1949 United States classifications (Baldwin, Kellogg, and Thorp 1938; Thorp and Smith 1949). This category has been introduced not only to provide for another needed level of generalization but also to provide opportunity to define central concepts of our great groups and intergrades (and extragrades—transitional to a "not-soil" population) in order to recognize and express more clearly in our taxonomy that soils are a continuum, with gradational changes in many properties.

The nature of the kinds of differentiating characteristics employed in the various categorical levels of the United States Comprehensive Soil Classification System is given in Table 14.9. Two types of comments and explanations should be made to aid in understanding this table and the general operating procedure in application of differentiating characteristics of the Comprehensive System to the soil population. One is that in soil classification practice, at the highest categorical level, one tends to look at the entire population of soils to the extent they are known, and place them into the ten broad groupings—the soil orders. Once this is accomplished, the practice is then to consider the nature and properties of only the soils within a given order and determine the suborder, or suborders, within a *specific order* into which the soils may be placed. The differentiating characteristics themselves for the suborders, and for great groups within a suborder, etc., have been determined and selected in this fashion. That is, determinations were made of the types of differentia needed to

TABLE 14.9 ✤ **Nature of differentiating characteristics of the categories of the United States Comprehensive Soil Classification System**

Category	Number of Taxa	Nature of Differentiating Characteristics
Order	10	Soil-forming processes as indicated by presence or absence of major diagnostic horizons.
Suborder	47	Genetic homogeneity. Subdivision of orders according to presence or absence of properties associated with wetness, soil moisture regimes, major parent material, and vegetational effects as indicated by key properties; organic fiber decomposition stage in Histosols.
Great group	206 (approximate)	Subdivision of suborders according to similar kind, arrangement, and degree of expression of horizons, with emphasis on upper sequum; base status; soil temperature and moisture regimes; presence or absence of diagnostic layers (plinthite, fragipan, duripan).
Subgroup		Central concept taxa for great group and properties indicating intergradations to other great groups, suborders, and orders; extragradation to "not soil."
Family		Properties important for plant root growth; broad soil textural classes averaged over control section or solum; mineralogical classes for dominant mineralogy of solum; soil temperature classes [based on mean annual soil temperature at 50 cm (20 in.) depth].
Series	10,000 (approximate) in United States	Kind and arrangement of horizons; color, texture, structure, consistence, and reaction of horizons; chemical and mineralogical properties of the horizons.

produce the desired suborders, great groups, etc., in an empirical fashion. This brings us to the second comment—differentiating characteristics are not uniformly applied, or applicable, to all soils at a given categorical level. Uniform application is often considered the optimum or ideal one according to classification theory, but in such a diverse and complex population as soils this is not feasible or practical. That is, certain types of differentiating characteristics are applied only to certain taxa (of the level above which one is considering) to produce the desired taxa at the level with which one is dealing. For example, the degree of wetness and soil moisture regimes are excellent, useful kinds of differentia at the suborder level of most orders, but they have no relevance or significance in the Aridisols, the soils of the deserts. Hence, another type of differentiating characteristic is applied to produce useful and relevant suborder taxa within the Aridisols—the presence or absence of an argillic horizon or of a natric horizon close to the soil surface which separates those soils on the "old" desert landscapes from those of the younger desert landforms. An even more striking example is found in the Histosol order. Differentiating characteristics used for taxa of levels below the order in mineral soils generally have no relevance in organic soils. Therefore, degree of decomposition of the organic fibers, which carries a large number of accessory and covariant characteristics, is used as a differentiating characteristic within the order Histosols only. This we call the *principle of limited applicability of differentiating characteristics*—that they may reasonably be applied only to portions of the entire population at any given categorical level.

Names of the orders, the formative element in the order name used as an identifier at lower levels, derivation or source of the formative element, and the mnemonic or memory device for each are listed in Table 14.10 (after Soil Survey Staff 1960). In this table, we have listed the orders in the same sequence of occurrence as they appear in the Soil Key (Soil Survey Staff 1967). We have prepared a partial key which may be used as a general semiquantitative guide to placement of soils in the orders.

Simplified Key to Soil Orders,
Comprehensive Soil Classification System

If soil has:	Order
> 30% clay to 1 meter (40 in.) or to lithic/paralithic contact gilgai or slickensides or wedge-shaped aggregates	Vertisols
No diagnostic horizon other than ochric, anthropic, albic, argic	Entisols
No spodic, argillic, natric, oxic, petrocalcic, plinthite; but has cambic or histic	Inceptisols
Ochric or argillic but no oxic or spodic and usually dry	Aridisols
Spodic	Spodosols
Mean annual soil temperature > 8 C (47 F), properties not placing it in one of above, percentage base saturation < 35 @ 1.25 m (50 in.) below top of argillic or 1.8 m (72 in.) below surface	Ultisols
Mollic but no oxic	Mollisols
All other mineral soils w/o oxic	Alfisols
Oxic horizon	Oxisols
> 30% organic matter to a depth of 40 cm (16 in.)	Histosols

TABLE 14.10 ❧ **Soil order names and their formative elements**

Order	Formative Element	Derivation	Mnemonicon
Vertisol	ert	L.—*verto,* turn	in*vert*
Entisol	ent	(nonsense syllable)	rec*ent*
Inceptisol	ept	L.—*inceptum,* beginning	ince*pt*ion
Aridisol	id	L.—*aridus,* dry	ar*id*
Spodosol	od	Gr.—*spodos,* wood ashes	P*od*zol; *odd*
Ultisol	ult	L.—*ultimus,* last	*ult*imate
Mollisol	oll	L.—*mollis,* soft	m*oll*ify
Alfisol	alf	(nonsense syllable)	Ped*alf*er
Oxisol	ox	Fr.—*oxide,* oxide	*ox*ide
Histosol	ist	Gr.—*histos,* tissue	h*ist*ology

SOURCE: Soil Survey Staff 1960.

It is very important to follow through this key, and especially the complete one (Soil Survey Staff 1967), in order, from the beginning, in attempting to "key out" the identification and name of an unknown soil, or confusion and misclassification will result. That is, one must start at the beginning or top of the key to check differentiating characteristics for placement purposes. (For example, we have noted students and inexperienced persons, upon noting a soil has a mollic epipedon, immediately assuming the soil is a Mollisol, which *could* be incorrect, as some soils in other orders *may* have mollic epipedons.)

We shall not list all the suborders similarly because there are too many. However, in Table 14.11 we list the formative elements that identify the suborder to which taxa of lower levels belong, source or derivation of the name, the memory device (mnemonicon), and the general meaning of each (Soil Survey Staff 1967).

Likewise, for the great groups we have listed in Table 14.12 the formative elements by means of which taxa at lower levels are tagged as to the great group to which they belong, and the derivation and memory devices of each.

All great groups presently recognized are listed as to formative element in Table 14.13. This table can be used to generate the full name of any presently recognized taxa down to and including the great group. Taking the entry in a column as the first part (usually one syllable) of the name (great group formative element), one then adds the suborder formative element found at the top of each column for the second part of the name. Then the order formative element from the row border at the left of the table is added at the end to complete the name. As an example, let us say that we are interested in the name of a soil with a fragipan which by reason of its soil moisture regime has been determined to fall in an "ud" suborder and by reason of its argillic horizon and low base status to fall in the Ultisol order. Putting these together, we arrive at the name Fragiudult. This table also is a useful checklist of all suborders and great groups established as of the time of writing (1970).

Subgroups are identified with a binomial nomenclature. That is, their names carry the great group name as a noun with the main differentiating

TABLE 14.11 ❧ Suborder names in the United States Comprehensive System—formative elements and meaning

Formative Element	Derivation	Mnemonicon	Meaning or Connotation
alb	L.—*albus*, white	*alb*ino	Presence of albic horizon (a bleached eluvial horizon)
and	Modified from *Ando*	*Ando*	Andolike
aqu	L.—*aqua*, water	*aqu*arium	Characteristics associated with wetness
ar	L.—*arare*, to plow	*ar*able	Mixed horizons
arg	Modified from argillic horizon; L.—*argilla*, white clay	*arg*illite	Presence of argillic horizon (a horizon with illuvial clay)
bor	Gr.—*boreas*, northern	*bor*eal	Cool
ferr	L.—*ferrum*, iron	*ferr*uginous	Presence of iron
fibr	L.—*fibra*, fiber	*fibr*ous	Least decomposed stage
fluv	L.—*fluvius*, river	*fluv*ial	Flood plains
hem	Gr.—*hemi*, half	*hem*isphere	Intermediate state of decomposition
hum	L.—*humus*, earth	*hum*us	Presence of organic matter
lept	Gr.—*leptos*, thin	*lept*ometer	Thin horizon
ochr	Gr.—base of *ochros*, pale	*ochr*er	Presence of ochric epipedon (a light-colored surface)
orth	Gr.—*orthos*, true	*orth*ophonic	The common ones
plag	Modified from Ger. *Plaggen*, sod		Presence of plaggen epipedon
psamm	Gr.—*psammos*, sand	*psamm*ite	Sand textures
rend	Modified from Rendzina	*Rend*zina	Rendzinalike
sapr	Gr.—*sapros*, rotten	*sapr*ophyte	Most decomposed stage
torr	L.—*torridus*, hot, dry	*torr*id	Usually dry
trop	Modified from Gr.—*tropikos*, of the solstice	*trop*ical	Continually warm
ud	L.—*udus*, humid	*ud*ometer	Of humid climates
umbr	L.—*umbra*, shade	*umbr*ella	Presence of umbric epipedon (a dark-colored surface)
ust	L.—*ustus*, burnt	comb*ust*ion	Of dry climates, usually hot in summer
xer	Gr.—*xeros*, dry	*xer*ophyte	Annual dry season

SOURCE: Soil Survey Staff 1967.

characteristic of the subgroup used as a modifying adjective(s) (Smith 1968; Soil Survey Staff 1967). A "typic" subgroup is defined as a taxon representing the "central concept" of the great group. In this case, the adjective typic becomes part of the name of the subgroup as a separate modifying word, along with the great group name. Other taxa are defined according to their possession of one or more characteristics of other great groups to which they intergrade. Certain subgroup taxa possess one or more characteristics not typical or representative of that great group to which they are assigned but not present in other great groups and therefore not intergradational. In these cases, an adjective term is used to denote this extragrade or aberrant property or feature. Rules have been devised for selection of names when the intergradation is to another order, which we shall not elaborate here (Soil Survey Staff 1967). As examples, let us say that we are dealing with a Paleudult which well typifies that great group. We name it Typic Paleudult. Now assume we are considering another soil in this same great group that possesses incipient plinthite. We designate this soil as a Plinthic Paleudult.

TABLE 14.12 ❧ **Great group names in the United States Comprehensive System—formative elements and meaning**

Formative Element	Derivation	Mnemonicon	Meaning or Connotation
acr	Modified from Gr.—*Akros*, at the end	*acr*olith	Extreme weathering
agr	L.—*ager*, field	*agr*iculture	An agric horizon
alb	L.—*albus*, white	*alb*ino	An albic horizon
and	Modified from *Ando*	*Ando*	Andolike
anthr	Gr.—*anthropos*, man	*anthr*opology	An anthropic epipedon
aqu	L.—*aqua*, water	*aqu*arium	Characteristic associated with wetness
arg	Modified from argillic horizon; L.—*argilla*, white clay	*arg*illite	An argillic horizon
calc	L.—*calcis*, lime	*calc*ium	A calcic horizon
camb	L.—*cambiare*, to exchange	change	A cambic horizon
chrom	Gr.—*chroma*, color	*chrom*a	High chroma
cry	Gr.—*Kryos*, cold	*cry*stal	Cold
dur	L.—*durus*, hard	*dur*able	A duripan
dystr	Modified from Gr.—*dys*, ill; *dystrophic*, infertile	*dystr*ophic	Low base saturation
eutr eu	Modified from Gr.—*eu*, good; *eutrophic*, fertile	*eutr*ophic	High base saturation
ferr	L.—*ferrum*, iron	*ferr*ic	Presence of iron
frag	Modified from L.—*fragillis*, brittle	*frag*ile	Presence of fragipan
fragloss	Compound of *fra(g)* and *gloss*		See the formative elements *frag* and *gloss*
gibbs	Modified from *gibbsite*	*gibbs*ite	Presence of gibbsite
gloss	Gr.—*glossa*, tongue	*gloss*ary	Tongued
hal	Gr.—*hals*, salt	*hal*ophyte	Salty
hapl	Gr.—*haplous*, simple	*hapl*oid	Minimum horizon
hum	L.—*humus*, earth	*hum*us	Presence of humus
hydr	Gr.—*hydro*, water	*hydr*ophobia	Presence of water
hyp	Gr.—*hypnon*, moss	*hyp*num	Presence of hypnum moss
luo, lu	Gr.—*louo*, to wash	ab*lu*tion	Illuvial
moll	L.—*mollis*, soft	*moll*ify	Presence of mollic epipedon
nadur	Compound of *na(tr)* and *dur*		
natr	Modified from *natrium*, sodium		Presence of natric horizon
ochr	Gr.—base of *ochros*, pale	*och*er	Presence of ochric epipedon (a light-colored surface)
pale	Gr.—*paleos*, old	*pale*osol	Old development
pell	Gr.—*pellos*, dusky		Low chroma
plac	Gr.—base of *plax*, flat stone		Presence of a thin pan
plag	Modified from Ger.—*Plaggen*, sod		Presence of plaggen horizon
plinth	Gr.—*plinthos*, brick		Presence of plinthite
quartz	Ger.—*quarz*, quartz	*quartz*	High quartz content
rend	Modified from Rendzina	*Rend*zina	Rendzinalike
rhod	Gr.—base of *rhodon*, rose	*rhod*odendron	Dark red colors
sal	L.—base of *sal*, salt	*sal*ine	Presence of salic horizon
sider	Gr.—*sideros*, iron	*sider*ite	Presence of free iron oxides
sombr	Fr.—*sombre*, dark	*somb*er	A dark horizon
sphagno	Gr.—*sphagnos*, bog	*sphagn*um moss	Presence of sphagnum moss
torr	L.—*torridus*, hot and dry	*torr*id	Usually dry
trop	Modified from Gr.—*tropikos*, of the solstice	*trop*ical	Continually warm
ud	L.—*udus*, humid	*ud*ometer	Of humid climates
umbr	L.—base of *umbra*, shade	*umbr*ella	Presence of umbric epipedon
ust	L.—base of *ustus*, burnt	comb*ust*ion	Dry climate, usually hot in summer
verm	L.—base of *vermes*, worm	*verm*iform	Wormy, or mixed by animals
vitr	L.—*vitrum*, glass	*vitr*eous	Presence of glass
xer	Gr.—*xeros*, dry	*xer*ophyte	Annual dry season

SOURCE: Soil Survey Staff 1967.

TABLE 14.13 & Great group formative elements arrayed according to suborder (columns) and orders (rows)

Order Formative Element	alb	and	aqu	arg	bor	fibr	fluv	hum	hem	ochr	orth	psamm	sapr	trop	ud	umbr	ust	xer	Sub-order Only†
ert (Vertisol)															Pell Chrom		Pell Chrom	Pell Chrom	Torr
ent (Entisol)			Hydr* Cry Trop Hapl				Cryo Torri Usti Xero Udi Tropo				Cry Torri Ust Xer Ud Trop	Cryo Quartzi Torri Usti Xero Udi							Ar
ept (Inceptisols)		Cry Dur Hydr Eutr Dystr Vitr	Cry Plac And Trop Fragi Hal Hum Hapl							Fragi Dur Cry Ust Xer Eutr Dys				Us(t)‡ Eu(tr)‡ Dys(tr)‡ Humi		Anthr Fragi Cry Hapl Xer			Plagg
id (Aridisols)				Nadur Dur Natr Pale Hapl							Dur Sal Pale Calci Camb								
od (Spodosols)			Fragi Cry Dur Plac Trop Hapl Sider					Placo Tropo Fragi Cryo Haplo			Plac Fragi Cry Hapl								Ferr
ult (Ultisols)			Plinth Fragi Trop Ochr Umbr Pale					Pale Tropo Haplo							Fragi Plinth Pale Rhod Trop Hapl		Plinth Pale Rhod Trop Hapl	Pale Haplo	

Order								
oll (Mollisols)	Natr Argi	Cry Dur Natr Calci Argi Hapl	Pale Cryo Natri Argi Vermi Calci Haplo		Pale Argi Verm Hapl	Dur Natr Calci Pale Argi Verm Hapl	Duri Natri Calci Pale Argi Haplo	Rend
alf (Alfisols)	Natr Trop Fragi Gloss Alb Ochr Umbr	Pale Fragi Natri Cryo Eutro Glosso		Agr Fragi Natr Trop Ferr Gloss Pale Hapl		Plinth Dur Natr Pale Rhod Hapl	Plintho Duri Natri Rhodo Pale Haplo	
ox (Oxisols)	Gibbsi Plinth Ochr Umbr	Sombri Gibbsi Haplo Acro	Gibbsi Acr Eutr Umbri Hapl			Acr Eutr Hapl	Torr	
ist (Histosols)	Cryo Sphagno Boro Medi Luvi	Cryo Boro Tropo Medi Luvi	Cryo Boro Tropo Medi					Fol

* Great group listing in order of the soil key.
† Suborders in this column have no great groups presently recognized.
‡ Letters in ( ) excluded in forming name.

205

Now for the soil family nomenclature. As indicated in Table 14.9, family criteria are mainly of three kinds: broad textural classes, mineralogy classes, and soil temperature classes. The appropriate class name for each of the three types of criteria is presently used as an adjective following the subgroup binomial name, set off by commas. For example, assume we are considering a Typic Hapludult which is rich in clay that is predominantly kaolinitic and is located in North Carolina in the thermic soil temperature belt. We then designate this soil as Typic Hapludult, clayey, kaolinitic, thermic.

PROPERTIES AND DIFFERENTIATING CHARACTERISTICS OF THE ORDERS. The ten taxa of the orders replace the Zonal, Intrazonal, and Azonal orders of the 1938 system and 1949 revisions, and are roughly equivalent to the suborders of that classification with respect to level of abstraction (Smith 1968). A listing of these ten orders with respect to the great groups of the 1938 system which fall in each of the orders is presented in Table 14.14.

The main properties of each of the orders and their associated suborders and great groups are presented in the following chapters.

NUMERICAL CLASSIFICATION OF SOILS. The concept of the universe as a continuum, as discussed for example by Whitehead (see Chapter 1), has led to experiments in arranging or ordinating soil taxa on a numerical basis. Hole and Hironaka (1960) followed the example of plant ecologists at the University of Wisconsin in applying a simple ordination method to the Miami family and catena and to twenty-five representative soil profiles of the world, described in the *7th Approximation* (Soil Survey Staff 1960). They built three-dimensional models that approximated the multidimensional product. The electronic computer has made possible a dramatic development of what Sneath and Sokal (1962) describe as the new or numerical taxonomy, which they define as "the numerical evaluation of the affinity

TABLE 14.14 ❧ **Soil orders of United States Comprehensive System in relation to great groups of 1938 classification**

Order	1938 Great Groups Included
Entisols	Azonal soils, some Low Humic Gley
Vertisols	Grumusols
Inceptisols	Ando, Sol Brun Acide, some Brown Forest, Low Humic Gley, Humic Gley
Aridisols	Desert, Reddish Desert, Serozem, Solonchak, some Brown and Reddish Brown soils, associated Solonetz
Mollisols	Chestnut, Chernozem, Brunizem, Rendzina, some Brown, Brown Forest, associated Humic Gley, and Solonetz
Spodosols	Podzols, Brown Podzolic, Groundwater Podzols
Alfisols	Gray-Brown Podzolic, Gray Wooded, Noncalcic Brown, Degraded Chernozem, associated Planosols and Half-Bogs
Ultisols	Red-Yellow Podzolic, Reddish-Brown Lateritic, associated Planosols, and some Half-Bogs
Oxisols	Laterite soils, Latosols
Histosols	Bog soils

SOURCE: Soil Survey Staff 1960.

or similarity between taxonomic units and the ordering of these units into taxa on the basis of their affinities."

Numerical taxonomy draws on Adansonian principles rather than Linnean principles of taxonomy (Sneath and Sokal 1962). These two approaches are compared:

Principles of Classification

Adansonian	Linnean
1. Ideal natural taxonomy is one in which taxa have greatest content of information	1. Distinct kinds of things without intermediates
2. Every natural feature is of equal weight in constructing a classification	2. Best method of classifying is to determine "essential nature" of a population and subdivide according to modifications
3. Affinity is a function of proportion of features in common	3. Most important characters for classification are those of greatest physiological importance (plant and animal taxonomy especially)
4. Affinity is independent of phylogeny (evolution of the species or genetic theory)	4. Species are basic constituents of higher groups

Some of the disadvantages or problems of numerical taxonomy have been discussed by Simpson (1964). These can be summarized:

1. Single measure of similarity involves an enormous loss of information.

2. Selection, measurement, and coding of multiple characters are highly subjective.

3. Many different kinds of characters must enter into a taxonomic classification. Can they all be considered and written into a computer program?

Progress in the numerical classification of soils tends to support the potential cited by Whitehead (1925): "Classification is necessary. But unless you can progress from classification to mathematics, your reasoning will not take you very far." Other advantages to computer use in soil classification include data storage and retrieval and long-term continuity of effort. In a sense the popularity of Aristotelian logic has retarded the advancement of physical biological sciences not only throughout the Middle Ages but also in recent times. The Darwinian theory of evolution, the Linnean classification of organisms, and soil classification up to this point have not been mathematically derived (Bidwell and Hole 1964). The new taxonomy, on the other hand, has a really quantitative basis.

Taxonomic dendrograms like the one in Figure 14.1 (Arkley 1968) can be prepared from indices of similarity with the use of a desk calculating machine. Sarkar, Bidwell, and Marcus (1966) used computer analysis to reduce the number of properties selected for a soil classification to a thoroughly correlated few. Arkley (1968) weighted factors and variables according to their communality with other properties, using cumulative communality cluster analysis. Jenny, Salem, and Wallis (1968) produced the first quantification of the general state factor equation for soil genesis. They found that certain properties (including productivity of a crop in a greenhouse) of the soils with which they worked correlated with the state factors of the original soil-sampling sites, regardless of presence or absence of presumed paleosolic features.

Although numerical taxonomy of soil is in its infancy, the current availability of high speed computers and increasing interest in quantification of soil data can be expected to lead to further work in this field.

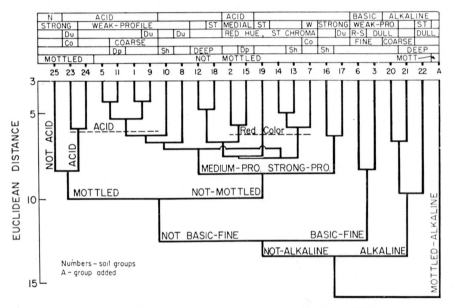

FIG. 14.1. Taxonomic dendrogram. (After Arkley 1968.)

LITERATURE CITED

Arkley, R. J. 1968. Statistical methods in soil classification. Trans. 9th Intern. Congr. Soil Sci. (Adelaide, Australia) 4:187–92.

Aubert, G. 1965. Soil classification. Tables used by the pedology section of ORSTOM (1965) for soil classes, subclasses, groups, and subgroups. Can. Pedol. ORSTOM 3, pp. 269–88.

———. 1968. Classification des Sols Utilisee par les Pedologues Francais, pp. 78–94. In World soil resources report 32, FAO, Rome, Italy.

Avery, B. W. 1956. A classification of British soils. Trans. 6th Intern. Congr. Soil Sci. (Paris, France) E:279–85.

Baldwin, M., C. E. Kellogg, and J. Thorp. 1938. Soil classification, pp. 979–1001. In Soils and men, yearbook of agriculture. U.S. Dept. Agr. U.S. Govt. Printing Office, Washington.

Ball, D. F. 1964. Loss-on-ignition as an estimate of organic matter and organic carbon in non-calcareous soils. J. Soil Sci. 15:84–92.

Bennema, J., and M. Camargo. 1964. Segundo esboco parcial de classificaco de solos Brasileiros. Ministeria Agricultura, Rio de Janeiro, Brazil.

Bidwell, O. W., and F. D. Hole. 1964. Numerical taxonomy and soil classification. Soil Sci. 97:58–62.

Costa de Lemos, R. 1968. The main tropical soils of Brazil, pp. 95–106. In World soil resources report 32. FAO, Rome, Italy.

Dudal, R. 1968a. Definitions of soil units for the soil map of the world. In World soil resources report 33. FAO, Rome, Italy.

———. 1968b. Problems of international soil correlation, pp. 137–43. In World soil resources report 32. FAO, Rome, Italy.

———. 1969. About the legend of the FAO/UNESCO soil map of the world. Technical Work-Planning Conf., National Cooperative Soil Survey, Charleston, S.C. Jan. 1969.

Gerasimov, I. P. 1968. World soil maps compiled by Soviet soil scientists, pp. 25–77. In World soil resources report 32. FAO, Rome, Italy.

Gerasimov, I. P., and E. N. Ivanova. 1959. Comparison of three scientific trends in resolving general questions of soil classification. Soviet Soil Sci., 1959, no. 11, pp. 1190–1205. (Transl. 1960 from Pochvovedeniye, 1959, 11:1–18.)

Gerasimov, I. P., A. A. Zavalishin, and E. N. Ivanova. 1939. A new scheme of a general soil classification for the USSR (Novaya skhema obshchei klassifikatsii pochva SSSR). Pochvovedeniye, 1939, 7:10–43.

Heller, J. L. 1963. The nomenclature of soils, or what's in a name? Soil Sci. Soc. Am. Proc. 27:216–20.

Hole, F. D., and M. Hironaka. 1960. An experiment in ordination of some soil profiles. Soil Sci. Soc. Am. Proc. 24:309–12.

Ivanova, E. N. 1956. An attempt at a general classification of soils. (Transl. from Russian.) Pochvovedeniye, 1956, 6:82–102.

Jenny, H., A. E. Salem, and J. R. Wallis. 1968. Interplay of soil organic matter and soil fertility with state factors and soil properties, pp. 5–37. In Study week on organic matter and soil fertility. John Wiley & Sons, New York.

Jensen, H. I. 1914. The soils of New South Wales. Govt. Printer, Sydney.

Kellogg, C. E. 1963. Why a new system of soil classification? Soil Sci. 96:1–5.

Kovda, V. A., B. G. Rozanov, and Ye. M. Samoylova. 1969. Soil map of the world. Soviet Soil Sci., 1969, no. 1, pp. 1–9. (Transl. by W. M. Johnson from Priroda, 1968, 12:2–7.)

Kowalinski, S. 1966. Attempt at a new classification of the soils of Europe. Soviet Soil Sci., 1966, no. 3, pp. 76–79. (Transl. 1966 from Pochvovedeniye, 1966, 3:37–41.)

Kubiena, W. L. 1948. Entwicklungslehre des Bodens. Springer-Verlag, Wien.

———. 1953. The soils of Europe. T. Murby & Co., London.

Leahey, A. 1961. The soils of Canada from a pedological viewpoint, pp. 147–57. In R. F. Legget (ed.), Soils in Canada. Rev. ed., 1965. Univ. Toronto Press, Toronto.

———. 1963. The Canadian system of soil classification and the seventh approximation. Soil Sci. Soc. Am. Proc. 27:224–25.

———. 1968. The Canadian taxonomic soil classification, pp. 37–52. In World soil resources report 32. FAO, Rome, Italy.

Leeper, G. W. 1950. Climates of Australia, pp. 23–24. In The Australian environment. CSIRO, Melbourne, Australia.

Liverosky, Yu. A. 1969. Some unresolved problems in classification and systematization of USSR soils. Soviet Soil Sci., 1969, no. 1, pp. 106–16. (Transl. 1969 from Pochvovedeniye, 1969, 2:119–30.)

Northcote, K. H. 1960. A factual key for the recognition of Australian soils. CSIRO. Australia Div. Soils, Divisional Rept. 4/60.

Prasolov, L. I. 1931. A unified nomenclature and the fundamentals of genetical soil classification. (O edinoi nomenklature i osnovakh genetischkoi klassifikatsii pochva.) Pochvovedeniye, 1931, no. 8.

Prescott, J. A. 1931. The soils of Australia in relation to vegetation and climate. Council for Sci. Ind. Res., Australia Bull. 52.

Ramann, E. 1911. Bodenkunde. J. Springer, Berlin.

———. 1918. Bodenbildung und Bodeneinteilung. J. Springer, Berlin.

Rozov, N. N., and E. N. Ivanova. 1968. Soil classification and nomenclature used in Soviet pedology, agriculture, and forestry, pp. 53–77. In World soil resources report 32. FAO, Rome, Italy.

Sarkar, P. K., O. W. Bidwell, and L. F. Marcus. 1966. Selection of characteristics for numerical classification of soils. Soil Sci. Soc. Am. Proc. 30:269–72.

Sigmond, A. A. J. de. 1938. The principles of soil science. (Transl. from Hungarian by A. B. Yolland.) Edited by J. V. Jacks. T. Murby & Co., London.

Simpson, G. G. 1964. Numerical taxonomy and biological classification. Science 144:712–13.

Smith, G. D. 1968. Soil classification in the United States, pp. 6–24. In World soil resources report 32. FAO, Rome, Italy.

Sneath, P. H. A., and R. R. Sokal. 1962. Numerical taxonomy. Nature 193:855–60.

Soil Survey Staff. 1960. Soil classification, a comprehensive system—7th approximation. U.S. Dept. Agr. U.S. Govt. Printing Office, Washington.

———. 1964. Supplement to Soil classification, a comprehensive system—7th approximation. U.S. Dept. Agr. U.S. Govt. Printing Office, Washington.

———. 1967. Supplement to Soil classification, a comprehensive system—7th approximation. U.S. Dept. Agr. U.S. Govt. Printing Office, Washington.

Stephens, C. G. 1962. A manual of Australian soils, 3rd ed. CSIRO, Melbourne, Australia.

Stobbe, P. C. 1962. Classification of Canadian soils. Intern. Soil Conf., New Zealand, pp. 318–24.

Tavernier, R., and R. Marechal. 1962. Soil survey and soil classification in Belgium. Intern. Soil Conf., New Zealand, pp. 298–307.

Thorp, J., and G. D. Smith. 1949. Higher categories of soil classification: Order, suborder, and great soil groups. Soil Sci. 67:117–26.

Whitehead, A. N. 1925. Science of the modern world. Macmillan, New York. (Memorial ed. Harvard Univ. Lowell Lectures. Mentor, 1962.)

Whitney, Milton. 1909. Soils of the United States. U.S. Dept. Agr. Bur. Soils Bull. 55. U.S. Govt. Printing Office, Washington.

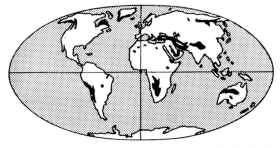

CHAPTER 15

Entisols: Recently Formed Soils ♒

ENTISOLS include soils of such slight and recent development that only an ochric epipedon or simple man-made horizons have formed.

Interestingly, the full evolutionary sequence of a well-developed soil from bare rock is paralleled by a sequence of increasing awareness of soil on the part of landscape painters in the Western world. The American soil geneticist, Hans Jenny, has pointed out (1968) that painters of the fourteenth century showed whole landscapes of bare rock outcrops. Artists of succeeding centuries came to portray first a vegetative cover over rock and, in more recent times, a layer of soil (with horizons) between vegetation and substratum. We are concerned in this chapter with the simpler end of the soil continuum, recognizing, however, that no soil is really simple.

SETTING. Of first importance are the factors limiting soil horizon development in wetlands, alluvial lands, sandy lands, higher-lying rocky lands, and various unconsolidated deposits such as loess and mudflows. The global distribution of Entisols (Fig. 15.1) indicates that many factors are involved, including the following, which may operate in various combinations:

1. Xeric macroclimates (Fig. 18.3) and microclimates, either cold or warm, may limit the amount and duration of water movement in the soil and biotic influence on the soil. Arctic, subarctic, antarctic, and temperate and intertropical desert zones provide these conditions.

2. Mass wasting and other forms of erosion may remove surficial material from the site as fast as or faster than most pedogenic horizons can form. Steep slopes favor the operation of this factor. Solifluction or creep can operate on slopes as low as 2%.

3. Cumulization may add new material to the surface of the soil as fast as or faster than the new material can be assimilated into a pedogenic horizon. This process is at work on alluvial floodplains (Riecken and Poetsch 1960), deltas, footslopes, land around volcanoes, areas of deposition of loess and dune sand, and shorelands bordering lagoons and estuaries.

Removal of vegetation, whether by natural or anthropic means, from part of a landscape accelerates erosion and related deposition.

4. Immobilization of soil plasma in inert materials, in carbonate-rich flocculated materials, and in some highly siliceous sediments inhibits profile differentiation by illuviation.

5. Exceptional resistance to weathering (pedologic inertia) of some initial materials, such as quartzite bedrock, prolongs the period of undistinguished horizonation.

6. Infertility and toxicity of some initial materials to plant growth limits biogenetic differentiation of the soil profile. In serpentine barrens, Entisols may be associated with bedrock outcrops and lithic subgroups of other soil orders.

7. Saturation with water or even submergence of the soil for long enough periods inhibits horizon development.

8. Shortness of time since exposure of initial materials to the active factors of soil formation limits profile development. Fresh lava flows, marine or lacustrine flats newly exposed by uplift of land or by lake drainage, and avalanche and landslip scars provide sites for very young soils, as do situations mentioned under item 3 above. Man's manipulation of soil and geologic materials with bulldozers and other machines leaves fresh initial materials exposed to active pedogenic factors.

9. A recent drastic change in the biotic factor may initiate formation of a different soil profile in an old one, which serves as initial material. For example, human activities may substitute plant communities that do not favor development of Spodosol (Podzol) soil profiles for plant communities that do. Hemlock forests may be replaced by aspen and bracken. This may allow degradation of the spodic horizon to the point that the soil no longer qualifies as Spodosol and becomes an Entisol, all within a period of less than a century.

The world map, Figure 15.1 (p. 214), shows Entisols in mountainous, desert, and sandy regions. They occupy possibly 20% of the land area of this planet.

Tedrow (1968) found wet mineral soils (Aquents) associated with polar Desert soils (Aridisols) in Greenland. Tedrow and Ugolini (1966) observed protorankers and bird-affected (ornithogenic) soils (Orthents) associated with Antarctic cold Desert soils. Rickert and Tedrow (1967) noted wet sands (Aquents) among sand dunes (Psamments) in Alaska.

Topolithosequences of Entisols may occur in floodplains of rivers, ranging from sandy soils (Psamments) of islands, bars, and river banks to Alluvial soils of finer texture (Fluvents) on the natural levees and wet silty soils (Aquents) in the backwater lowlands. Of the 590 million hectares of Alluvial soils (Alluvial as defined in the 1938 classification) in the world (Kellogg and Orvedal 1969) that support about one-third of the human population, perhaps less than half are Entisols, the remainder being Vertisols, Mollisols, and Inceptisols.

NATURE OF PEDOGENIC PROCESSES. The emphasis of the previous section has not been on the nature of the processes, but on the curtailment of their operation in Entisols. This can amount to a complete absence of some processes at certain sites. For example, oxidation may be excluded from saturated or ponded soils; cryoturbation from climatic zones in which there is no freezing; cutan formation from sandy soils lacking plasma. In this respect, Entisols are no different from soils of some other orders.

The full range of kinds of processes of soil formation listed in Table 6.1 commonly operate to some degree in Entisols of the world. But their impact has not been great enough to produce soil features recognized as diagnostic for other orders. For example, melanization is usually active to a limited degree in all Entisols, but soils with mollic epipedons are excluded by definition from the Entisol order. Fragmentation of diagnostic horizons by human manipulation of soils can create profiles with no diagnostic horizons. This is a basis for definition of the suborder Arents. Buried horizons, such as ochric epipedons, are observed in profiles of many cumulative Entisols. Some lithic Entisols (Orthents, in particular) may be unstable soils that are prevented by erosion from developing more distinct profiles. As such they may be stable soils in equilibrium with the environment (Whittaker et al. 1968) and represent one of several kinds of "climax" communities that constitute the complete soilscape of the region.

USES. Entisols present engineering problems in many regions. Erosion by water, wind, and mass wasting is important in steep and hilly to mountainous areas, where runoff or infiltration is rapid. Rocky, shaly, and sandy lands pose these hazards in different ways. Flooding and deposition must be reckoned with in lowlands, particularly in river floodplains and tidal flats. The boundaries of areas endangered by flooding are usually the boundaries of certain alluvial and shoreland soil bodies, Entisols included. Drainage of wetlands involves notable areas of Aquents, including some of the "cat clays" (kattekleien of the Dutch) of tidal marshes, which become extremely acid upon drainage through oxidation of sulfides.

Inhospitable lands, including terrains of Entisols and associated salt flats and rock outcrops, may be assigned a value as wildlife preserve or at least as wilderness areas with aesthetic qualities. The Psamment-Aquent-Udult soilscape[1] of Figure 15.2 is largely wooded and set aside for wildlife (Horton 1967). Military testing operations by various nations of the world have been concentrated to a notable extent on Entisols. Cattle and sheep range in subhumid, semiarid, and arid regions covers vast areas of soils of this order, including parts of the Sand Hills of Nebraska. In the Arctic region, caribou herds move over landscapes containing Entisols.

Millions of people support themselves on fertile and workable Entisols developed from alluvium and shore deposits in association, in many places, with soils of other orders. Entisols formed by active cumulization of fertile soil material are highly prized in primitive agriculture because of their ability to grow crops without fertilization. Flooding of these soils often limits construction of more permanent forms of agricultural practices unless protected by dams or levees. Changes of water table levels resulting

1. The term "soilscape" is a contraction of "soil landscape." See Chapter 25 for a more detailed discussion.

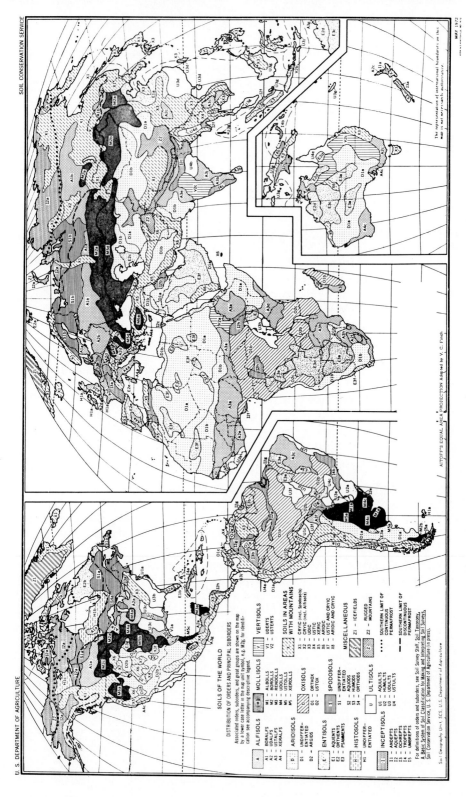

FIG. 15.1. Soils of the world. Probable occurrence of orders and suborders. (Courtesy of U.S. Soil Conservation Service.)

A. ALFISOLS. Soils with subsurface horizons of clay accumulation and medium to high base supply; either usually moist or moist for 90 consecutive days during a period when temperature is suitable for plant growth.

   A1—Boralfs: cool.
       A1a—with Histosols, cryic temperature regimes common.
       A1b—with Spodosols, cryic temperature regimes.
   A2—Udalfs: temperate to hot, usually moist.
       A2a—with Aqualfs.
       A2b—with Aquolls.
       A2c—with Hapludults.
       A2d—with Ochrepts.
       A2e—with Troporthents.
       A2f—with Udorthents.
   A3—Ustalfs: temperate to hot, dry more than 90 cumulative days during periods when temperature is suitable for plant growth.
       A3a—with Tropepts.
       A3b—with Troporthents.
       A3c—with Tropustults.
       A3d—with Usterts.
       A3e—with Ustochrepts.
       A3f—with Ustolls.
       A3g—with Ustorthents.
       A3h—with Ustox.
       A3j—Plinthustalfs with Ustorthents.
   A4—Xeralfs: temperate or warm, moist in winter and dry more than 60 consecutive days in summer.
       A4a—with Xerochrepts.
       A4b—with Xerorthents.
       A4c—with Xerults.

D. ARIDISOLS. Soils with pedogenic horizons; usually dry in all horizons and never moist as long as 90 consecutive days during a period when temperature is suitable for plant growth.

   D1—Aridisols: undifferentiated.
       D1a—with Orthents.
       D1b—with Psamments.
       D1c—with Ustalfs.
   D2—Argids: with horizons of clay accumulation.
       D2a—with Fluvents.
       D2b—with Torriorthents.

E. ENTISOLS. Soils without pedogenic horizons; either usually wet, usually moist, or usually dry.

   E1—Aquents: seasonally or perennially wet.
       E1a—Haplaquents with Udifluvents.
       E1b—Psammaquents with Haplaquents.
       E1c—Tropaquents with Hydraquents.
   E2—Orthents: loamy or clayey textures, many shallow to rock.
       E2a—Cryorthents.
       E2b—Cryorthents with Orthods.
       E2c—Torriorthents with Aridisols.
       E2d—Torriorthents with Ustalfs.
       E2e—Xerorthents with Xeralfs.
   E3—Psamments: sand or loamy sand textures.
       E3a—with Aridisols.
       E3b—with Orthox.

E3c—with Torriorthents.
E3d—with Ustalfs.
E3e—with Ustox.
E3f—of shifting sands.
E3g—Ustipsamments with Ustolls.

H. **HISTOSOLS.** Organic soils.

H1—Histosols: undifferentiated.
H1a—with Aquods.
H1b—with Boralfs.
H1c—with Cryaquepts.

I. **INCEPTISOLS.** Soils with pedogenic horizons of alteration or concentration but without accumulations of translocated materials other than carbonates or silica; usually moist or moist for 90 consecutive days during a period when temperature is suitable for plant growth.

I1—Andepts: amorphous clay or vitric volcanic ash or pumice.
I1a—Dystrandepts with Ochrepts.
I2—Aquepts: seasonally wet.
I2a—Cryaquepts with Orthents.
I2b—Halaquepts with Salorthids.
I2c—Haplaquepts with Humaquepts.
I2d—Haplaquepts with Ochraqualfs.
I2e—Humaquepts with Psamments.
I2f—Tropaquepts with Hydraquents.
I2g—Tropaquepts with Plinthaquults.
I2h—Tropaquepts with Tropaquents.
I2j—Tropaquepts with Tropudults.
I3—Ochrepts: thin, light-colored surface horizons and little organic matter.
I3a—Dystrochrepts with Fragiochrepts.
I3b—Dystrochrepts with Orthox.
I3c—Xerochrepts with Xerolls.
I4—Tropepts: continuously warm or hot.
I4a—with Ustalfs.
I4b—with Tropudults.
I4c—with Ustox.
I5—Umbrepts: dark-colored surface horizons with medium to low base supply.
I5a—with Aqualfs.

M. **MOLLISOLS.** Soils with nearly black organic-rich surface horizons and high base supply; either usually moist or usually dry.

M1—Albolls: light gray subsurface horizon over slowly permeable horizon, seasonally wet.
M1a—with Aquepts.
M2—Borolls: cool or cold.
M2a—with Aquolls.
M2b—with Orthids.
M2c—with Torriorthents.
M3—Rendolls: subsurface horizons have much calcium carbonate but no accumulation of clay.
M3a—with Usterts.
M4—Udolls: temperate or warm, usually moist.
M4a—with Aquolls.
M4b—with Eutochrepts.
M4c—with Humaquepts.
M5—Ustolls: temperate to hot, dry more than 90 cumulative days in the year.
M5a—with Argialbolls.
M5b—with Ustalfs.

M5c—with Usterts.
M5d—with Ustochrepts.
M6—Xerolls: cool to warm, moist in winter and dry more than 60 consecutive days in summer.
M6a—with Xerorthents.

O. OXISOLS. Soils with pedogenic horizons that are mixtures principally of kaolin, hydrated oxides, and quartz, and are low in weatherable mineral content.
O1—Orthox: hot, nearly always moist.
O1a—with Plinthaquults.
O1b—with Tropudults.
O2—Ustox: warm or hot, dry for long periods but moist more than 90 consecutive days in the year.
O2a—with Plinthaquults.
O2b—with Tropustults.
O2c—with Ustalfs.

S. SPODOSOLS. Soils with accumulation of amorphous materials in subsurface horizons; usually moist or wet.
S1—Spodosols: undifferentiated.
S1a—with Boralfs, cryic temperature regimes.
S1b—with Histosols, cryic temperature regimes.
S2—Aquods: seasonally wet.
S2a—Haplaquods with Quartzipsamments.
S3—Humods: with accumulations of organic matter in subsurface horizons.
S3a—with Hapludalfs.
S4—Orthods: with accumulations of organic matter, iron, and aluminum in subsurface horizons.
S4—Haplorthods with Boralfs.

U. ULTISOLS. Soils with subsurface horizons of clay accumulation and low base supply, usually moist or moist for 90 consecutive days during a period when temperature is suitable for plant growth.
U1—Aquults: seasonally wet.
U1a—Ochraquults with Udults.
U1b—Plinthaquults with Ordox.
U1c—Plinthaquults with Plinthaquox.
U1d—Plinthaquults with Tropaquepts.
U2—Humults: temperate or warm and moist all of year, high content of organic matter.
U2a—with Umbrepts.
U3—Udults: temperate to hot, never dry more than 90 cumulative days in the year.
U3a—with Andepts.
U3b—with Dystochrepts.
U3c—with Udalfs.
U3d—Hapludults with Dystochrepts.
U3e—Rhodudults with Udalfs.
U3f—Tropudults with Aquults.
U3g—Tropudults with Hydraquents.
U3h—Tropudults with Orthox.
U3j—Tropudults with Tropepts.
U3k—Tropudults with Tropudalfs.
U4—Ustults: warm or hot, dry more than 90 cumulative days in the year.
U4a—with Ustochrepts.
U4b—Plinthustults with Ustorthents.
U4c—Rhodustults with Ustalfs.
U4d—Tropustults with Tropaquepts.
U4e—Tropustults with Ustalfs.

V. VERTISOLS. Soils with high content of swelling clays; deep, wide cracks develop during dry periods.

  V1—Uderts: usually moist in some part in most years, cracks open less than 90 cumulative days in the year.
    V1a—with Usterts.

  V2—Usterts: cracks open more than 90 cumulative days in the year.
    V2a—with Tropaquepts.
    V2b—with Tropofluvents.
    V2c—with Ustalfs.

X. SOILS IN AREAS WITH MOUNTAINS. Soils with various moisture and temperature regimes; many steep slopes, relief and total elevation vary greatly from place to place. Soils vary greatly within short distances and with changes in altitude; vertical zonation common.

  X1—Cryic great groups of Entisols, Inceptisols, and Spodosols.

  X2—Boralfs and cryic great groups of Entisols and Inceptisols.

  X3—Udic great groups of Alfisols, Entisols, and Ultisols; Inceptisols.

  X4—Ustic great groups of Alfisols, Inceptisols, Mollisols, and Ultisols.

  X5—Xeric great groups of Alfisols, Entisols, Inceptisols, Mollisols, and Ultisols.

  X6—Torric great groups of Entisols; Aridisols.

  X7—Ustic and cryic great groups of Alfisols, Entisols, Inceptisols, and Mollisols; ustic great groups of Ultisols, cryic great groups of Spodosols.

  X8—Aridisols, torric and cryic great groups of Entisols, and cryic great groups of Spodosols and Inceptisols.

Z. MISCELLANEOUS.

  Z1—Icefields.

  Z2—Rugged mountains: mostly devoid of soil (includes glaciers, permanent snow fields, and, in some places, areas of soil).

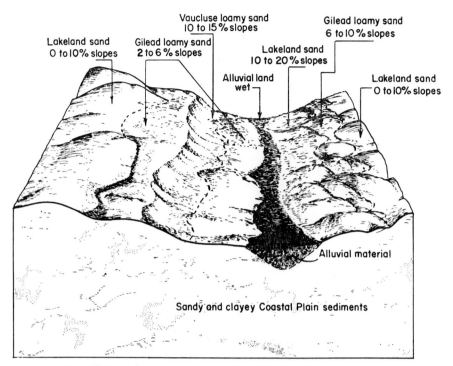

FIG. 15.2. Soilscape pattern of a Typic Quartzipsamment (Lakeland) and Aquents (alluvial wetland) with a Typic Fragiudult (Vaucluse) and Typic Hapludult (Gilead) in Scotland County in south central North Carolina. (From Horton 1967.)

from flood management works may affect soil productivity drastically, and even the classification of the soils. General farming, with and without irrigation, and specialized farming, such as for alfalfa production, are practiced on level, fertile Entisols. Citrus groves are on Psamments in Florida, and some truck farms are on lithic silty "marl land" Orthents. Rice paddies and other forms of cultivation requiring control of surface water are facilitated by the level terrain associated with the water-deposited Entisols (Fluvents). The presence of unpredictable lithologic discontinuities in Fluvents, created by irregular flood intensities, requires careful observation of subsoil during mapping and makes interpretation of mapping units difficult.

CLASSIFICATION. The ochric epipedon must be present to differentiate the Entisol from fresh initial material. Entisols may also have anthropic, albic, and agric horizons. Accumulation of salts, iron oxides, etc., may be present but usually below a depth of one meter.

   Figure 15.3 arranges the five suborders to show relationships to some factors and properties. These suborders may be defined briefly as follows:

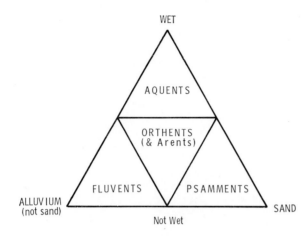

FIG. 15.3. Diagram showing some relationships between suborders of Entisols.

1. Aquents are permanently or seasonally wet (saturated) and even if artificially drained, they display bluish gray (gleyed) or very mottled horizons.

2. Arents are Entisols that are better drained than Aquents (lacking their gleying and mottling) and that exhibit fragments of diagnostic horizons below the Ap horizon. The great disturbance of the deeper horizons has in most cases been by deep plowing one or more times, but not often enough to produce a homogenized deep Ap horizon.

3. Fluvents (Fig. 22.2) are for loamy and clayey (finer in texture than loamy fine sand) Alluvial soils with very simple profiles. Irregularity of content of organic matter with depth is diagnostic. Stratification is common in alluvium and in soils derived from it.

4. Orthents (Fig. 19.2) are loamy and clayey Entisols (better drained than Aquents), with a regular decrease in content of organic matter with depth.

5. Psamments (Figs. 15.2, 19.2, 22.2) have textures of loamy fine sand or coarser and are better drained than Aquents.

Except for Arents, the suborders are subdivided into 23 great groups (Table 15.1) on the basis of several factors: mean annual soil temperature and range of soil temperature, sandiness and quartz content, periods of drying of the soil for two months or more, and low bearing capacity (Soil Survey Staff 1967).

LITERATURE CITED

Horton, R. E. 1967. Soil survey of Scotland County, North Carolina. Soil Conserv. Serv., U.S. Dept. Agr. U.S. Govt. Printing Office, Washington.

Jenny, H. 1968. The image of soil in landscape art, old and new. Pontificiae Academiae Scientarum Scripta Varia 32:947–79.

Kellogg, C. E., and A. C. Orvedal. 1969. Potentially arable soils of the world, and critical measures for their uses. Adv. Agron. 21:109–70.

Rickert, D. A., and J. C. F. Tedrow. 1967. Pedologic investigations on some aeolian deposits of northern Alaska. Soil Sci. 104:250–62.

Riecken, F. E., and E. Poetsch. 1960. Genesis and classification considerations of some prairie-formed soil profiles from local alluvium in Adair County, Iowa. Proc. Iowa Acad. Sci. 67:268–76.

Soil Survey Staff. 1967. Supplement to Soil classification, a comprehensive system—7th approximation. U.S. Dept. Agr. U.S. Govt. Printing Office, Washington.

Tedrow, J. C. F. 1968. Soil investigation in Ingdield land, Greenland. Arctic Inst. North Am. AROD Proj. 6325:1-EN.

Tedrow, J. C. F., and F. C. Ugolini. 1966. Antarctic soils. Publ. Natl. Acad. Sci., Natl. Res. Council, Washington, No. 1418, pp. 161–77.

Whittaker, R. H., S. W. Buol, W. A. Niering, and Y. H. Havens. 1968. A soil and vegetative pattern in the Santa Catalina Mountains, Arizona. Soil Sci. 105:440–50.

❧

TABLE 15.1 ❧ **Suborders and great groups in the Entisol order**

Suborder	Great Group
Aquents:	Hydraquents—very soft; low bearing capacity; N value is more than 0.5. At 20–50 cm (8–20 in.) clay content exceeds 8% and organic matter content is more than 3%
	Cryaquents—mean annual soil temperature is <8 C (47 F) and mean summer soil temperature is <15 C (59 F) at 50 cm (20 in.) or at a shallower lithic or paralithic contact
	Tropaquents—mean summer and mean winter temperatures at 50 cm (20 in.) differ by less than 5 C (9 F)
	Psammaquents—sands and loamy sands
	Haplaquents—are not distinguished by features diagnostic to the other Aquents
Psamments:	Cryopsamments—the soil is cold (see Cryaquents above)
	Quartzipsamments—more than 95% of the sand fraction by weight is quartz, zircon, tourmaline, rutile, and other resistant minerals
	Torripsamments—usually dry in most years at 18–50 cm (7–20 in.)
	Ustipsamments—usually moist, but the soil dries out for considerable periods (see Ustifluvents)
	Xeropsamments—soil is dry for long periods (see Xerofluvents)
	Udipsamments—usually moist (see Udifluvents)
Fluvents:	Cryofluvents—the soil is cold (see Cryaquents above)
	Torrifluvents—dry in most years at 18–50 cm (7–20 in.)
	Ustifluvents—soil is usually moist, but in most years is dry for more than 90 cumulative days in some subhorizon at 18–50 cm (7–20 in.) and is dry for less than 60 consecutive days in more than 7 out of 10 years, in that depth range
	Xerofluvents—are usually moist, but soil is dry for more than 60 consecutive days in more than 7 out of 10 years at 18–50 cm (7–20 in.) under a "Mediterranean climate"
	Udifluvents—are usually moist; mean range of soil temperature is more than 5 C (9 F) at 50 cm (20 in.)
	Tropofluvents—mean annual soil temperature range is slight (see Tropaquents above)
Orthents:	Cryorthents—the soil is cold (see Cryaquents above)
	Torriorthents—are usually dry (see Torrifluvents above)
	Ustorthents—are usually moist, but dry out for considerable periods (see Ustifluvents above)
	Xerorthents—are usually moist but soil is dry for long periods (see Xerofluvents above)
	Udorthents—are usually moist (see Udifluvents above)
	Troporthents—mean annual soil temperature range is slight (see Tropaquents above)
Arents:	No great group

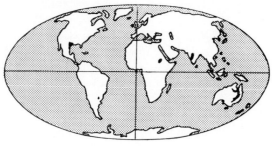

Vertisols: Shrinking and Swelling Dark Clay Soils

A BOUT 2,350,000 km² of dark clay soils are found distributed on every continent except Antarctica (Dudal 1963). Distribution is between 45° N and 45° S latitude. Extensive areas of these soils are located in Australia, the Deccan plateau of India, Sudan, Chad, Ghana, Cuba, Puerto Rico, and Taiwan (Hagenzieker 1964). Uderts (Grumusols) have been reported in Uruguay (Troeh 1969).

In the United States the most notable area of Vertisol formation is in Texas, where the Houston Black clay has been described as the type Grumusol (Templin, Mowery, and Kunze 1956). Over 7 million acres of swelling, clayey soils (which can be considered to be Vertisols) have been reported in western United States (Johnson, Cady, and James 1962; Western Regional Soil Survey Work Group 1964). In general, the Vertisols in western United States are developed in areas where the average annual precipitation ranges from 254 to 1,000 mm (10 to 40 in.), mean annual temperatures are 4 to 18 C (40 to 65 F), and altitudes are 90 to 2,250 m (300 to 7,500 ft) above sea level. Hallsworth, Robertson, and Gibbons (1955) reported "gilgai" in New South Wales under annual rainfall conditions of 180 to 1,520 mm (7 to 60 in.).

SETTING. The single commonest feature in the Vertisol environments is a seasonal drying of the soil profile. Rainfall patterns associated with Vertisols are varied. Although a dry season is a necessary feature, the duration of the dry season is highly variable. The modal situation for the Vertisols involves an annual wet-dry, monsoon type climate, illustrated in Figure 16.1. The more arid Vertisol areas remain dry for most of the year, with only a month or two of wetness. On the other end of the Vertisol range, soils are commonly wet, with moisture deficiency present for only a few weeks, often at irregular intervals, during the year.

A basic (alkaline) reaction is a feature common to the various Vertisol parent materials. These include calcareous sedimentary rocks, basic igneous rocks, basalt, ash, and alluvium from these materials. A variety of parent materials ranging from vitreous basalt to coarse-grained dolerite and argillaceous limestone are reported for the Vertisols of India by Simonson (1954). He summarized the available data for the Regur (Vertisol) soils in India as follows: profile pH values range from 7.2 to 8.5; organic matter content is low; carbonate content ranges from 0.5 to 8%. On the

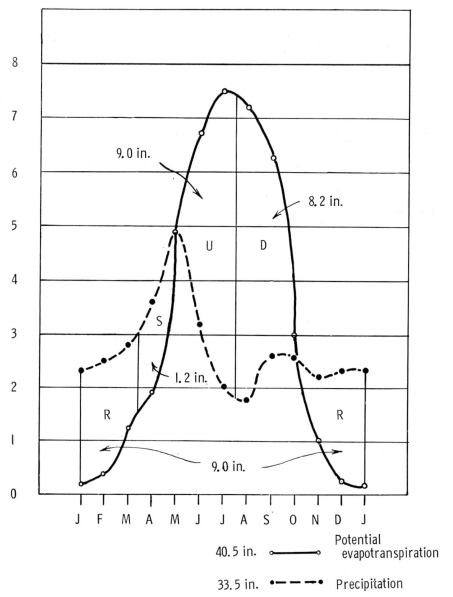

FIG. 16.1.   Average water budget of soils at Ferris, Texas.

Deccan plateau of India, Simonson found the Regur soils mostly confined to landscapes with slopes from 1 to 8%. Although some Vertisols were present on steeper slopes, their frequency of occurrence was low on rolling landscapes and they were largely absent in hilly areas.

A peculiar pedogenic land form occurs on at least 50% of the terrain occupied by such soils (Thorp 1957). The entire landscape may be crumpled into a complex microtopography of mounds and depressions. This is commonly called "gilgai" topography, but is also referred to as "crabhole," "Bay of Biscay," "hushabye," or "polygonal" topography. Gilgai topography may be lattice, dendritic, or wavey in pattern, depending on whether the depressions are more or less extensive (Hagenzieker 1963). Six forms of gilgai, normal, lattice, wavy, tank, stony, and melon-hole are described by Hallsworth, Robertson, and Gibbons (1955) and Hallsworth and Beckmann (1969).

NATURE OF PROFILE PROCESSES. Although there are several processes active in the formation of Vertisols, the predominant process seems to be haploidization by argillipedoturbation. To fully consider the development of the Vertisol profiles, one must first account for the high content of clay ($> 35\%$ by definition) and the predominance of 2:1 expanding clay. It is not difficult to explain the presence of the necessary clay where the soils are developed in argillaceous limestones, marine clays (Fig. 16.2), or

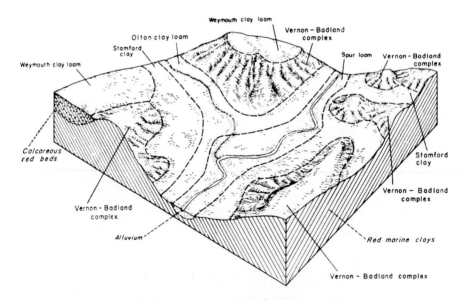

FIG. 16.2.   Soilscape pattern of Typic Chromusterts (Stamford) with Typic Ustochrepts (Vernon, Weymouth), a Typic Paleustoll (Olton) and a Fluventic Haplustoll (Spur) in Runnels County in central Texas. (From Wiedenfold, Barnhill, and Novosad 1970.)

shales. It would appear that those Vertisols developed on basalt, however, require a fairly extensive weathering period unless the solum is developed from volcanic ash deposited over the basalt. The weathering environment of the profile must be such that the 2:1 expandable clays are not further weathered or interlayered to the extent their expanding properties are destroyed. Once the requisite content of clay and dominant 2:1 expanding clay have been achieved, a sequence of events operates in the profiles in the following manner (Fig. 16.3). During the dry season, the soil cracks to the surface, due to the shrinkage of the 2:1 expanding clays. The cracks usually extend to a depth of one meter or more. While the cracks are open, surface soil material falls into them. The surface material can be dislodged by several mechanisms such as animal activity, wind, or at the onset of the rainy season by water. The clays hydrate and expand on rewetting. As expansion takes place, the cracks close, but because of the "extra" material now present in the lower parts of the profile, a greater volume is required and the expanding material presses and slides the aggregates against each other, developing a "lentil" angular blocky structure with slickenside features on the ped faces (Krishna and Perumal 1948). This expansion buckles the landscape, forming the microrelief called gilgai. The microbasins contain 0.3% more organic matter than the microridges in some Vertisols in Texas (Templin, Mowery, and Kunze 1956). This probably results from admixtures of subsurface material into the microridge area and slight erosion of organic-rich fines from the "ridges" to the "basins."

It is difficult to assign the Vertisols a place in a genetic scheme of soil classification. One may consider that they are old soils, the end product of a developmental sequence involving soils whose B horizons became so

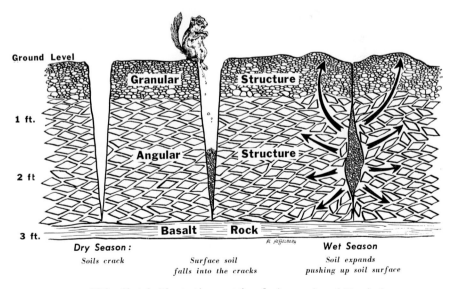

FIG. 16.3. Sketch illustrating wetting-drying cycles of Vertisols.

clayey (argillic) that shrink-swell cycles developed and eventually "swallowed" the A horizon. It is possible that the high content of fine clay in some of these soils (76% of $< 0.2\mu$ clay in a Texan soil studied by Kunze and Templin 1956), and a high fine clay/coarse clay ratio, were produced by lessivage on a large scale.

The fate of a Vertisol may be to undergo alteration of the 2:1 clays to nonexpanding types of clay. The profile would then cease to churn as intensively and eluviation processes would dominate. This interpretation would suggest that Vertisols are relatively young soils.

A third interpretation is that Vertisols are in equilibrium with their environment and that the 2:1 expanding lattice clays are stable and will persist, barring a climatic change. The Vertisols can then be considered diagnostic of environments in which the parent material is basic and gives rise to the formation of 2:1 expanding lattice silicates under the influence of wet-dry climates.

USE OF VERTISOLS. In general, the high content of expanding lattice clay is of primary concern in the management of these soils. Agronomic uses of Vertisols vary widely, depending on the climate. The high clay content and associated slow permeability of these soils when wet makes them desirable for cultures that require retention of surface water. Few, if any, commercial forests are found on Vertisols.

The largest acreages of Vertisols are used for pasture. It is rumored that calves can be hurt or even lost in the cracks during a dry season in Texas. The cracks are wide enough to present hazardous footing for animals. Runoff from initial rains after a dry period is almost entirely infiltrated through the wide cracks. After the soil has become saturated and the cracks closed, runoff may approach 100%.

Many engineering problems are associated with these soils. Structural failures are common. Highways, buildings, fences, pipelines, and utility lines are moved about and distorted by the shrinking and swelling of the soil. These soils will not accept sewage effluent. Percolation tests can be misleading and indicate too high a percolation rate if conducted during the dry season when cracks are open. It is necessary to keep these soils wet for several days to fully saturate and expand the clays to determine valid percolation values.

CLASSIFICATION OF VERTISOLS. Soils over 50 cm (20 in.) deep to lithic or paralithic contact, having at least 30% or more clay in all horizons above the lithic or paralithic contact or to a depth of 1 meter (40 in.), whichever is shallower, and having cracks at least 1 cm wide at a depth of 50 cm (20 in.) during part of most years are classified as Vertisols (Soil Survey Staff 1967). In addition, these soils usually have gilgai microrelief and slickensides on wedge-shaped or parallelepiped natural structural aggregates at some depth below the surface.

Such soils have been called Grumusols (Oakes and Thorp 1950; Templin, Mowery, and Kunze 1956), Tropical Black clays or Regur (Kossovich

1912), and Tirs (Villar 1944), as well as several other names (Simonson 1954). It is also probable that many of these soils have been referred to as Rendzinas. Oakes and Thorp have listed the following profile characteristics for Grumusols and they can be considered as typical properties of Vertisols although not all the properties are likely to be present in any given tessera.

1. Clay texture
2. No evidence of eluviation or illuviation
3. Strong granular structure in upper 15 to 50 cm (6 to 20 in.)
4. Calcareous or with neutral reaction
5. High coefficient of expansion
6. Gilgai relief
7. Extremely plastic wet consistency
8. Ca or Ca and Mg as dominant exchangeable cations
9. Montmorillonite as the dominant clay mineral
10. Calcareous clay parent material
11. Sola thicker than 25 cm (10 in.); usually over 76 cm (30 in.)
12. Dark colors of low chroma
13. One to 3% organic matter content
14. Little weathering
15. Tall grass or savanna vegetation

Four suborders are presently recognized in the Vertisol order (Soil Survey Staff 1966). They are determined primarily on the length of time the cracks are open to the surface (Table 16.1) (Fig. 16.4).

Torrerts are Vertisols that are usually dry in all parts of the solum and the cracks are open to the surface in most years unless irrigated.

Usterts are Vertisols in which the cracks open and close more than once a year but remain open for a total of 90 or more days a year but not all year.

Xererts are Vertisols in which the cracks open and close once each year and remain open for 60 or more consecutive days each year.

Uderts are Vertisols that are usually moist but in which cracks open at some time during the year but do not remain open for as many as 90 days each year.

Vertisol great groups are determined on the basis of chroma in the upper 30 cm (12 in.) of the solum.

TABLE 16.1 ℘ **Suborders and great groups in the Vertisol order**

Suborder	Great Group
Torrerts:	No great groups recognized at this time
Usterts:	Pellusterts—moist chromas less than 1.5 throughout the upper 30 cm (12 in.) Chromusterts—other Usterts
Xererts:	Pelloxererts—moist chromas less than 1.5 throughout the upper 30 cm (12 in.) Chromoxererts—other Xererts
Uderts:	Pelluderts—moist chromas less than 1.5 throughout the upper 30 cm (12 in.) Chromuderts—other Uderts

Number of Days per Year That Cracks Are Open

<90 cumulative <60 consecutive	>90 cumulative	>60·consecutive	All year in most yrs
UDERTS	USTERTS	XERERTS	TORRERTS

FIG. 16.4. Diagram showing some relationships between suborders of the Vertisols.

A number of Inceptisols, Mollisols, Alfisols, Ultisols, Aridisols, and Entisols intergrade at the subgroup level to Vertisols. These soils have significant cracking but not enough to be Vertisols, a coefficient of linear extensibility (COLE) of 0.09 or more, and a potential linear extensibility (COLE × thickness of layer involved) of at least 6 cm in the upper 1 m (40 in.) of the profile (DeMent and Bartelli 1969).

LITERATURE CITED

DeMent, J. A., and L. J. Bartelli. 1969. The role of vertic subgroups in the comprehensive soil classification system. Soil Sci. Soc. Am. Proc. 33:129–31.

Dudal, R. 1963. Dark clay soils of tropical and subtropical regions. Soil Sci. 95: 264–70.

Hagenzieker, F. 1963. Contribution to the agropedological knowledge of the tropical black earths of southern Ghana. Neth. J. Agr. Sci. 11:54–65.

———. 1964. The tropical grey and black earths of the Accra Plains, Ghana; their environment, characteristics and air-photograph mapping. Neth. J. Agr. Sci. 12:123–31.

Hallsworth, E. G., and G. G. Beckmann. 1969. Gilgai in the Quaternary. Soil Sci. 107:409–20.

Hallsworth, E. G., G. K. Robertson, and F. R. Gibbons. 1955. Studies in pedogenesis in New South Wales. VII. The "Gilgai" soils. J. Soil Sci. 6:1–31.

Johnson, W. M., J. G. Cady, and M. S. James. 1962. Characteristics of some Brown Grumusols of Arizona. Soil Sci. Soc. Am. Proc. 26:389–93.

Kossovich, P. S. 1912. "Dic Schwarzerde" (Tschernosion) Verl. f. Fachliteratur, GMBH, Berlin.

Krishna, P. C., and S. Perumal. 1948. Structure in Black Cotton soils of the Nizamsager project area, Hyderabad state, India. Soil Sci. 66:29–38.

Kunze, G. W., and E. H. Templin. 1956. Houston Black Clay, the type Grumusol. II. Mineralogical and chemical characterization. Soil Sci. Soc. Am. Proc. 20: 91–96.

Oakes, H., and J. Thorp. 1950. Dark-clay soils of warm regions variously called Rendzina, Black Cotton soils, Regur, and Tirs. Soil Sci. Soc. Am. Proc. 15: 347–54.

Simonson, R. W. 1954. Morphology and classification of the Regur soils of India. J. Soil Sci. 5:275–88.

Soil Survey Staff. 1966. Second supplement to Soil classification, a comprehensive system—7th approximation. U.S. Dept. Agr. U.S. Govt. Printing Office, Washington.

————. 1967. Supplement to Soil classification, a comprehensive system—7th approximation. U.S. Dept. Agr. U.S. Govt. Printing Office. Washington.

Templin, E. H., I. C. Mowery, and G. W. Kunze. 1956. Houston Black clay, the type Grumusol. I. Field morphology and geography. Soil Sci. Soc. Am. Proc. 20:88–90.

Thorp, J. 1957. Report on a field study of soils in Australia. Sci. Bull. 1. Earlham College, Richmond, Ind.

Troeh, F. R. 1969. Noteworthy features of Uruguayan soils. Soil Sci. Soc. Am. Proc. 33:125–28.

Villar, E. H. del. 1944. The Tirs of Morocco. Soil Sci. 57:313–39.

Western Regional Soil Survey Work Group. 1964. Soils of the western United States. Washington State Univ., Pullman.

Wiedenfeld, C. C., L. J. Barnhill, and C. J. Novosad. 1970. Soil survey of Runnels County, Texas. Soil Conserv. Serv., U.S. Dept. Agr. U.S. Govt. Printing Office, Washington.

❧

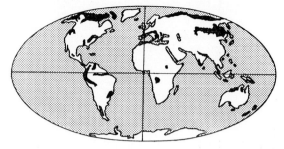

*Inceptisols: Embryonic Soils with
Few Diagnostic Features* ⚘

INCEPTISOLS are soils that have not developed features diagnostic for other orders but have some features in addition to the ochric epipedon and albic horizons permitted in the Entisols.

SETTING. Inceptisols are immature soils having profile features more weakly expressed than mature soils and retaining close resemblances to the parent material. No description of the settings can be made that is truly representative of all Inceptisols, but these are salient features: (1) highly resistant parent material; (2) abundance of volcanic ash; (3) extreme landscape positions, that is, steep lands and depressions; and (4) geomorphic surfaces so young as to limit soil development.

Although widely distributed over the world, under an astonishing range of environmental regimes and across a complete spectrum of parent material, Inceptisols have features that indicate pedologic immaturity. In many cases the direction of soil development is evident and one can predict that certain Inceptisols will ultimately become Ultisols, or Alfisols, etc. Some Inceptisols are in equilibrium with their environment and will not "mature" until that environment changes. The present Inceptisol on steep slopes will no longer be there when geologic erosion has leveled the slope to a more normal relief; but the Inceptisol is in steady state with respect to the present environment and the process of geologic erosion that is taking place under present conditions.

Many of the world's Inceptisols are formed from volcanic material. Such soils have been referred to as volcanic ash soils, or more recently as Ando soils. Volcanic forest soils are reported on the Kamchatka Peninsula of Siberia, USSR (Sokolov and Karayeva 1965). Andosols and Ando soils have been reported in Indonesia (Tan and Schuylenborgh 1961), South America and Africa (D'Hoore 1964; Wright 1964), in New Zealand (Taylor and Cox 1956), and in Japan (Aomine and Yoshinaga 1955; Kanno 1956). Several authors contributed papers concerned with volcanic ash soils in Latin America at a conference at Turrialba, Costa Rica (Training and Research Center of the IAIAS 1969). Cryandepts have been reported to cover most of the well-drained, nonmountainous areas of the Aleutian Islands in the Alaskan Peninsula (Simonson and Rieger 1967). To the north of the Aleutian Islands on the Kenai Peninsula of Alaska, Andepts are reported to be in association with Spodosols in landscapes in which invasion of grassland by spruce forest seems to be converting Andepts into Orthods.

PEDOGENIC PROCESSES. No single pedogenic process operates in all members of the order except leaching. It is probably more correct to say that virtually all of the pedogenic processes are active to some extent in Inceptisol profiles but none predominates. The Aquept suborder is characterized by active gleization that produces colors of low chroma. The Aquepts are commonly located in depressional parts of the landscape where leaching may be more extensive than in other landscape positions, but the process of lessivage and thus argillic horizon formation is somewhat retarded, probably because the soils do not undergo desiccation. In areas of acid rocks, soils formed in landscape depressions tend to be more leached and somewhat lower in base content and higher in exchangeable aluminum than soils in surrounding areas. This is especially true in regions of Ultisols where the surrounding better drained soils have low base status and some exchangeable aluminum. In landscapes of high base status soils, the Aquepts are commonly higher in base status than surrounding soils. This can be attributed to the enrichment of the low-lying parts of the landscape by lateral moving groundwater.

In situations where Inceptisols occupy upland positions on young geomorphic surfaces, both initial and secondary minerals are present in the soil profile. McCracken, Shanks, and Clebsch (1962) concluded that feldspar, present in the C horizon but not in the solum of some Dystrochrepts on steep, sloping areas of southeastern United States, had altered to intergradational vermiculite-chlorite and even to kaolinite and gibbsite in the A and B horizons.

Many of the Inceptisols of the world are in unusually resistant parent materials and hence are not developed to the same extent as associated soils. The nature of the parent materials slows weathering processes and limits the amount of clay produced. In other situations, the overpowering influence of leaching has systematically removed colloidal material from the soil profile, leaving an Inceptisol at the land surface.

Jackson (1964) has pointed out that allophane, a common product of weathering of volcanic ash, combines with humus to form resistant, dark-colored material. Soils containing large quantities of this material have low bulk densities and do not conform to classic categories of Zonal soils. In the main these are the Ando soils or the Andepts of the new classification system. These soils result from the formation of stable complexes from volcanic ash material and humus.

Quartzose parent materials, susceptible to spodic horizon formation under podzolization (see Chapter 20), are suited to genesis of Inceptisols elsewhere. In the case of quartz-rich coarse silt and sand, there is simply too little fine-textured material, both actually and potentially, to permit formation of an argillic horizon. Quartzite and siliceous sandstones provide conditions for Inceptisol formation.

USES OF INCEPTISOLS. The interpretations of Inceptisols for agricultural and nonfarming uses are necessarily diverse. Steep areas are best suited to woodland, recreation, or wildlife. Poorly drained Inceptisols can be extensively used for cultivated field crops, provided artificial drainage is feasible. In some landscapes, Inceptisols are more productive than associated soils of other orders. For example, associated soils with thick, rather impermeable argillic horizons may be less productive; as are associated sandy, droughty Entisols.

The Pullman soil of Figure 17.1 is used for cotton, grain sorghum, and wheat, while the Ustochrepts and other soils, in short, mid, and tall grasses, are used for range.

Since Inceptisols are developed in a variety of climates, excluding arid regions, the temperature and rainfall distribution cannot be discussed with specific reference to the order.

Liming and phosphate fertilization problems associated with allophane or amorphous material are probably acute in many soils of this order, especially in the Andepts. The high amount of pH-dependent charge on the amorphous clay particles requires considerable liming for maintenance of a soil reaction favorable to growth of agricultural crops. Soils containing

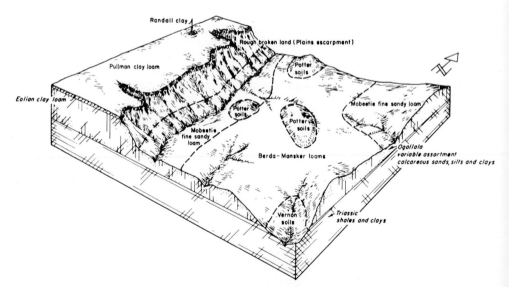

FIG. 17.1.   Soilscape pattern of some Typic Ustochrepts
(Berda, Mobeetie, Vernon), with Ustollic Calciorthid (Potter),
Typic Calciustoll (Mansker), Pachic Paleustoll (Pullman),
and Udic Pellustert (Randall) in Dickens County in
northwestern Texas. (After Girdner and Richardson 1970.)

amorphous material have been shown to be extremely strong fixers of phosphate.

CLASSIFICATION OF INCEPTISOLS. Inceptisols are mineral soils that exhibit profile development sufficient to exclude them from Entisols. A cambic horizon is usually present, but spodic, argillic, natric, and oxic horizons are lacking. Inceptisols are excluded from the Aridisol order by soil climate and from the Vertisol order by lack of argillipedoturbative features. Soils with umbric, histic, plaggen, or mollic epipedons are considered Inceptisols when one of the following properties is present: (1) bulk density of the fine earth is less than 0.85 g/cc; (2) base saturation is less than 50% at a depth of 1.8 m (72 in.); (3) mean summer and winter temperatures differ by less than 5 C (9 F) and the clay fraction is composed of more than 35% montmorillonite with the epipedon resting on material containing less than 40% calcium carbonate equivalent. Criteria of wetness, mineralogy, and temperature are used to separate the order into six suborders (Fig. 17.2).

Aquepts (Figs. 23.1, 24.3) are saturated with water at some period in the year unless drained by man.

Plaggepts have dark brown or black plaggen epipedons.

Andepts have a fine earth fraction bulk density less than 0.85 g/cc in the epipedon and/or cambic horizon, and amorphous material is the dominant source of the exchange complex. Vitric volcanic ash, cinders, or other vitric pyroclastic materials constitute 60% or more of the silt, sand, and gravel fractions.

Tropepts are the Inceptisols of the tropical regions. They have mean annual soil temperatures of 8 C (47 F) or more and have less than 5 C (9 F) difference in mean summer and mean winter soil temperatures.

Umbrepts have umbric, mollic, or anthropic epipedons. A mollic epipedon is rarely present in undisturbed Umbrepts but is readily formed by the rather common management practice of adding lime. To avoid reclassification of the soil in this instance, the low base status (<50%) in the

FIG. 17.2. Diagram showing some relationships between suborders of the Inceptisols.

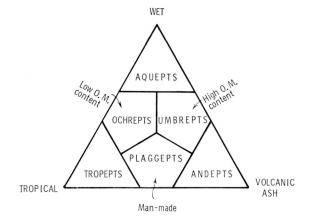

underlying cambic horizon is used to justify retention of the soil in the Inceptisol order. In view of the slow translocation of calcium in a profile, such a soil will probably qualify indefinitely as an Umbrept.

Orchrepts (Figs. 21.4, 22.2) include all other Inceptisols. Table 17.1 lists the great groups recognized in the Inceptisol order.

TABLE 17.1 ❧ **Suborders and great groups in the Inceptisol order**

Suborder	Great Group
Aquepts:	Cryaquepts—mean annual temperature less than 8 C (47 F)
	Plinthaquepts—plinthite within 1.25 m (50 in.) of the surface
	Andaquepts—bulk density less than 0.85 g/cc
	Tropaquepts—less than 5 C (9 F) difference in mean summer and mean winter temperatures
	Fragiaquepts—presence of a fragipan
	Halaquepts—sodium saturation >15% in some subsoil horizons
	Humaquepts—presence of a mollic, umbric, or histic epipedon
	Haplaquepts—other Aquepts with an ochric epipedon
Plaggepts:	No great groups presently recognized
Andepts:	Cryandepts—mean annual temperature less than 8 C (47 F)
	Durandepts—presence of a duripan within 1 m (40 in.) of the surface
	Hydrandepts—irreversible dehydrating clays
	Eutrandepts—have mollic epipedons and are thixotropic in some horizons
	Dystrandepts—have an umbric or ochric epipedon and are thixotropic in some horizons
	Vitrandepts—other Andepts
Tropepts:	Ustropepts—base saturation >50% in epipedon and cambic horizon and are dry for 90 cumulative days per year or have free lime within 1.5 m (60 in.) of the surface
	Eutropepts—base saturation >50% in epipedon and cambic horizon
	Dystropepts—mean annual soil temperature > 22 C (72 F) and base saturation < 50%
	Humitropepts—other Tropepts
Umbrepts:	Anthrumbrepts—presence of an anthropic epipedon
	Fragiumbrepts—presence of a fragipan
	Cryumbrepts—mean annual soil temperature <8 C (47 F)
	Haplumbrepts—not dry for as much as 60 consecutive days between 18 and 50 cm (7–20 in.)
	Xerumbrepts—other Umbrepts
Ochrepts:	Fragiochrepts—presence of a fragipan
	Durochrepts—presence of a duripan within 1 m (40 in.) of surface
	Cryochrepts—mean annual soil temperature at 50 cm (20 in.) <15 C (59 F) if cultivated or <8 C (47 F) if O horizon present
	Ustochrepts—dry for 90 cumulative days in most years but not as much as 60 consecutive days in 7 of 10 years unless mean annual soil temperature is 22 C (72 F) and/or mean summer and winter temperatures differ <5 C (9 F)
	Xerochrepts—dry for 60 consecutive days or more in 7 of 10 years
	Eutrochrepts—carbonates in cambic horizon or base saturation (by NH₄ OAc) of 60% or more within 75 cm (30 in.) of surface
	Dystrochrepts—other Ochrepts

LITERATURE CITED

Aomine, S., and N. Yoshinaga. 1955. Clay minerals of some well-drained volcanic ash soils in Japan. Soil Sci. 79:349–58.

D'Hoore, J. L. 1964. Soil map of Africa—scale 1:5,000,000—explanatory monograph. Commission for Technical Cooperation in Africa. Lagos, Nigeria. Publ. 93.

Girdner, C. L., Jr., and W. E. Richardson. 1970. Soil survey of Dickens County, Texas. Soil Conserv. Serv., U.S. Dept. Agr. U.S. Govt. Printing Office, Washington.

Jackson, M. L. 1964. Chemical composition of soils, pp. 71–141. In F. E. Bear (ed.), Chemistry of the soil, 2nd ed. Reinhold, New York.

Kanno, I. 1956. A pedological investigation of Japanese volcanic-ash soils. Bull. Kyushu Agr. Exp. Sta. 4:81–84.

McCracken, R. J., R. E. Shanks, and E. E. C. Clebsch. 1962. Soil morphology and genesis at higher elevations of the Great Smoky Mountains. Soil Sci. Soc. Am. Proc. 26:384–88.

Simonson, R. W., and S. Rieger. 1967. Soils of the Andept suborder in Alaska. Soil Sci. Soc. Am. Proc. 31:692–99.

Sokolov, I. A., and Z. S. Karayeva. 1965. Migration of humus and some elements in the profile of volcanic forest soils of Kamchatka. Soviet Soil Sci., 1965, No. 3, pp. 467–75. (Transl. 1966 from Pochvovedeniye, 1965, 5:2–22, by Am. Inst. Biolog. Sci., Washington.)

Tan, K. H., and J. van Schuylenborgh. 1961. On the classification and genesis of soils developed over acid volcanic materials under humid tropical conditions. II. Neth. J. Agr. Sci. 9:41–54.

Taylor, N. H., and J. E. Cox. 1956. The soil pattern of New Zealand. New Zealand Soil Bur. Publ. 113.

Training and Research Center of the IAIAS. 1969. Panel on volcanic ash soils in Latin America. Inter-American Institute of Agricultural Sciences of the OAS. Turrialba, Costa Rica.

Wright, A. C. S. 1964. The "Andosols" or "Humic Allophane" soils of South America. Report on the meeting on the Classification and Correlation of Soils from Volcanic Ash, Tokyo, Japan. World soil resources report 14. FAO, Rome, Italy, 1965, pp. 9–22.

❧

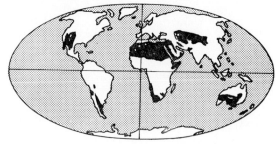

Aridisols: Soils of Arid Regions

IF DEFINED on the basis of climate, the arid regions occupy 36% of the earth's land surface; 35% if defined on the basis of vegetation (Shantz 1956). These include and center around the Kalahari-Namib, Sahara, Somali-Chalbi, Arabian, Iranian, Thar, Turkestan, Takla-Makan, Gobi, Monte-Patagonian, Atacama-Peruvian, Australian, and North American deserts (Dregne 1967). The regions of soils classified in the Aridisol order do not conform to all the parameters of either climatic or vegetative zones, but are associated with arid and semiarid climates and with desert vegetation.

It is a common, though erroneous, concept that soils in dry regions are sandy, rocky, and devoid of vegetation. Such a concept is easily obtained when a person drives through desert areas and sees the almost barren mountain ranges and dry sandy river beds. Also, the typical movie setting of Arabian Nights in the sand dunes has helped foster this concept. Sand dunes and other masses of constantly shifting sand, often referred to as "true desert," are interspersed throughout many desert areas and are less extensive than is commonly believed. Nevertheless, Entisols are common inclusions in Aridisol regions.

The areas of Aridisol development are associated with Mollisols in surrounding cooler or more moist regions. Along a transect from arid into adjacent semiarid regions, the ochric epipedons darken with organic matter as the mollic epipedons first appear on the fine-textured and/or calcareous parent materials.

In the more humid parts of the semiarid zones, Aridisols are commonly bordered by forested Alfisols. Vertisols are sometimes found associated with the Aridisols on bodies of lithogenetic or pedogenetic 2:1 expanding clays.

The climatic regime in which Aridisols form can be characterized as one in which the potential evapotranspiration greatly exceeds the precipitation during most of the year and no water percolates through the soil. A characteristic water balance for the dry region is shown in Figure 18.1 (Buol 1964).

Figure 18.1 shows that about 1.25 cm (1/2 in.) of water is stored (S) in the soil during December, January, and February. The stored water is utilized (*U*) during March, and the soils are deficient (*D*) in water throughout most of the year.

Evidence of leaching below the average depth of water storage is often observed in Aridisols and ascribed to humid paleoclimates (Smith 1965).

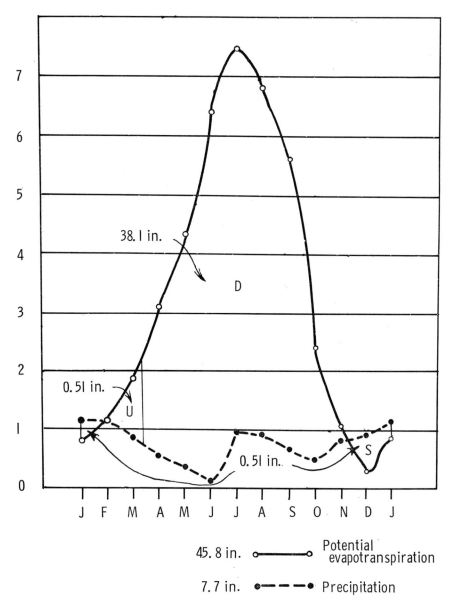

FIG. 18.1. Average water budget for soils at Phoenix, Arizona.

However, another explanation is that the typically erratic distribution of the rainfall causes occasional periods of relatively high precipitation in the winter months during which deeper leaching takes place. In this case the depth of leaching may reflect the rainfall of the extreme years rather than that of average years (Chapter 10).

The dry lands are characterized by a sparse growth of plants. Extensive barren areas are not common. Various species of cactus *(Cactaceae),* mesquite *(Prosopis),* creosotebush *(Larrea),* yucca *(Yucca),* sagebrush *(Artemisia),* shadscale *(Atriplex),* hopsage *(Grayia),* and muhly grass *(Muhlenbergia)* are common. Microbial populations are usually low. The organic matter content of the surface soil is low. Low carbon/nitrogen ratios are probably brought about by action of nitrifying bacteria and/or nitrogen-fixing blue-green algae which form a crust on some of these soils (Fuller, Cameron, and Raica 1960; Martin and Fletcher 1943).

The above considerations refer to deserts of temperate and tropical zones. In addition, there are vast areas of desert soils in arctic and antarctic zones (Tedrow 1968).

PROFILE PROCESSES. The chemical and physical reactions which are important in Aridisol profiles are essentially the same as those in humid regions. However, because of the limited amount of water available in the profile, these reactions are relatively less intense. As a result, the soils inherit much of their morphology from the parent material (Buol 1965). The lack of leaching has left the profiles with a high base status. The clay fraction of many Aridisols has been reported to be largely illite (Buol and Yesilsoy 1964; Hseung and Jackson 1952; Jackson et al. 1948; Merwe and Heystek 1955). It is probable that the illite is directly related to mica of the parent rock (Jackson et al. 1948). In some Aridisols more montmorillonite is present in the lower calcareous and least weathered part of the profile than in the upper horizons. Since weathering is minimal and most active in the upper horizons, the montmorillonite in the subsoil may have formed by crystallization from dilute solutions in the presence of abundant calcium. This phenomenon has been demonstrated in the laboratory (Henin 1955).

Aridisol profiles are well oxidized as evidenced by their low organic matter content and lack of "free" iron oxide movement (Rozanov 1951). Aridisol profiles on the older, more stable land forms (Fig. 18.2) have accumulated clay to form argillic horizons. The process by which argillic horizons form in Aridisols has been the subject of much controversy. The lack of an eluvial horizon and the apparent perennial dryness of the surface layer caused Nikiforoff (1937) to conclude that the clay in the argillic horizon has formed in situ. The theory of in situ clay formation has been supported by Rozanov (1951) in Asia where he observed a weathered condition of pebbles in the B horizon but not in overlying and underlying horizons. Illuviation argillans (clay skins) are not well developed in argillic horizons of Aridisols.

It is probable that the formation of argillic horizons involves both

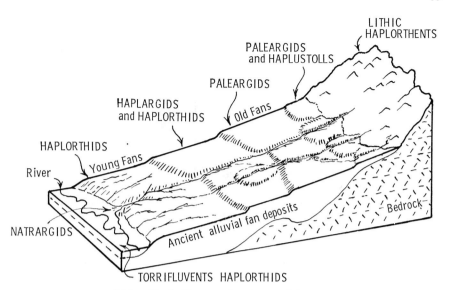

FIG. 18.2. Block diagram showing positions of some major kinds of Aridisols and their associates.

the formation of clay in situ and lessivage. Index minerals and molecular ratio methods to determine relative weathering intensities have indicated that some weathering has taken place to form clay both in the surface and argillic horizons but that the surface layers are more intensively weathered than the lower horizons (Barshad 1955; Smith and Buol 1968). It is probable that illuviation argillans have been destroyed by shrink-swell pedoturbation during cycles of extreme desiccation and seasonal rainfall common to many of these soils (Buol and Yesilsoy 1964; Nettleton, Flach, and Borst 1968).

A surface "pavement" of gravel is present over many of the Aridisols. It is probable that much of this has been produced by deflation (Kellogg 1953; Merwe 1954; Rozanov 1951). However, it has been demonstrated that this gravel layer could be formed by repeated wetting and drying. The gravel may be moved up to the surface by action of entrapped air when the soil is wet by rain. In this case a gravel-free layer of vesicular structure directly underlies the surface pavement (Springer 1958). Upper surfaces of stones in the gravel layer are commonly stained black with manganese and iron oxides, often called desert varnish.

The most striking feature observed in many soils of the dry regions is the layer of carbonate accumulation or "caliche." The genesis of the caliche layer is the subject of much controversy that in part stems from the lack of a clear, quantitative definition of caliche. The amount of materials other than carbonates, namely, silica and other salts, varies from one caliche to another. The general assumption is that calcium bicarbonate moves down the profile, and that $CaCO_3$ is precipitated when the percolation stream stops and is lost through transpiration (Chapter 6). Unconsolidated caliche layers, called calcic horizons when they contain more than 15% calcium carbonate equivalent, generally are found at increasing depths with increasing rainfall. On the other hand, indurated or "petrocalcic" layers are commonly found at depths that do not correlate directly with rainfall. One possible explanation is that successive truncations and burials have taken place. Many petrocalcic layers, because of their resistance to deterioration, may be relics of older land forms and not related to present profiles.

Common salt features in soils of dry regions are salic horizons (salt accumulations) and natric horizons (sodium-affected). Their formation is favored by periodic accumulation of water as in or near broad seasonal lakes known as playas or locally in areas of seepage similar to spring sites in more humid areas. Many of the salt and sodium concentrations are caused by man through his application of irrigation water.

Aridisols of the polar areas are normally frozen for 9 or 10 months of the year (Tedrow 1966). While frozen, the soils are chemically inactive and mineral decomposition is retarded.

USES OF ARIDISOLS. Use of Aridisols for agriculture is limited chiefly by the lack of water. The soils shown in Figure 18.3 are used for range except

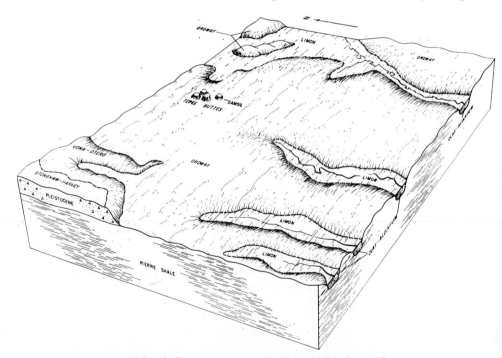

FIG. 18.3.   Soilscape patterns of a Ustertic Camborthid (Ordway), Mollic Haplargids (Vona Stoneham), and a Mollic Calciorthid (Harvey), with Typic Ustorthent (Otero, Samsil) and Vertic Usthorthent (Limon) in Crowley County in southeastern Colorado. (From Larsen, Martin, and Mayhugh 1968.)

for level areas on which irrigation is practiced. However, many Aridisols present some problem with respect to irrigation, and in any case are commonly less well situated for watering than are the associated Fluvents. Where irrigation of Aridisols is anticipated, it is usually necessary to level the land. This practice may expose calcic, petrocalcic, natric, or argillic horizons, or duripans. Only soils with adequate internal permeability should be selected for irrigation to avoid problems of salinization and alkalization arising from the irrigation water itself.

Although low in content of nitrogen, Aridisols have surprising amounts of the other fertility elements, especially potash from feldspars and mica. Supplies of micronutrients are usually abundant, although they may not be available because of the high pH. Foliar applications of iron and trace elements may be necessary for satisfactory crop production.

Without irrigation the Aridisols are useful to only a limited extent for seasonal grazing, most of which is concentrated on the associated Fluvents. Few engineering problems are encountered on the Aridisols. Flash floods are a hazard along drainageways. Petrocalcic layers impede excavation in building and landscaping.

Where food and services, particularly refrigeration, can be brought in from other areas, the dry regions provide excellent conditions for urbanization.

CLASSIFICATION OF ARIDISOLS. Aridisols are defined as having ochric epipedons and one or more of the following subsurface horizons: argillic, cambic, natric, gypsic, calcic, petrocalcic, or duripan. They are separated from the Inceptisols by the circumstance that unless irrigated they are usually dry or have a saturated extract conductivity of more than 2 mmhos per cm at 25 C in the 18- to 50-cm (7- to 20-in.) layer or above a lithic or paralithic contact, whichever is shallower.

The occurrence of an argillic horizon is important to the theory of Aridisol formation. This horizon relates to age and position on the landscape as well as to soil management. The presence or absence of an argillic horizon is used as the criterion for dividing the order into two suborders, Argids (with argillic horizons) and Orthids (without argillic horizons).

The Argids have formed on the oldest geomorphic surfaces, as on the crests of dissected alluvial fans. Orthids are found on geologically younger sideslopes and surfaces of intermediate age (Figs. 17.1, 18.2). Soils of the youngest surfaces in the dry region, both the steep mountain slopes and recent alluvial bottoms, have not developed any diagnostic subsurface horizons and are classified as Entisols (Figs. 18.3 and 18.4).

The other diagnostic horizons and features associated with the Aridisol order are used as criteria at the great group level (Table 18.1). These features all must be within 1 m (40 in.) of the surface to be diagnostic. Horizons within the depth influence land use and contain most of the features of importance to soil classification and genesis, with two exceptions. One is that land leveling may expose horizons lying below a depth of 1 m. The other is that deep-lying layers such as duripans may restrict water movement, limiting drainage and imposing problems for irrigation.

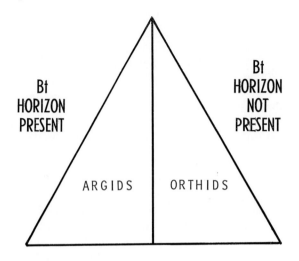

FIG. 18.4. Diagram showing some relationships between suborders of Aridisols.

This casts doubt on the wisdom of including a 1-m (40-in.) depth limitation in the definition of great groups of Aridisols.

TABLE 18.1 ❧ **Suborders and great groups in the Aridisol order**

Suborder	Great Group
Orthids:	Durorthids—duripan within 1 m (40 in.) of the surface Salorthids—salic horizon within 1 m (40 in.) of the surface Paleorthids—petrocalcic horizon within 1 m (40 in.) of the surface Calciorthids—calcic or gypsic horizon within 1 m (40 in.) of the surface Camborthids—only a cambic horizon
Argids:	Nadurargids—duripan within 1 m (40 in.) of surface with an overlying natric horizon with columnar structure Durargids—duripan within 1 m (40 in.) of surface with an overlying argillic or natric horizon without columnar structure Natrargids—columnar structured natric horizon within 1 m (40 in.) of the surface Paleargids—petrocalcic horizon within 1 m (40 in.) of surface or argillic horizon with >35% clay Haplargid—argillic horizon having <35% clay

LITERATURE CITED

Barshad, I. 1955. Soil development, pp. 1–52. In F. E. Bear (ed.), Chemistry of the soil, 1st ed. Reinhold, New York.

Buol, S. W. 1964. Calculated actual and potential evapotranspiration in Arizona. Univ. Ariz. Agr. Exp. Sta. Tech. Bull. 162.

——. 1965. Present soil-forming factors and processes in arid and semiarid regions. Soil Sci. 99:45–49.

Buol, S. W., and M. S. Yesilsoy. 1964. A genesis study of a Mohave sandy loam profile. Soil Sci. Soc. Am. Proc. 28:254–56.

Dregne, H. E. 1967. Inventory of research on surface materials of desert environments. Chapter V. Office of Arid Lands Studies, Univ. Ariz., Tucson.

Fuller, W. H., R. E. Cameron, and N. Raica, Jr. 1960. Fixation of nitrogen in desert soils by algae. Trans. 7th Intern. Congr. Soil Sci. (Madison, Wis.) 2: 617–24.

Henin, S. 1955. Synthesis of clay minerals at low temperatures, pp. 54–60. In Ada Swineford (ed.), Clay and clay minerals. Proc. 4th Natl. Conf. Natl. Acad. Sci. Natl. Res. Council, Washington, 1956.

Hseung, Y., and M. L. Jackson. 1952. Mineral composition of the clay fraction. III. Some main soil groups of China. Soil Sci. Soc. Am. Proc. 16:294–97.

Jackson, M. L., S. A. Tyler, A. L. Willis, G A. Bourbeau, and R. P. Pennington. 1948. Weathering sequence in clay-size minerals in soils and sediments. I. Fundamental generalizations. J. Phys. Colloid Chem. 52:1237–60.

Kellogg, C. E. 1953. Potentialities and problems of arid soils. Desert Research, Proc. Intern. Symp., 1952. Israel Research Council, Jerusalem, pp. 19–40.

Larsen, R. J., D. R. Martin, and R. E. Mayhugh. 1968. Soil survey of Crowley County, Colorado. Soil Conserv. Serv., U.S. Dept. Agr. U.S. Govt. Printing Office, Washington.

Martin, W. P., and J. E. Fletcher. 1943. Vertical zonation of great soil groups on Mt. Graham, Arizona, as correlated with climate, vegetation and profile characteristics. Univ. Ariz. Agr. Exp. Sta. Tech. Bull. 99.

Merwe, C. R. van der. 1954. The soils of the desert and arid regions of South Africa. Proc. 2nd Interafrican Soils Conf. (Leopoldville) 2:827–34.

Merwe, C. R. van der, and H. Heystek. 1955. Clay minerals of South African soils groups. III. Soils of the desert and adjoining semiarid regions. Soil Sci. 80: 479–94.

Nettleton, W. D., K. W. Flach, and G. Borst. 1968. A toposequence of soils in tonalite grus in the southern California peninsular range. Soil Surv. Invest. Rept. 21. Soil Conserv. Serv., U.S. Dept. Agr.

Nikiforoff, C. C. 1937. General trends of desert type of soil formation. Soil Sci. 43:105–31.

Rozanov, A. N. 1951. The Serozems of central Asia. (Transl. by A. Gourevich.) Israel Prog. for Sci. Trans., Jerusalem, 1961. Available from U.S. Dept. Commerce, Washington.

Shantz, H. L. 1956. History and problems of arid lands development, pp. 3–5. In G. F. White (ed.), The future of arid lands. Am. Assoc. Adv. Sci. 43.

Smith, B. R., and S. W. Buol. 1968. Genesis and relative weathering intensity studies in three semiarid soils. Soil Sci. Soc. Am. Proc. 32:261–65.

Smith, G. D. 1965. Lectures on soil classification. Pedologie Special Issue 4. Belgian Soil Sci. Soc. Rozier 6, Gent, Belgium.

Springer, M. E. 1958. Desert pavement and vesicular layer of some soils of the desert of the Lahontan basin, Nevada. Soil Sci. Soc. Am. Proc. 22:63–66.

Tedrow, J. C. F. 1966. Polar desert soils. Soil Sci. Soc. Am. Proc. 30:381–87.

———. 1968. Soil investigations in Inglefield Land, Greenland. Final Report for U.S. Army Research Office, Durham. DA-ARO-D-31-124-G820. DA Project No. 2M014501B52B ARO—D Project No. 6325—EN. The Arctic Institute of North America.

❧

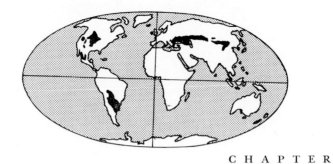

Mollisols: Grassland Soils of Steppes and Prairies

NEARLY all soils with deep, dark, relatively fertile topsoil (mollic epipedon) are formed under grassland vegetation. Exceptions include poorly drained Mollisols of lowland hardwood forests, some well-drained Brown Forest soils (Udolls), and a few soils of other orders (see below).

The grasslands of the middle latitudes occupy about 13,000,000 km², of which approximately 5,500,000 km² are in tall-grass prairies and 7,500,000 km² are in short-grass steppes. Weaver (1954) has described in detail the vegetation of the grasslands of North America. There is evidence that some components migrated from the north and others from the south to form the complex community of plants, including many Compositae, some legumes, and cacti.

Weaver and Albertson (1956) state that 55 million years ago in Eocene times a warm temperate zone forest grew where the Great Plains now are. The rise of the Rocky Mountains cut off the supply of moisture from the Pacific Ocean, with the result that the forest was replaced by grasses and herbs. Midgrasses (*Stipa comata* and others) and grasslike sedges migrated into the region from the north. Short grasses (*Buchloe dactyloides*) and side-oats grama moved in from the southwest. Bluestems (*Andropogon gerardi*) and other tall grasses came from the south.

About 45,000 and again about 27,000 years ago paleohunters, ancestors of the Plains Indians of North America, crossed the land bridge at the Bering Straits (Müller-Beck 1966). Through the millenia they hunted mastodons, bison, ground sloths, camels, tapirs, horses, and hares on the grasslands of North America. These hunters were cut off from communication with Eurasia during the period 23,000 to about 13,000 B.P. by glacial ice lobes which coalesced across the continent from the Atlantic Ocean to the Pacific Ocean.

SETTING. The short-grass grassland, the steppe, resembles a pastured meadow extending monotonously to the horizon. Only in unusually wet years do patches of taller grasses develop enough to give the vegetative cover an uneven appearance. Where sagebrush is dominant, the land has a shrubby appearance.

The tall-grass prairie is a grassland of relatively luxurious growth of vegetation that stands 1 to 3.3 m (3 to 10 ft) high at maturity. The natural stands of the Argentine Pampa were so tall that a man riding on horseback could disappear from sight.

The short grasses stand 13 to 30 cm (5 to 12 in.) high; the midgrasses, 30 to 100 cm (1 to 3 ft) high; the tall grasses, 1 to 3.3 m (3 to 10 ft) high.

The versatility of the grassland community is such that in a prolonged succession of dry years cacti, particularly prickly pear *(Opunti camanchica)*, spread and in some situations protect the grasses from overgrazing by animals. A single patch of the cactus might reach a size of 10 m (30 ft) across. In some dry seasons on parts of the Great Plains, the cactus plants leave insufficient room for cattle to lie down (Weaver 1954). In wet years, the grasses and herbs respond and crowd back the cacti to sandy terrain. Legumes in the grassland community contribute considerable nitrogen to the soil.

Sagebrush *(Artimisia* spp.) is a major component of the drier steppe flora (Fosberg 1965); *Stipa* is the main species in the western Russian steppes. Small soapweed *(Yucca glauca)* is prominent in places. Its roots spread as much as 2 m (6 ft) vertically and 10 m (32 ft) laterally. Western wheat grass *(Agropyron smithii)* and buffalo grass *(Buchloe dactyloides)* are found on Chestnut and Reddish Chestnut soils. Buffalo-grass sod was used by pioneers for building houses. Blue grama grass *(Bouteloua gracillis)* is common on the drier Brown soils (Thorp 1948).

Big and little bluestem grasses *(Andropogon gerardi* and *scoparius)* are among the tall grasses found on Brunizem and Chernozem soils (now called Mollisols). Thorp (1948) estimates that 113 to 409 kg (250 to 900 lb), dry weight, of raw organic matter are added annually to a Mollisol (Chernozem) soil profile per acre. Possibly 136 to 500 kg (300 to 1,100 lb) of organic matter are added annually to a Mollisol of the tall-grass prairies (Brunizem) soil profile.

The boundaries between dry land and steppe, steppe and prairie, prairie and forest land are irregular and with outliers and inliers as a result of complexities in the geographical distribution of initial material, topography, climatic zones, and communities of organisms.

On the borders of drier zones, sandy soils permit extensions of grasslands into the drier regions. On the borders of more humid zones, sandy materials favor forest growth in the prairie lands, as in the case of the cross-timbers of Texas.

The natural grassland is, depending on the season, a continuous green or brown landscape with strips of forest along valleys and with scattered trees in ravines. The monotony of the prairie to some early settlers seemed, "except [for] the absence of sand, like crossing the desert. . . . We hoped to descry the distant woods" (Engel and Hopkins 1956).

Under cultivation, soils of the prairies appear very dark brown (moist color), with black bodies in depressions in humid to subhumid regions and light-colored, salt-affected soils in depressions in subhumid to semiarid regions. Shallow soils on steeper slopes plow up light colored where a calcareous Cca is near the surface or where salts are concentrated by evaporation.

Borchert's analysis (1950) of the climate of the prairie triangle of the Great Plains of North America lists the essential ingredients of the climates of grasslands of the middle latitudes. These features occur in different proportions over these lands: (1) severe, dry winters with much wind and relatively slight accumulations of snow; (2) relatively moist springs in most years; (3) droughty summers with some thunderstorms and tornadoes. The short-grass steppe has a drier climate than does the tall-grass prairie.

Conditions on middle-latitude steppes are represented by the mean annual temperatures and precipitation of -2 C (28 F) and 20 cm (8 in.) at Urga, Mongolia; 14 C (58 F) and 25 cm (10 in.) at Quetta, Pakistan; and 4 C (39 F) and 36 cm (14 in.) at Williston, N.C. (Finch et al. 1957). Prairies develop under relatively moist conditions: 8 C (46 F) and 81 cm (32 in.) in eastern Iowa and 16 C (60 F) and 76 cm (30 in.) in central Oklahoma. Major shifts of climatic boundaries occur from year to year in the grasslands (Horn and Bryson 1960; James 1959; Kendall 1935).

Barshad (1946) noted that prairie soils (Argiudolls) occur in California under moist winters and dry summers. It is evident, therefore, that different but equivalent climates can produce similar soils.

The grasslands cover a wide range of land forms, from low mountains to extensive plains. At the edges of the grasslands, as for example at the boundary of deciduous forest in Wisconsin, extensions of prairies by fire have formed preferentially on topography over which fire moves easily, namely, ridge tops and some windward slopes. The wide variety of soils of the Great Plains is shown on the map prepared by Aandahl (1972).

Aandahl (1948) correlated variations in depth of noncalcareous soil and nitrogen profile with slope position in prairie soils of western Iowa. Soils on lower slopes were found to be more leached and to contain more nitrogen in the first 61 cm (24 in.) than soils on higher slopes.

Thorp (1947) described catenary and noncatenary complexes in Chernozem landscape and in zones on mountain slopes where grassland soils grade into forest soils.

Ruhe (1969) reported evidence of forests in Iowa before the spread of prairie over the state. Curtis (1959) refers to the alto-thermal period (approximately 5,000 to 2,000 b.p.) as the great period of prairie expansion. About 5,000 years ago significant portions of the Lake Michigan basin were occupied by soils, the lake (called Lake Chippewa) being small and standing at 75 m (230 ft) above sea level.

Thorp (1948) proposed the designation of three degrees of development of grassland soils: minimal, medial, and maximal. Matelski (1959) considered Chestnut soils under the same three categories, which may have significance as chronosequences.

Smith, Allaway, and Riecken (1950) suggested that with time a Prairie (Brunizem; Argiudoll) soil will develop into a Planosol (Albaquoll).

Thorp (1957) suggested that Reddish Chestnut soils are older than Chernozem and Chestnut soils. He noted that Chernozem soils are no older than Sangamonian and occur on calcareous or basic initial materials. In Australia, Chernozems probably are destined to become Reddish Chestnut soils (Thorp 1965).

Some Russian soil scientists suggest that the Chestnut soils evolved since glaciation as a flooded condition gave way to a dry condition. The soils became alkaline, and then dealkalized (solodized) and were left with the present carbonate-rich condition.

PROFILE PROCESSES. Melanization, the process of darkening of the soil by addition of organic matter, is of course the dominant process in Mollisols. It is the process by which the mollic epipedon or dark surface horizon extends down into the profile. The process is a bundle of several specific processes (Hole and Nielsen 1968): (1) extension of roots of prairie vegetation into the soil profile; (2) partial decay of organic materials in the soil, producing some relatively stable, dark compounds; (3) reworking of the soil and organic matter by earthworms, ants, cicada nymphs, moles, and rodents with formation of dark soil-organic matter complexes and mixtures, krotovinas (filled burrows), and mounds; (4) eluviation and illuviation of organic colloids, along with some mineral colloids, as along cracks between peds, the surfaces of which become coated with dark cutans; (5) formation of resistant "ligno-protein" residues, such as give a black color even to long-cultivated Chernozem soils.

Eluviation and illuviation, apart from their involvement in melanization, accomplish the downward movement and precipitation of clay with iron oxides and organic matter in argillans, and iron oxides in concretions. Formation of concretions is particularly evident in the A2 horizons of prairie "Planosols" (Albolls of the new system). Milfred (1966) reports occurrence of iron and manganese concretions, called glaebules, in a well-drained prairie soil. These exhibit a wide range of distinctness, from mere stains in the soil to hard pellets. Free iron and clay apparently move together in well-drained prairie soils but not in wet ones (Simonson, Prill, and Riecken 1957). Studies by Swenson and Riecken (1955) indicate that in Argiudolls and Hapludolls free iron moves with weathering to the coarse clay and fine silt fraction in the A and B1 horizons; iron is retained in clay minerals of the very fine clay fraction in the B2 and C horizons.

Percolation of water is influenced by systems of cracks, krotovinas, and open channels made by roots and fauna. Weaver and Albertson (1956) report that soon after a 3-inch rain had fallen on a dry grassland soil in which drought cracks about 2 inches wide were spaced about 7 feet apart, borders of green vegetation developed 12 inches wide, paralleling the cracks.

Coatings of complexes of clay and organic matter (organo-argillans) on ped surfaces in the B horizon indicate that blocky and prismatic soil structures are prerequisite to the development of the coatings. The coatings in turn may favor the maintenance of the structure (Wittmuss and Mazurak 1958). Buntley and Westin (1965) noted that the soil profile development within peds lags behind that in ped coatings. As a climate becomes more moist, or a soil becomes older, coating may be moved from vertical ped faces in the upper B to those in the lower B. Coatings on horizontal ped faces may escape movement for long periods and constitute a clue to the former condition of the profile. The absence of iron oxide coatings from quartz grains in the Al horizon of Chernozems may result from a temporary reducing condition in the spring when the frozen B horizon creates a perched water table, followed by downward movement of the water with mobilized iron.

Robinson (1950) estimated the ages of the two soils of adjoining forest and prairie lands at 20,000 and 27,000 years, respectively, in part on the basis of depth to carbonates. Yet these soils occur so close together that such a great difference in age seems incredible. If one assumes that they are the same age, then the difference in depth to carbonates might be explained by hypothesizing (1) that the compact textural B horizon of the forest soil holds back percolating waters for evapotranspiration more than the corresponding horizon does in the prairie soil, or (2) that small increments of loess during soil formation have amounted to more on the prairie than in the forest, or (3) that the more porous prairie soil profile, which Baxter and Hole (1967) suggest was influenced by ants and worms and prairie vegetation over a 3,000-year period, following a previous forest episode, has expanded an extra foot during its development.

The presence in many Mollisols of a clay content in the A horizon nearly equal to that in the B horizon in medium-textured soils of good to moderate drainage, and a progressively increasing ratio of content of clay in the B horizon to that in the A with increasing degree of impedance of drainage, suggest several possibilities. One is that somehow clay is translocated from the B horizon up to the A horizon, particularly at well-drained sites. Another is that clay forms rapidly in the A horizon of well-drained prairie soils, as compared with the A horizons of forest soils and of poorly drained soils. A third possibility is that clay and other colloids eluviate very slowly in grassland soils, because of the complexing of mineral and organic colloids and because of the effects of the intimate binding of soil and rapid absorption of water by plant roots.

Observations and calculations by Baxter and Hole (1967) substantiate the suggestion that clay is translocated from the B horizon to the A horizon. The agency is a common prairie ant, *Formica cinerea,* which builds mounds as much as a foot high and at least as wide (Fig. 19.1). Clay content in the mounds is equivalent to that in the B horizon. Argillans characteristic of B and C horizons have been reported from thin sections of ant-mound material. Contents of available phosphorus and potassium are extremely high in the mounds, possibly in part because the mounds are largely composed of yellowish B horizon material, and in part because the ants concentrate organic materials from aphids, vegetation, and from their own bodies in the mounds.

FIG. 19.1. Cross section through two mounds of the
common western mound-building ant.
(After Baxter and Hole 1967.)

Biologic activity in grassland soils is greater than in forest soils. Evidence for this has been reported by various workers who observed rates of disintegration of (1) in situ cores of puddled soil (Telfair, Garner, and Miars 1957) and (2) pieces of cloth (Cooke 1955) and cord (Richard 1945).

Activity of earthworms is considerable in Brunizems and some Chernozems. Baxter and Hole (1967) noted earthworms and their casts even in active ant mounds. Milfred (1966) observed aggregation and tubules made of fecal pellets in the A1 horizon of a Tama silt loam, a Brunizem (now Mollisol). Buntley and Papendick (1960) used the term Vermisol for Chernozems which had been thoroughly worked to a depth of 2 feet or so by earthworms. These soils showed no textural change in the solum, had B horizons with granular rather than prismatic structure, a dispersed Cca horizon, and an unusually thick A1 horizon. It is possible that earthworms bring some lime to the surface in Brown and Chestnut soils (Vermustolls).

Working of soil profiles by cicada nymphs is considerable in Brown and Sierozem soils of Idaho, Utah, and Nevada (Hugie and Passey 1963). Medium-textured soil profiles may consist largely of "cylindrical blocky" peds formed by fillings of krotovinas. Both krotovinas and fillings are the work of cicada nymphs in the area they studied.

Thorp (1949) and Matelski (1959) have calculated that rodents annually bring to the surface 20 to 40 tons of subsoil material (air-dry weight) per acre. Laycock (1958) noted that gophers in Wyoming burrow in snow and may fill the tunnels in snow with soil, which is left on the surface of the ground after the snow melts. Curtis (1959) judged that the upper 2 feet of a prairie (Brunizem; Hapludoll) soil were turned over by ants, worms, and rodents once each century.

Some soil scientists working in the short-grass steppes of North America have the opinion that small increments of loess are added to the A1 horizon of Chestnut soils annually. As a result, the A1 horizon is continually growing upward. The horizons of calcite and gypsum accumulations also shift upward, producing a Bca above the older Cca. Loess currently accumulates deepest along fence rows where grass and tumbleweeds trap the silt. Processes of wetting and drying produce the striking prismatic structure in clayey sola of Chestnut soils. Rupture of roots may occur because of swelling and shrinking of soil (Thorp 1948) during wetting and drying and freezing and thawing.

Severe frost action (congellipedoturbation) and erosion by runoff waters affected steppe soils during periglacial climatic periods, producing mounds and stone stripes, cutting drainage ways into B horizons in places and burying soils in others (Fosberg 1965).

USES OF MOLLISOLS. To a notable extent, Mollisols are utilized by man for food production. These soils have been only slightly leached and the base status remains high. The lack of sufficient moisture is critical in the drier regions of Mollisol occurrence, and periodic flooding is a hazard in some lowlands. Early farmers were quick to realize that these soils would be productive once the tough sod could be broken by the plow. Clearing could easily be accomplished by fire. Mollisols were first farmed with little or no additions of fertilizer. But, for modern high yields, significant quantities of complete fertilizer are required. In the United States today, corn and wheat have replaced native grasses. Virgin areas of Mollisols are exceedingly rare, and relatively undisturbed areas are available for study only in corners of cemeteries and along little-used railroad tracks and rough lands such as the Flint Hills of Kansas. The USSR has preserved several areas for scientific study.

CLASSIFICATION OF MOLLISOLS. In general, Mollisols (Fig. 19.2) are thought of as those soils that have mollic epipedons. While it is true that all Mollisols have mollic epipedons, the presence of a mollic epipedon does not automatically qualify a soil as a Mollisol. Mollic epipedons are present in some Vertisols, in which case the plastic, churning nature of the clay has been felt to be more significant than the mollic epipedon. Also, mollic epipedons are found in the Inceptisols, especially where volcanic glass and acid cambic horizons more significantly influence the profile than does the mollic epipedon. It is also noted (Smith 1965) that epipedons that are made to meet the mollic criteria by the common practice of agricultural liming are excluded from criteria when placing a soil in the Mollisol order. Mollisols may or may not have albic, argillic, calcic, cambic, and natric horizons.

Seven suborders are recognized as the major subdivisions of Mollisols (Fig. 19.3). Albolls are Mollisols with an albic horizon. In general, these soils are in that part of the Mollisol region that borders on the Alfisols or Spodosols.

Aquolls are the Mollisols with characteristics of wetness.

Borolls are Mollisols with mean annual soil temperatures less than 8 C (47 F), excluding Aquolls.

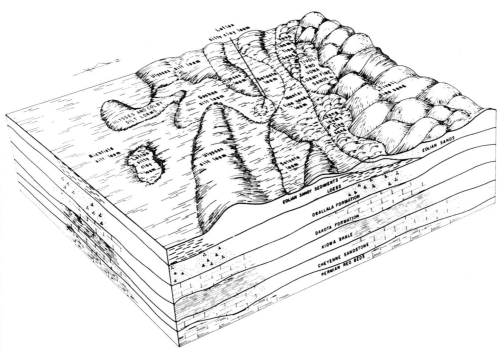

FIG. 19.2.   Soilscape patterns of a Typic Argiustoll (Richfield), Pachic Argiustoll (Goshen), Vertic Argiustoll (Lofton), with Typic Haplustolls (Ulysses, Manter), Mollic Haplargid (Vona), Typic Ustorthent (Colby), and Typic Ustipsamment (Tivoli) in Grant County in southwestern Kansas. (After Hamilton et al. 1969.)

FIG. 19.3.   Diagram showing some relationships between suborders of Mollisols.

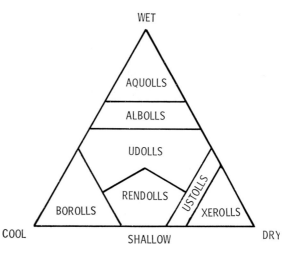

Rendolls have no argillic or calcic horizons. Their epipedons are less than 50 cm (20 in.) thick and overlie coarse fragments of calcareous rock and stones.

Three suborders are recognized on the basis of soil moisture conditions, thus separating the order from the dry to the more humid range of its occurrence.

Udolls (Fig. 22.3) are not dry for as much as 90 days per year or 60 consecutive days per year.

Ustolls (Figs. 16.2, 17.1, 19.2) are dry for more than 90 days per year, but not as much as 60 consecutive days per year.

Xerolls are dry for more than 60 consecutive days in most years.

In summary, the suborders reflect characteristics of both genesis and use of Mollisols. The occurrence of albic horizons in Albolls reflects the operation of podzolization. Ambient moisture balance and temperature,

TABLE 19.1 ❧ Suborders and great groups in the Mollisol order

Suborder	Great Group
Rendolls:	No great groups presently recognized
Albolls:	Natralbolls—presence of a natric horizon Argialbolls—presence of an argillic horizon
Aquolls:	Cryaquolls—mean annual temperature less than 8 C (47 F) Duraquolls—presence of duripan within 1 m (40 in.) Natraquolls—presence of a natric horizon Calciaquolls—presence of calcic horizon within 40 cm (16 in.) Argiaquolls—presence of an argillic horizon Haplaquolls—other Aquolls
Borolls:	Paleborolls—upper boundary of argillic horizon deeper than 60 cm (24 in.) Cryoborolls—mean summer temperature less than 8 C (47 F) when O horizon present Natriborolls—presence of a natric horizon Argiborolls—presence of an argillic horizon Vermiborolls—mollic epipedon over 50 cm (20 in.) thick with >25% of underlying horizon consisting of worm casts or filled animal burrows Calciborolls—presence of a calcic or gypsic horizon within 1 m (40 in.) Haploborolls—other Borolls
Ustolls:	Durustolls—presence of duripan within 1 m (40 in.) Natrustolls—presence of a natric horizon Calciustolls—presence of a calcic horizon within 1 m (40 in.) or a petrocalcic horizon within 1.5 m (60 in.) Paleustolls—presence of a petrocalcic horizon within 1.5 m (60 in.) or an argillic horizon where the clay content does not decrease 20% within 1.5 m (60 in.) of the surface Argiustolls—presence of an argillic horizon thinner than in Paleustolls Vermustolls—have >50% of the mollic epipedon and >25% of the underlying horizon composed of worm casts or filled animal burrows Haplustolls—other Ustolls
Xerolls:	Subgroups the same as for Ustolls except no vermixerolls are identified
Udolls:	Paleudolls—presence of an argillic horizon that does not have a clay content decrease by as much as 20% within 1.5 m (60 in.) of the surface Argiudolls—presence of an argillic horizon thinner than in Paleudolls Vermudolls—>50% of the mollic epipedon and >25% of the underlying horizon consists of worm casts or filled animal burrows Hapludolls—other Udolls

TABLE 19.2 Estimates of some properties of modal concepts of some Mollisols

	A1 Horizon				Structure	Devel. of Bt & Prisms	Devel. of Organs & Argillans	pH of B Horizon	O.M. content of horizons		Cca Horizon	Depth to Carbonates
	Thickness	pH	C/N	B.S.					Ap	B		
	(cm)			(%)					(%)	(%)		(cm)
Brunizems, Reddish Prairie soils (Argi-, Hapl-, & Verm-udolls)	30	5.1–6.5	10–11	70–90	granular	weak to none	mod.	5.3–6.7	4.5–2.0	1.5–1.0	absent	76–200
Chernozems (Agri-, Haplo-, Natri-, Vermiborolls)	20	6.0–7.5	11–12	85	granular	none to weak	mod.	6.5–7.5	4	2–3	usually present	38–60
Chestnuts, Reddish Chestnuts (Argi-, Hapl-, Verm-, Durustolls)	20	6.8–7.5	10–13	90	gran. to blocky and prismatic	mod. to strong	weak to strong	7.5	3	2–3	present	30–90
Brown soils (Argi-, Haplustolls; Argi-, Dur-Xerolls)	10	6.9–7.5	11	95	platy to gran.	weak to mod.	very weak	7.5	2	1.5	present	20–60
Sierozems (Haploxeroll)	7	7.5	8	100	fine platy, vesicular	none to very weak to mod.	very weak	9.1	1.3	1.3	present	17–30

NOTE: The purpose of this table is to identify in a general way the central concept of each of the traditional grassland soil groups of well-drained sites.

of primary consideration in farming, are reflected in the Ustolls, Udolls, Xerolls, and Borolls. The Aquolls reflect local topography as well as management limitations imposed by their need for drainage, and the shallow rocky nature of the Rendolls reflects genesis and use parameters.

Table 19.1 lists further separation at the great group level in the sequence used to key out the order. Table 19.2 lists some estimates of Mollisol properties.

LITERATURE CITED

Aandahl, A. R. 1948. The characterization of slope positions and their influence on the total nitrogen content of a few virgin soils of western Iowa. Soil Sci. Soc. Am. Proc. 13:449–54.

———. 1972. Soils of the Great Plains. 1:2,500,000 Map. Soil Conserv. Serv., U.S. Dept. Agr., Lincoln, Nebr.

Barshad, I. 1946. A pedologic study of California prairie soils. Soil Sci. 61:523–42.

Baxter, F. P., and F. D. Hole. 1967. Ant *(Formica cinerea)* pedoturbation in a Prairie soil. Soil Sci. Soc. Am. Proc. 31:425–28.

Borchert, J. R. 1950. The climate of the central North American grassland. Annals Assoc. Am. Geog. 40:1–39.

Buntley, G. J., and R. I. Papendick. 1960. Worm-worked soils of eastern South Dakota, their morphology and classification. Soil Sci. Soc. Am. Proc. 24:128–32.

Buntley, G. J., and F. C. Westin. 1965. A comparative study of developmental color in a Chestnut-Chernozem-Brunizem soil climosequence. Soil Sci. Soc. Am. Proc. 29:579–82.

Cooke, W. B. 1955. Fungi, lichens and mosses in relation to vascular plant communities in eastern Washington and adjacent Idaho. Ecol. Monographs 25:119–80.

Curtis, J. T. 1959. Vegetation of Wisconsin: An ordination of plant communities. Univ. Wis. Press, Madison.

Engel, Martha S., and A. W. Hopkins. 1956. The prairie and its people. Univ. Wis. Agr. Exp. Sta. Bull. 520.

Finch, V. C., G. T. Trewartha, A. H. Robinson, and E. H. Hammond. 1957. Elements of geography, 4th ed. McGraw-Hill, New York.

Fosberg, M. A. 1965. Characteristics and genesis of patterned ground in Wisconsin time in a Chestnut soil zone of southern Idaho. Soil Sci. 99:30–37.

Hamilton, V. L., Q. L. Markley, W. R. Swafford, and H. P. Dickey. 1969. Soil Survey of Grant County, Kansas. Soil Conserv. Serv., U.S. Dept. Agr. U.S. Govt. Printing Office, Washington.

Hole, F. D., and G. A. Nielsen. 1968. Some processes of soil genesis under prairie. Proc. symp. on prairie and prairie restoration. Knox College, Galesburg, Ill., pp. 28–34.

Horn, L. H., and R. A. Bryson. 1960. Harmonic analysis of the annual march of precipitation over the United States. Annals Assoc. Am. Geog. 50:157–71.

Hugie, V. K., and H. B. Passey. 1963. Cicadas and their effect upon soil genesis in certain soils in southern Idaho, northern Utah and northeastern Nevada. Soil Sci. Soc. Am. Proc. 27:78–82.

James, P. E. 1959. A geography of man, 2nd ed. Ginn, New York.

Kendall, H. M. 1935. Notes on climatic boundaries in the eastern United States. Geograph. Rev. 25:117–24.

Laycock, W. A. 1958. The initial pattern of revegetation of pocket gopher mounds. Ecology 39:346–51.

Matelski, R. P. 1959. Great soil groups of Nebraska. Soil Sci. 88:228–39.

Milfred, C. J. 1966. Pedography of three soil profiles of Wisconsin representing the Fayette, Tama and Underhill series. Ph.D. thesis. Univ. Wis., Madison.

Müller-Beck, H. 1966. Paleohunters in America: Origins and diffusion. Science 152:1191–1210.

Richard, F. 1945. The biological decomposition of cellulose and protein test cord in soil under grass and forest association. Amst. Forstl. Vorschswensen. 24: 297–397.

Robinson, G. H. 1950. Soil carbonate and clay contents as criteria of rate and stage of soil genesis. Ph.D. thesis. Univ. Wis., Madison.

Ruhe, R. V. 1969. Quaternary landscapes of Iowa. Iowa State Univ. Press, Ames.

Simonson, G. H., R. C. Prill, and F. F. Riecken. 1957. Free iron distribution in some poorly drained prairie soils in Iowa. Proc. Iowa Acad. Sci. 64:385–92.

Smith, G. D. 1965. Lectures on soils classification. Pedologic Special Issue 4, Belgian Soil Science Soc. Rozier 6, Gent, Belgium.

Smith, G. D., W. H. Allaway, and F. F. Riecken. 1950. Prairie soils of the upper Mississippi valley. Adv. Agron. 2:157–205.

Swenson, R. M., and F. F. Riecken. 1955. Movement of iron in the development of loess-derived Brunizem soils. Soil Sci. 79:177–86.

Telfair, D., M. R. Garner, and D. Miars. 1957. The restoration of a structurally degenerated soil. Soil Sci. Soc. Am. Proc. 21:131–34.

Thorp, James. 1947. Practical problems in soil taxonomy and soil mapping in great plains states. Soil Sci. Soc. Am. Proc. 12:445–48.

———. 1948. How soils develop under grass, pp. 55–66. In Yearbook of Agriculture, U.S. Dept. Agr. U.S. Govt. Printing Office, Washington.

———. 1949. Effects of certain animals that live in the soil. Sci. Monthly 68:180–91.

———. 1957. Report on a field study of soils of Australia. Sci. Bull. No. 1, Earlham College, Richmond, Ind.

———. 1965. The nature of the pedological record in the Quaternary. Soil Sci. 99:1–8.

Weaver, J. E. 1954. North American prairie. Johnsen, Lincoln, Nebr.

Weaver, J. E., and F. W. Albertson. 1956. Grasslands of the great plains: Their nature and use. Johnsen, Lincoln, Nebr.

Wittmuss, H. D., and A. P. Mazurak. 1958. Physical and chemical properties of soil aggregates in a Brunizem soil. Soil Sci. Soc. Am. Proc. 22:1–5.

❧

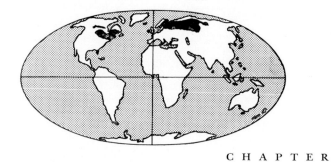

Spodosols: Soils with Subsoil Accumulations of Sesquioxide and Humus ❧

THE "WHITE EARTHS," the Spodosols, contrast sharply with the "black earths," the Borolls (Chernozems) (see Chapter 19). Spodosols encompass most but not all soils named Podzol—from the Russian terms *pod* (beneath) and *zol* (ash).

Spodosols are widely known as acid ashy gray sands over dark sandy loams. These two contrasting horizons, with an abrupt boundary between them, place these soils among the most eye-catching and photogenic in the world. Many scientists take pleasure in studying Spodosols, which respond quickly to changes in vegetation, and which have the appearance of a chromatographic column. The development of these soils in quartzose sand is favored by the presence of a vegetative cover under which acid litter accumulates. Moved by percolating water, organic compounds from the litter clean the quartz grains in the first horizon and coat them with a dark mixture of iron oxide and humus in the second. The great variety of Spodosols, and their distribution in very different climatic zones, suggest that this first impression of these soils is far too simple.

SETTING. Not every "white" A2 (albic) horizon that one sees signifies a Spodosol. The wide application of the term "podzolization" to many soils with bleached surface or subsurface horizons is not followed here. Albaquolls and Albaqualfs (salt-affected soils and Planosols), Eutroboralfs (Gray Wooded soils), and many Hapludalfs (Gray-Brown Podzolics) are loci of active podzolization but are not Spodosols. Neither is the pseudopodzol of Madagascar, the albic horizon of which is a 1- to 30-cm-thick accumulation of opal phytoliths under acacias and bamboos (Riquier 1960). Thus, it is fortunate that the albic horizon is not considered a diagnostic horizon, whereas the spodic horizon is. This black to strong brown subsoil horizon is the essential feature by which a Spodosol is identified, whether the topsoil is a dark plow layer or ashy gray sand.

The setting in which such a soil is produced is one in which the combination of factors of soil formation yields the necessary conditions—

the accumulation of iron, aluminum, and/or organic matter in a subsoil horizon. Many kinds of vegetation, including grasses, can yield certain organic compounds that speed podzolization under laboratory conditions, but litter from a certain few species of plants is particularly capable of fostering such accumulations. Among these plants are the hemlock tree *(Tsuga canadensis)* in forests of northern latitudes, the kauri tree *(Agathis australis)* of New Zealand (see Chapter 11), and heath *(Calluna vulgaris* and *Erica* sp.) of northern Europe. Individual Spodosol soil bodies may be correlated with single kauri trees or groups of hemlocks or communities of heath plants. Spodosols do occur under a wide variety of trees *(Picea, Pinus, Larix, Thuya, Populus, Quercus, Betula)* and understory plants (including *Vaccinium)* in climatic zones that are cool (McFee and Stone 1965). This includes the taiga and similar plant communities of mountainous lands that together may occupy more than 38 million square kilometers of land in the world. Apparently the moisture regime in these soils is propitious for the formation of requisite organic compounds in the O horizon and their delivery, under a flushing regime, to the subsoil.

On the coastal plain of southeastern United States extensive areas of Aquods (Groundwater Podzols) are developed under *Quercus, Pinus,* and associated understory plants.

The giant Podzol of Surinam (Eyk 1957) has a 25-cm-thick sandy Bhir horizon on schist-derived clay under nearly 3 m of gray to white sand. The mean annual temperature is 26 C (79 F) and mean annual rainfall is 2,300 mm (92 in.), which includes a relatively dry period in October and November when monthly precipitation is about 75 mm (3 in.). The present savanna vegetation may be quite different from that under which this soil first formed.

A 2-m-deep Typic Sideraquod (Groundwater Podzol with iron pan) is in ecological equilibrium with a pigmy forest 1.5 to 3 m tall in the Fort Bragg area of coastal California (Jenny, Arkley, and Schultz 1969). Dwarf Spodosols (less than 45 cm deep) are reported from Alaska (Kubota and Whittig 1960) and heath balds of the Great Smoky Mountains (McCracken, Shanks, and Clebsch 1962).

The great "Russian generalization" that the climate of the boreal zone can be expected to impress, via organisms, the O-A2-Bhir horizon sequence on any mineral material, including calcareous clay, has validity because of the youthfulness of landscapes in this zone. But because of the precise definition of properties of the Spodosol order, emphasizing the spodic horizon, the zonal pattern of Podzols is only partially realized as Spodosols, and is quite discontinuous.

Tree-tip mounds, sometimes called cradle knolls, are numerous in some Spodosol polypedons. They are formed by uprooting of trees in storms. They may have a relief of as much as 1 m and length of 3 m. Tall trees with strong trunks that will not snap in a wind, and a somewhat shallow root system, produce these features during storms (Baxter, in Milfred, Olson, and Hole 1967). Cradle knolls occupy about 20% of the area of the forest floor in parts of the Great Lakes region of North America.

Macrotopography is variable. Spodosols are extensive on nearly level to undulating sand plains, but some are reported on slopes approaching 90% in mountainous areas (Bouma et al. 1969), with best solum development on slopes facing away from the equator.

It is generally conceded that Spodosols may form relatively quickly. Estimates range from a few hundred years (Soil Survey Staff 1967) to several thousand years (Franzmeier and Whiteside 1963a). The movement of organic and mineral colloids from surface soil to subsoil in coarse sandy material to the point of equilibrium with the environment should not take an inordinate amount of time and energy. Burges and Drover (1953) reported evidence that 200 years were required to leach calcite from beach sand in New South Wales, 2,000 years to produce an iron-Podzol, and 3,000 years to produce an iron-humus Podzol with pH as low as 4.5. A Podzol buried under peat in northern Ireland was dated by artifacts and paleobotany as being formed between 3,000 and 2,000 B.C. (Proudfoot 1958). Franzmeier and Whiteside (1963a, b) studied a 10,000-year-long chronosequence of Spodosols in Michigan and concluded that between 3,000 and 8,000 years were necessary for the formation of a Spodosol. The age of a Typic Sideraquod of coastal California has been estimated at about 1 million years (Jenny, Arkley, and Schultz 1969).

NATURE OF PEDOGENIC PROCESSES. Podzolization is a bundle of processes (Gerasimov 1960; Ponomareva 1964; Stobbe and Wright 1959) which brings about translocation, under the influence of the hydrogen ion (Hallsworth, Costin, and Gibbons 1953) and organic compounds (Bloomfield 1953a, b; 1954), of organic matter, iron, and aluminum (and a small amount of phosphorus) from the upper part of the mineral solum to the lower part. If clay is also transported in suspension, this may be considered atypical as a pedogenic process in this order, although it is actually a fairly common accompaniment of podzolization. This is evidenced by clay accumulation in spodic horizons (Franzmeier and Whiteside 1963a) and in argillic horizons below them (Milfred, Olson, and Hole 1967). Free iron: clay ratios are not constant with depth as they are in Alfisols and Ultisols (Soil Survey Staff 1967). "Purest" Spodosols may be expected to have formed in initial material containing little or no clay and receiving little if any by wind action during pedogenesis.

Processes of Spodosol formation may be considered under the following headings: accumulation of organic matter; leaching and acidification; weathering; translocation of Fe and Al (with some P, Mn, and clay) from the A to B horizon; immobilization of humic and fulvic acids (and some clay) in the B; pelleting of humus coatings; reduction in bulk density; cementation.

Accumulation of organic matter in the solum of a Spodosol is concentrated in the O and Bhir on Bh (Spodic) horizons. In dry weight per acre, the total organic matter in the Spodosol sequum increased steadily to 30 tons over a period of 10,000 years, in a chronosequence studied by Franzmeier and Whiteside (1963b). In well-developed, well-drained Spodosols, proliferation of tree roots tends to be in both the O and spodic horizons.

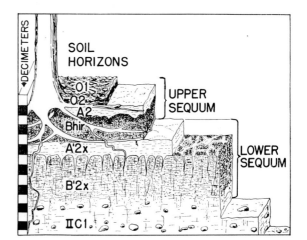

FIG. 20.1. Diagrammatic cross section of a Spodosol in Wisconsin that is a bisequal soil.

The O horizons of a Spodosol constitute a mor layer (Fig. 20.1) containing an abundance of fungi, the hyphae of which permeate the material at certain seasons. The decomposition of this acid mat is slowed by freezing, drought, low summer temperatures, acidity, resistant nature of the plant debris, and possibly by antibiotics produced by fungi, actinomycetes, and bacteria. The mor may be thicker than 1 m adjacent to a kauri tree trunk (where it resembles a Folist). *Collembola* (springtails) and *Acarina* (mites) are numerous in these horizons, along with many other Arthropods and some Annelidae (Eaton and Chandler 1942; Macfadyen 1963). Although total oxygen consumption of the biomass in mor may be only about a third of that in mull (Kevan 1955), its respiration rate per gram of biomass is nearly three times as fast. This is presumably because the microfauna in mor are smaller and individually more intensely active than those in mull. In mor, the ratio of volume of living organisms to volume of litter is about 1:30,000 in the litter layer (O1) and 1:15,000 in the fermenting and humus layer (O2).

Leaching of carbonates and significant replacement of the exchangeable cations, Ca^{++}, Mg^{++}, K^+, and Na^+, by H^+ and Al^{+++} in the A horizon are prerequisite to mobilization of organic matter, and with it, of Fe^{+++}, Al, and about 1/20,000 as much P, and some Mn. Thorp[1] observed a Rendzinalike soil under a kauri tree growing in unleached material. Bailey, Whiteside, and Erickson (1957) noted collapse of sola developed in initial materials containing more than 12% of acid-soluble mineral matter.

The amount of Fe and Al accumulated in a Spodosol may be far more than has been biocycled in the history of the soil (Franzmeier and Whiteside 1963b). The source of the bulk of these two elements in the spodic horizon, therefore, has come from weathering of ferromagnesian minerals, feldspars, illite, and chlorite in the A horizon, where pH usually drops below 5.0, and in one case went as low as 2.8 (Jenny, Arkley, and Schultz 1969). The particles of weatherable minerals may not by any means have

---

1. Personal communication, 1962.

all been present in the initial material. If iron is not present in appreciable amounts, the spodic horizon is composed largely of quartz sand coated with humus, as in Humods (Edelman, 1950).

The albic horizon of a Spodosol is the site of accumulation of both (1) resistant minerals and of (2) difficultly soluble products of decomposition. The resistant minerals include quartz with small amounts of montmorillonite and beidellite (Bouma et al. 1969). The second category of materials includes silica in various forms such as dehydrated silicic acid or secondary quartz with small amounts of titanium oxide and barium sulfate in some soils. The albic horizon is pink (5YR 6/3, moist) under forest cover and gray (10YR 6/2) under heath (Schuylenborgh 1962). "Bleicherde" is the German term for the albic horizon.

Illuviation of clay, a process called lessivage (Duchaufour 1958) or illmerization (Fridland 1957), is sometimes viewed as a precursor of podzolization. After clay is eluviated from surficial horizons, albic and spodic horizons can form in the coarser residue. Tonguing of Spodosol sola into older degrading horizons is not uncommon. In some Spodosols lessivage may, however, be concurrent with podzolization. Franzmeier, Whiteside, and Mortland (1963) observed in a spodic horizon very thin birefringent free-grain argillans coated with thick (1.08 mm) amorphous organs (organic cutans) with some included clay.

Chemical eluviation in these soils includes reduction and translocation of Fe (McKenzie, Whiteside, and Erickson 1960) by chelates (Atkinson and Wright 1957) and complexing compounds of branching structure, such as polyphenols (Bloomfield 1957). The maximum accumulation of Al is commonly below that of Fe in the spodic horizon. Aaltonen and Mattson (after Jenny 1941) found evidence that the Podzol B may form from the bottom up, the zone of maximum colloid accumulation shifting from a depth of about 40 cm to 15 cm over a period of 5,000 years. Burges and Drover (1953), on the contrary, observed that this boundary shifted downward with time.

Precipitation of illuviated sesquioxides in the spodic horizon, Orterde and Orstein, in German parlance (referring by corruption to "Ertz" or ore), may be mechanical, chemical, or biological. Insufficiency of percolating water, from individual storms, to carry the colloids and solutes farther down may force precipitation in the Bhir horizon. Sieving action can be mechanical. Colloids clog pores. When this process is important, the spodic horizon may grow upward into the albic horizon, a trend suggested by data of Franzmeier, Whiteside, and Mortland (1963). Bacteria may destroy the chelating and complexing organic compounds which mobilized the Fe and Al. Negative charges on thin clay films may immobilize the positively charged ions of Fe and Al. Polyphenols, organic acids, and reducing sugars were found to be major active components in mobilizing Fe and Al in a North Carolina study (Malcolm and McCracken 1968). In this study it was concluded that a major source of mobile organic matter for mobilization of Fe and Al in the podzolization process comes from tree canopy drip. They also concluded from their studies that precipitation and immobilization took place when sufficient Al and/or Fe was present to exceed the solubility product believed to prevail in incipient spodic horizons. Calcium was not found to affect mobilization and immobilization in this study. Kawaguchi and Matsuo (1960) also report studies showing that the movement of iron in the soil is regulated by the ratio of the amount of the mobilizing agent to the amount of iron to be mobilized, relating this to the immobilization of Fe in a Spodic horizon due to presence of excess iron. Presumably, similar relationships exist with respect to Al, based on data of Malcolm and McCracken (1968). This mechanism, then, of mobilization and immobilization seems to provide the basis for a reasonable process for "podzolization" and development of spodic horizons.

Humic acids may convert to fulvic acids in the albic horizon by increase in oxygen-containing functional groups. These acids move down into the spodic horizon where they are bonded to the Fe^{++} and Al^{+++} already accumulated there. The nearly black organic subhorizon now forms at the top of the spodic horizon. As the films of organic matter thicken around sand grains, the films become susceptible to cracking during droughts. This results in pelleting of the films. The amorphous pellets sift down into voids, increasing the available water-holding capacity of the soil. This may foster a forest succession from conifers to deciduous trees, with ultimate degradation of the Spodosol. Some clay is occluded in the amorphous films and pellets.

Much of what has been said about cheluviation (Swindale and Jackson 1956) in well-drained Spodosols can also be said of Aquods (Groundwater Podzols) in which a fluctuating water table is an important agent of translocation.

This whole process of formation of a Spodosol entails an overall decrease in bulk density of the mineral portion of the solum. A 13% increase in volume is reported for Michigan Spodosols (Bailey, Whiteside, and Erickson 1957). Reductions in bulk density from 1.67 g/cc (C horizon) to 1.21 g/cc (A2 horizon) in a sandy Spodosol and from 1.62 to 1.13 g/cc in a silty one were reported by Hole and Schmude (1959) from northern Wisconsin. McFee and Stone (1965) reported a bulk density of 0.14 g/cc for a mor weighing 108 T/a (87 T of organic matter) under a 70-T standing crop and over 71 T of organic matter in the mineral solum. The commonly observed tonguing of the albic horizon into the spodic horizon, and the latter into the C horizon, may have been formed by flow of soil and water down tree-root channels.

Cementation may be striking in a Spodosol solum. *Durinodes* (Soil Survey Staff 1967) are cemented portions of the albic horizon. The cementing material is largely silica. A thin (1 to 10 mm) wavy or involuted black to reddish brown iron pan within 50 cm (20 in.) of the soil surface is called the *placic horizon* (Soil Survey Staff 1967). This horizon has been observed in tropical as well as subarctic regions. Cementation of a thicker mass of the spodic horizon to a rocklike consistency is called an Ortstein. Pol'skii (1961) reports analyses from this material showing 5% to 21% Fe_2O_3 and 2% to 15% MnO. P. E. Müller distinguished several kinds of Ortsteins

(Muir 1961), including a black peatlike Ortstein, a brown humus Ortstein, and two concretionary Ortsteins, one in quartz sand converted to a pseudo-sandstone, and one in peat, called Raseneisenstein, that contains about 90% iron hydroxide and is porous and slaglike in appearance.

Under the spodic horizon of many Spodosols (Nygard, McMiller, and Hole 1952) is a reversibly and weakly cemented horizon called the fragipan (Fig. 20.1). It has been suggested that the invisible, reversible cement is silica, or alumina, or illite. This seems reasonable in the light of calculations of Rode (quoted by Russell 1956) that losses from a Spodosol during 10,000 years arranged the elements in the following order of diminishing quantity: $Si>Al>Fe>K>Mg>Ca>Na$. This seems to reflect the abundance of elements in the initial materials. The fabric of a fragipan is one of packing of finer particles in voids between larger ones to yield a bulk density (oven dry) as high as 1.92 g/cc (Hole et al. 1962). Vesicular, platy, and prismatic structures are present and may have been produced by frost action not involving cryopedoturbation. Repeated movement of a freezing front down through this horizon may have swept fines out of the horizon, increasing its close packing and fragile brittleness. Silty and clayey coatings on the upper surfaces of stones in the subjacent C horizon below support this idea. The fragipan is a barrier to root growth and downward water movement. Lateral movement of water over its surface is common on slopes.

Shifting of ecotones may enhance or diminish podzolization. In south central Alaska forests have encroached on grasslands, converting Cryandepts into Cryorthods (Rieger and DeMent 1965). Removal of hemlock from hemlock hardwood forests in northern Wisconsin results in fading of the spodic B horizon in the soils (Milfred, Olson, and Hole 1967). Lateral growth of sphagnum moss bogs in the taiga of northern Canada may bury Spodosols under Histosols.

USES. Spodosols are used for forestry, pasture, hay land, and cultivated crops. The soilscape shown in Figure 20.2 is covered for the most part with poor quality pine-oak woodland, but some level areas are used for truck crops (and the peat for cranberries). Three centuries of haying on sparsely stable-manure-fertilized alpine meadows have converted Spodosols to a "Brown" soil (Inceptisol) in parts of Switzerland (Bouma et al. 1969). In the north central region of the United States, crop rotations on Spodosols include silage corn, oats, rye, potatoes, red clover, flax, strawberries, and raspberries. Spodosols provide the major sites for commercial blueberry production in the coastal plain of North Carolina. Fertilization and cultivation raise the nutrient levels of these soils and lead to soil compaction, mixing of O and A2 horizons, and some degradation of spodic horizons by aeration and leaching, particularly in irrigated potato fields. Recreational activities are on the increase on these soils, as larger numbers of city people vacation in the forests near lakes and streams. Spodosol terrain includes extensive wildlife preserves.

CLASSIFICATION. Once the spodic horizon (or iron-cemented placic horizon over a fragipan) of a profile has been identified as meeting the

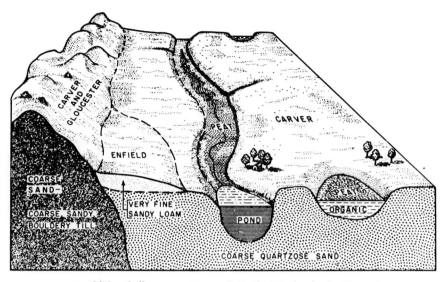

FIG. 20.2. Soilscape pattern of Entic Haplorthods (Carver, Gloucester, Enfield) with Histosols (Peat) in Plymouth County in eastern Massachusetts. (After Upham 1969.)

requirements,[2] the Spodosol may be classified into one of four suborders (see Fig. 20.3) which are briefly defined as follows:

1. Aquods are Spodosols that are commonly saturated with water (at least down to the placic horizon or duripan if such is present) or, if artificially drained, display such evidences of former wetness as (a) a histic epipedon, (b) mottling in the albic and upper spodic horizons, and (c) a duripan in the albic horizon.

---

2. Limits of the spodic horizon (see Soil Survey Staff 1967 for details) include: substantial amounts of amorphous material including carbon, iron, and aluminum; relatively little layer lattice clay; a thickness of at least 1 cm to this horizon; a 15-bar water content of less than 20% and less than 60% glassy volcanic ash (20 to 200μ); presence of amorphous cutans and pellets; hue of 10YR or redder and either value less than 3 or chroma of 3 or more.

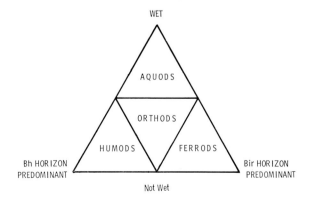

FIG. 20.3. Diagram showing some relationships between suborders of Spodosols.

TABLE 20.1 ❧ Suborders and great groups in the Spodosol order

Suborder	Great Group
Aquods	Fragiaquods—have a fragipan below the spodic horizon but no placic horizon
	Cryaquods—MAST* <8 C (47 F)
	Duraquods—indurated albic horizon in which dry peds will not slake in water
	Placaquods—have a placic horizon that rests on a spodic horizon and a fragipan
	Tropaquods—MAST >8 C (47 F) and <5 C (9 F) summer to winter temperature change
	Haplaquods—have spodic horizons where >50% of the horizon turns redder upon ignition
	Sideraquods—other Aquods
Humods:	Placohumods—have a placic horizon in the spodic horizon
	Tropohumods—MAST >8 C (47 F) and summer to winter temperature difference <5 C (9 F)
	Fragihumods—have fragipan below spodic horizon
	Cryohumods—MAST <8 C (47 F)
	Haplohumods—other Humods
Orthods:	Placorthods—have a placic horizon in or above the spodic horizon
	Fragiorthods—have a fragipan below the spodic horizon
	Cryorthods—MAST <8 C (47 F)
	Haplorthods—other Orthods
Ferrods:	No great groups

* MAST = mean annual soil temperature.

2. Ferrods are Spodosols that are not as wet as Aquods and have more than six times as much free (elemental) iron as carbon in the spodic horizon.

3. Humods are Spodosols that are not as wet as Aquods and have spodic horizons containing dispersed organic matter and aluminum but little free iron (less than 0.5% of the fine earth).

4. Orthods are Spodosols that are not as wet as Aquods and have in the spodic horizon a content of free iron not more than six times that of carbon, usually much less.

These suborders are divided into 16 great groups (Table 20.1) on the basis of several factors: mean annual soil temperature and range of soil temperature; degree of cementation of the albic horizon with silica; degree of accumulation of iron in the spodic horizon; degree of development of a fragipan; presence of an umbric epipedon.

No reference is made in this classification to buried (thaptic) Spodosols such as were observed by Bryson, Irving, and Larsen (1965) in a tension zone between tundra and taiga in Canada.

Some Haplorthods with ochric epipedons resting on the spodic horizon (without albic horizon) have been called Brown Podzolics in the past (Nygard, McMiller, and Hole 1952).

The terms maximal, medial, and minimal Spodosols have been used by some workers (Hole and Schmude 1959) to indicate degrees of profile development within a region.

LITERATURE CITED

Atkinson, H. J., and J. R. Wright. 1957. Chelation and the vertical movement of soil constituents. Soil Sci. 84:1–11.

Bailey, H. H., E. P. Whiteside, and A. E. Erickson. 1957. Mineralogical composition of glacial materials as a factor in the morphology and genesis of some Podzol, Gray Wooded, Gray-Brown Podzolic and Humic-Gley soils in Michigan. Soil Sci. Soc. Am. Proc. 21:433–41.

Bloomfield, C. 1953a. A study of podzolization. I. The mobilization of iron and aluminum by Scots pine needles. J. Soil Sci. 4:5–16.

———. 1953b. A study of podzolization. II. The mobilization of iron and aluminum by the leaves and bark of *Agathis australis* (Kauri). J. Soil Sci. 4:17–23.

———. 1954. A study of podzolization. III. The mobilization of iron and aluminum by Rimu *(Dacrydium cupressium)*. J. Soil Sci. 5:39–45.

———. 1957. The possible significance of polyphenols in soil formation. J. Sci. Food Agr. 8:389–92.

Bouma, J., J. Hoeks, L. Van der Plas, and B. van Scherrenburg. 1969. Genesis and morphology of some Alpine Podzol profiles. J. Soil Sci. 20:384–98.

Bryson, R. A., W. N. Irving, and J. A. Larsen. 1965. Radiocarbon and soil evidence of former forest in the southern Canadian Tundra. Science 147:46–48.

Burges, A., and D. P. Drover. 1953. The rate of Podzol development in sands of the Woy Woy district, N.S.W. Australian J. Botany 1:83–94.

Duchaufour, P. 1958. Dynamics of forest soils under Atlantic climate. Ecole National des Eaux et Forets. Nancy, France.

Eaton, T. H., Jr., and R. F. Chandler, Jr. 1942. The fauna of forest-humus layers in New York. Cornell Agr. Exp. Sta. Memoir 247.

Edelman, C. H. 1950. Soils of the Netherlands. North-Holland Pub. Co., Amsterdam, Holland.

Eyk, J. J. Van der. 1957. Reconnaissance soil survey in northern Surinam. Ph.D. thesis. Wageningen.

Franzmeier, D. P., and E. P. Whiteside. 1963a. A chronosequence of Podzols in northern Michigan. I. Ecology and description of pedons. Mich. State Univ. Agr. Exp. Sta. Quart. Bull. 46:2–20.

———. 1963b. A chronosequence of Podsols in northern Michigan. II. Physical and chemical properties. Mich. State Univ. Agr. Exp. Sta. Quart. Bull. 46:21–36.

Franzmeier, D. P., E. P. Whiteside, and M. M. Mortland. 1963. A chronosequence of Podsols in northern Michigan. III. Mineralogy, micromorphology, and net changes occurring during soil formation. Mich. State Univ. Agr. Exp. Sta. Quart. Bull. 46:37–57.

Fridland, V. M. 1957. Podzolization and illmerization. Dokl. Akad. Nauk. 115:1006–9.

Gerasimov, I. P. 1960. Gleyey pseudo-Podzols of central Europe and the formation of binary surface deposits. Soils Fertilizers 23:1–7.

Hallsworth, E. G., A. B. Costin, and F. R. Gibbons. 1953. Studies in pedogenesis in New South Wales. VI. On the classification of soils showing features of Podzol morphology. J. Soil Sci. 4:241–56.

Hole, F. D., and K. O. Schmude. 1959. Soil survey of Oneida County, Wisconsin. Univ. Wis. Geol. Nat. Hist. Survey Bull. 82.

Hole, F. D., G. W. Olson, K. O. Schmude, and C. J. Milfred. 1962. Soils survey of Florence County, Wisconsin. Univ. Wis. Geol. Nat. Hist. Survey Bull. 84.

Jenny, H. 1941. Factors of soil formation. McGraw-Hill, New York.

Jenny, H., R. J. Arkley, and A. M. Schultz. 1969. The pygmy forest-Podzol ecosystem and its dune associates in the Mendocina Coast. Madrono. 20:60–74.

Kawaguchi, K., and Y. Matsuo. 1960. The principle of mobilization and immobilization of iron oxide in soils and its application to the experimental production of podzolic soil profiles. Trans. 7th Intern. Congr. Soil Sci. (Madison, Wis.) 4:305–13.

Kevan, D. K. McE. (ed.). 1955. Soil zoology; proceedings of the University of Nottingham Second Easter School in Agricultural Science. Academic Press, New York.

Kubota, J., and L. D. Whittig. 1960. Podzols in the vicinity of Nelchina and Tazlina glaciers, Alaska. Soil Sci. Soc. Am. Proc. 24:133–36.

Malcolm, R. L., and R. J. McCracken. 1968. Canopy drip: A source of mobile soil organic matter for mobilization of iron and aluminum. Soil Sci. Soc. Am. Proc. 32:834–38.

McCracken, R. J., R. E. Shanks, and E. E. C. Clebsch. 1962. Soil morphology and genesis at higher elevations of the Great Smoky Mountains. Soil Sci. Soc. Am. Proc. 26:384–88.

Macfadyen, A. 1963. Animal ecology: Aims and methods. Pitman, New York.

McFee, W. W., and E. L. Stone. 1965. Quantity, distribution and variability of organic matter and nutrients in a forest Podzol in New York. Soil Sci. Soc. Am. Proc. 29:432–36.

McKenzie, L. J., E. P. Whiteside, and A. E. Erickson. 1960. Oxidation-reduction studies on the mechanism of B horizon formation in Podzols. Soil Sci. Soc. Am. Proc. 24:300–305.

Milfred, C. J., G. W. Olson, and F. D. Hole. 1967. Soil resources and forest ecology of Menominee County, Wisconsin. Soil Series 60. Univ. Wis. Geol. Nat. Hist. Survey Bull. 85.

Muir, A. 1961. The Podzol and Podzolic soils. Adv. Agron. 13:1–56.

Nygard, I. J., P. R. McMiller, and F. D. Hole. 1952. Characteristics of some Podzolic, Brown Forest, and Chernozem of the northern portion of the lake states. Soil Sci. Soc. Am. Proc. 16:123–29.

Pol'skii, B. N. 1961. The question of the chemistry of Ortsteins of sod-podzolic soils. Pochvovedeniye N.2:93–96.

Ponomareva, V. V. 1964. Theory of podzolization (Izdatel'stvo Nauka). (Transl. by A. Gourevitch.) Israel Prog. for Sci. Trans., Jerusalem, 1969. Available from U.S. Dept. Commerce, Springfield, Va.

Proudfoot, V. B. 1958. Problems of soil history. Podzol development at Goodland and Torr Towlands, County Antrim, Northern Ireland. J. Soil Sci. 9:186–98.

Rieger, S., and J. A. DeMent. 1965. Cryorthods of the Cook Inlet-Susitna Lowland, Alaska. Soil Sci. Soc. Am. Proc. 29:448–53.

Riquier, J. 1960. Les phytolithes de certains sols tropicaux et des Podzols. Trans. 7th Intern. Congr. Soil Sci. (Madison, Wis.) 4:425–31.

Russell, E. W. 1956. Soil conditions and plant growth. Longmans, Green and Co., London.

Schuylenborgh, J. van. 1962. On soil genesis in temperate humid climate. I. Some soil groups in the Netherlands. Neth. J. Agr. Sci. 10:127–44.

Soil Survey Staff. 1967. Supplement to Soil classification, a comprehensive system—7th approximation. U.S. Dept. Agr. U.S. Govt. Printing Office, Washington.

Stobbe, P. C., and J. R. Wright. 1959. Modern concepts of the genesis of Podzols. Soil Sci. Soc. Am. Proc. 23:161–64.

Swindale, L. D., and M. L. Jackson. 1956. Genetic processes in some residual podozolized soils of New Zealand. Trans. 6th Intern. Congr. Soil Sci., pp. 233–39.

Upham, C. W. 1969. Soil survey of Plymouth County, Massachusetts. Soil Conserv. Serv., U.S. Dept. Agr. U.S. Govt. Printing Office, Washington.

❧

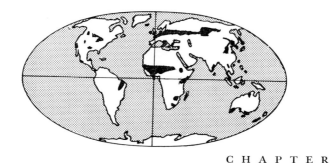

Alfisols: High Base Status Forest Soils   ✿

Many SOILS that have significantly more clay in the B horizon than in the A horizon are divided between this order and the more highly weathered Ultisol order. It is to be expected that the requisite soil properties could be produced under many conditions and although the modal Alfisols are in temperate regions, representatives of the order are also found in moderately weathered materials of tropical and subtropical regions.

SETTING. Two prerequisites are met by Alfisolic landscapes: (1) a moderate abundance of layer lattice clay, and (2) its accumulation in the subsoil enough to produce an argillic horizon. Translocation of clay has been under acid conditions or under sodic alkaline conditions. Equally important is the absence of conditions that favor formation of Spodic and Mollic and other diagnostic horizons that would take precedence in the classification of the soil. Although the Alfisols are formed in many climatic zones, they are most extensive in humid and subhumid temperate regions on young land surfaces that have been stable and relatively free of pedoturbations and erosion for at least several thousand years. Yet the soils are young enough to retain notable supplies of primary minerals, layer lattice clays, and available plant nutrients. Vast acreages of present or former deciduous forest lands and even prairie lands on glacial drift, loess, and in deserts, on old alluvial fans are occupied by Alfisols. Their presence in intertropical landscapes is commonly on younger geomorphic surfaces than those of associated Ultisols and Oxisols. Alfisols include loamy sands, loams, and clays in which illuvial clay has accumulated in the Bt horizon to form argillans.

Gardner and Whiteside (1952) proposed a textural lithosequence of Alfisols in glacial drift in the state of Michigan (Fig. 21.1). In the state of New York, Cline (1949) observed a chronolithosequence from unleached to highly leached soils: Brown Forest soil (Udoll) → Gray-Brown Podzolic (Udalf) → Brown Podzolic and Podzol (Haplorthod) (Fig. 8.1). Very clayey

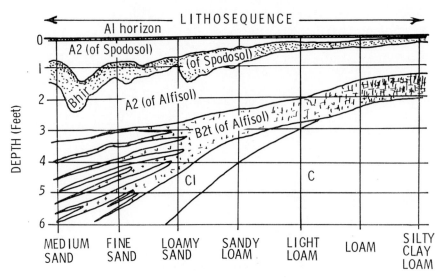

FIG. 21.1.   Lithosequence of bisequal soils observed in
Michigan. (After Gardner and Whiteside 1952.)

and reddish brown Bt horizons have formed over limestone and highly
calcareous sediments of glacial origin (Bartelli and Odell 1960a, b). Under
humid temperate climates, Alfisols may occupy most of the landscape, except on very steep slopes, alluvial floodplains, and very poorly drained
depressions. Under subhumid to arid conditions, Alfisols are commonly
confined to borders of depressions where slight concentrations of water
have favored sodium and clay migration to form the natric horizon. High
elevations or limited rainfall or concentrations of sodium in the parent
rock favor Alfisol formation in the tropics (Guerrero 1963).

Figure 15.1 shows that major bodies of Alfisols are in cool moist (A1,
A1s) and moderate moist (A2, A2s) regions, chiefly in the Northern Hemisphere; and in subhumid (A3, A3a) and Mediterranean (A4, A4s) regions
in all continents.

Alfisols develop on macrotopography that ranges from level to hilly.
They are frequently associated with soils of other orders where the patterns
of microclimate are related to aspect in landscapes of mature topography.
In eastern Ohio, a 90-cm (34-in.) solum of a Sol Brun Acide-Gray-Brown
Podzolic intergrading to Brown Podzolic soil (Alfic Ochrept) has developed
on south-facing slopes under a xeric oak *(Quercus)* plant community; and
on the north-facing slope under a mesophytic beech *(Fagus)*-maple *(Acer)*-
basswood *(Tilia)* forest a 1-m (40-in.) solum of a Brown Forest-Sol Brun
Acide soil (probably a Mollic Ochrept) is on upper slopes and a Gray-
Brown Podzolic-Sol Brun Acide (probably Alfic Ochrept) is present on the
footslopes (Finney, Holowaychuk, and Heddleson 1962).

Genesis of an Alfisol profile in eastern Iowa requires about 5,000 years
(Arnold and Riecken 1964) because of the slowness of the process of clay
accumulation to form the argillic horizon.

NATURE OF PEDOGENIC PROCESSES. Leaching of carbonates from the zone of the developing solum and braunification (see below) appear to be prerequisites to Alfisol development. The first process, by removing a powerful flocculant, namely, calcium carbonate (and bicarbonate), leaves the solum in a condition favorable to relatively free movement of plasma under the influence of percolating water. Soil reaction may be as acid as 4.5 in the Bt of Hapludalfs. Soil colloids move in an acid environment in these soils. In Natrustalfs the sodium ion is important in the dispersion and mobilization of clay. Movement of colloids by cheluviation does occur, as suggested by the presence of striking albic and argillic horizons in Eutroboralfs, which may be nearly neutral to the surface. The second process (braunification) releases a milder flocculant, iron, which fosters deposition of clay in the main B horizon, while carbonates of the C horizon flocculate colloids at the floor of the solum.

The O1 and O2 are dynamic horizons that exhibit an annual cycle of winter maximum accumulation and late summer minimum in some Hapludalfs under forest cover (Nielsen and Hole 1964) (Fig. 21.2). Under prairie vegetation the O horizon may become more bulky than in the forest, forming a mulch that is a barrier to evaporation of soil moisture and that in effect shortens the growing season by delaying warming of the soil in the spring.

Formation of an ochric epipedon (A1 horizon) has been virtually universal in these soils, whether presently under prairie or forest. Roots of plants are the chief agents in melanization in prairie lands. Where this process is not sufficient to form a mollic epipedon, Alfisols have formed under prairie cover. The trees deliver the bulk of their annual production of organic matter above ground, including the litter which is not as deeply incorporated as in grass-covered soils. In forest lands, additional work in the genesis of mull is done by earthworms (Nielsen and Hole 1964) and other fauna that mix leaf litter and humus with mineral soil to shallow depths (2 to 10 cm). Biocycling of nutrients from subsoil to O and A1 horizons is an important process in Udalfs. This is indicated by the presence of nearly neutral (pH 6.5 to 7.0) surface soil (A1) over quite acid (pH 4.8 to 5.8) subsoil. Residues of calcareous concretions from tissues of earthworms have been observed in mull of forested Alfisols. The changing

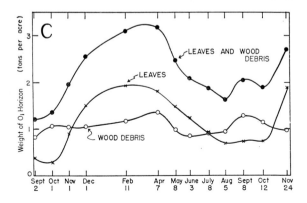

FIG. 21.2. Seasonal changes in dry weight of organic horizons of a Udalf under forest. (After Nielsen and Hole 1964.)

microtopography of the forest floor can be related to seasonal activities of animals, including moles and rodents. Bacterial gums are important in the formation of granular structure of ochric epipedons.

The effects of leaching and eluviation of clay are most obvious in the albic horizon, where melanization has been minimal. Platy structure is common in albic horizons of soils of the cooler regions and may result from expansion and contraction normal to the soil surface, in freeze-thaw and wetting-drying cycles, and from parallel orientation of planar silt particles. Movement of seasonally perched water table may be involved. Vesicular structure may result from degassing of water on freezing. Physical breakdown of fine sand and coarse silt may be most intense in the albic horizon (Arnaud and Whiteside 1964). Some albic horizons are relatively high in content of colorless or light-colored organic matter (Wilde 1950). Resistant minerals, such as quartz, are more concentrated in the A horizons, and the $SiO_2:R_2O_3$ ratio is higher than in the Bt (Allan and Hole 1968).

Tonguing of albic material down cracks in the Bt horizon is outstanding in Eutroboralfs and Glossudalfs (Gray Wooded soils) (Ranney and Beatty 1969). Grainy ped coatings observed in the upper part of the argillic horizon of degrading Udolls (Arnold and Riecken 1964) may represent a first stage in this process.

Presence of argillans showing optical orientation in thin section in the argillic horizon is considered proof of eluviation of clay from the A horizon and its illuviation into the B. More fine clay is moved than coarse clay. Mechanical movement of the clay by percolating water could account for these cutans, as Thorp, Strong, and Gamble (1957) have demonstrated, although movement of silica and alumina in solution, and later synthesis of them into clay in the Bt horizon, is not ruled out. In many Udalfs the bulk of the argillans occurs in the lower part of the solum (Buol and Hole 1959).

Deposition of clay accompanying sesquioxides and organic matter in the argillic horizon may be brought about by (1) depletion of percolating waters through soaking into peds, (2) swelling shut of voids and consequent slowing of percolating water, (3) sieve action of clogging fine pores, and (4) flocculation of the negatively charged clay by positively charged iron oxides in the Bt and by higher base saturation in the lower solum. Conditions for significant formation of argillans in Alfisols may be relatively rare. Intensely rainy periods following prolonged droughts may be especially favorable occasions.

It is possible that pedoturbation by clay swelling (Nettleton, Flach, and Brasher 1969) frost action, root growth, and faunal activity may have been instrumental in shattering argillans in upper subhorizons of the argillic horizon. Some argillans extend for many meters down cracks and joints in the substratum and have escaped fragmentation (Gooding, Thorp, and Gamble 1959). Some papules in upper B horizons may indeed be argillan fragments.

The activity of cicadas (Hugie and Passey 1963) has produced a special kind of "cylindrical" blocky structure in Bt horizons of certain Ustalfs and has very probably comminuted any previously existing argillans.

Floralpedoturbation is probably unimportant in the genesis of Alfisols. Tree-tip mounds are scarce in originally forested Alfisols as compared with Spodosols, probably because rooting is commonly deeper in the Alfisols.

Prismatic structure is most strikingly developed in the B horizons of Ustolls and Xerolls (Chestnut, Brown, Sierozem, and Solod soils) in which opening and closing of cracks during moist-dry cycles have been confined to the subsoil and have not led to pedoturbation.

In some argillic horizons the clay is largely inherited from initial material or is weathered in situ. Borchardt, Hole, and Jackson (1968) found evidence that about 20 cm (8 in.) of loess (about half clay and half silt) had been incorporated into the Bt horizon of a common Hapludalf in southeastern Wisconsin.

Where impeded drainage brings about seasonal reducing conditions, braunification and gleization apparently are coextensive in the zone of seasonal water table fluctuation, resulting in a mottled brown and gray pattern. Small iron concretions are frequently present. Brown and Thorp (1942) reported distributions of these features in the Miami catena and family (see Fig. 21.3). European pedologists report an evolutionary sequence of soils from Parabraunerde (Udalfs) to Pseudogleys and Stagnogleys as a clay pan forms, and mottling develops.

In summary, the following processes may contribute to the formation of Alfisols:

1. Concentration of clay in the Bt by mere removal of carbonates from the initial material of that horizon.

2. Eluviation of initial (and progressively added eolian) clay from the A horizon.

3. Illuviation of clay into the Bt horizon.

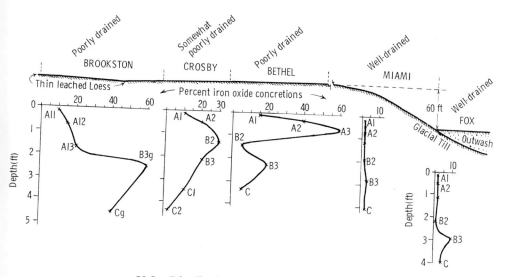

FIG. 21.3. Distribution of iron concretions in sand and fine gravel fractions in a Miami catena. (After Brown and Thorp 1942.)

4. Formation of clay in the B by weathering of feldspars, micas, and ferromagnesian minerals.

5. Differential loss by leaching of materials, so that the A is more depleted, particularly in clay, than is the Bt, but both are depleted.

6. Initial arrangement of a coarse material over a fine material, producing a pseudoprofile resembling that of a bona fide Alfisol.

USES. Alfisols are used for cultivated crops (with relatively short season), winter (hardy) hayland, pasture, range, and forest. The relatively high

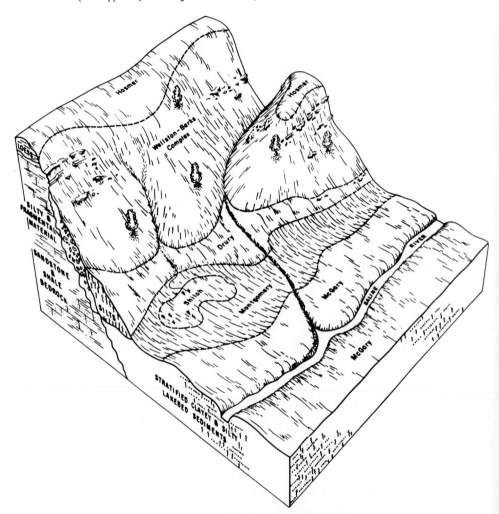

FIG. 21.4. Soilscape pattern of a Typic Fragiudalf (Hosmer) and Aeric Orchraqualf (McGary) with Ultic Hapludalf (Wellston), Typic Dystrochrept (Berks), Dystric Eutrochrept (Drury), Cumulic Haplaquoll (Shiloh), and Typic Haplaquoll (Montgomery) in Gallatin County in southeastern Illinois. (After Wallace and Fehrenbacher 1969.)

base saturation of the pedon and the presence of notable reserves of plant nutrients in the more highly base-saturated C horizon indicate the outstanding fertility of these soils. They support the eastern portion of the Corn Belt (in association with Aquolls; Fig. 21.4) and the western part of the wheat-producing areas of North America. Arable soils of Mediterranean climatic zones are to a considerable extent Alfisols. Scenery is a principal feature of some areas of Xeralfs.

Accelerated erosion presents a serious hazard to future productivity of Alfisols. The argillic horizon, being relatively high in clay content, is not a very desirable medium for seed germination and plant development. Infiltration may be decreased on an area from which ochric and albic horizons have been lost by erosion. Accelerated erosion not only fosters a droughty condition at the eroded site but also increases flood hazard in lower portions of the adjacent landscape.

The abundance of relatively unweathered clay minerals provides Alfisols with a high cation exchange capacity. Aluminum toxicity is not a problem. However, liming is needed for legumes. Potassium and ammonia fixation is an important consideration in Alfisols that have notable contents of slightly weathered illite components of the clay fraction.

CLASSIFICATION. Alfisols have either an argillic or natric horizon that is not under a spodic or oxic horizon. No mollic epipedon is present. The base saturation, by sum of cations, is more than 35% at a depth of 1.25 m (50 in.) below the top of the argillic horizon or 1.8 m (72 in.) below the surface. Where a lithic or paralithic contact is present within one of the aforementioned depths, a base saturation of more than 35% is required at that depth. A fragipan is permitted in the Alfisols only when clay skins (argillans) greater than 1 mm in thickness are present in the fragipan. Mean annual soil temperature of Alfisols is usually > 8 C (47 F). Below this temperature the base saturation criterion is waived and all soils meeting the other criteria for Alfisols are considered such, regardless of low base status.

Alfisols are divided into five suborders (Fig. 21.5) which are briefly defined as follows:

1. Aqualfs (Fig. 24.3) are seasonally saturated, or if artificially drained, display evidence of former wetness such as mottles, low chromas, and Fe-Mn concretions 2 mm or more in diameter. These soils are found in a wide range of climatic conditions and exhibit a variety of soil horizons and features.

2. Boralfs are not as wet as Aqualfs and are cool (mean annual temperature is < 8 C (47 F) and commonly exhibit an albic horizon that tongues into the argillic or natric horizon. Mean soil temperatures at 50 cm differ by > 5 C (9 F) between winter and summer.

3. Udalfs (Figs. 22.3, 24.3) are not as wet as Aqualfs nor as cool or glossic as Boralfs.

4. Ustalfs are annually dry (for less than 60 consecutive days) and are not as cool nor as glossic as Boralfs, and commonly have a carbonate ac-

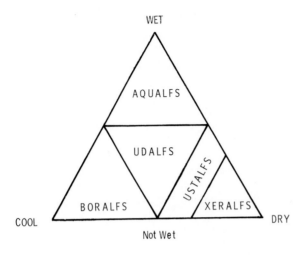

FIG. 21.5. Diagram showing some relationships between suborders of Alfisols.

cumulation at the base of the solum. Mean soil temperature at 50 cm differs by 5 C (9 F) or more between winter and summer.

5. Xeralfs are annually dry (for more than 60 consecutive days) and are not as cool nor as glossic as Boralfs. The mean soil temperature at 50 cm differs by > 5 C (9 F) between winter and summer.

The suborders are divided into 33 great groups (Table 21.1) on the basis of a number of factors: mean summer soil temperature (at 50 cm) and range of soil temperatures between summer and winter; tonguing of A2 into Bt; abruptness of A2/Bt transition; thickness of Bt; reddish hue of Bt; presence or absence of fragipan, duripan, natric, or petrocalcic horizon; presence or absence of iron nodules or plinthite; and percent base saturation in the Bt.

Subgroups in the Alfisol order reflect the relationship of these soils with several other orders that border the modal Alfisol ecological niche. Commonest of these are the intergrades to the Mollisol order: Mollic, Natraborollic, Lithic-Mollic, Udollic, etc. Also, intergradations to the Ultisol, Vertisol, Entisol, Inceptisol, and Aridisol are noted.

Drainage differences commonly noted in a drainage catena of Alfisols are reflected in the suborder and subgroup classification. For example, a well-drained soil would be the Typic Hapludalf; the moderately well-drained soil, an Aquic Hapludalf; the imperfectly or somewhat poorly drained soil is an Aeric Ochraqualf, and the poorly drained soil is a Typic Ochraqualf. A very poorly drained soil associated with the above catena would be a Typic Umbraqualf.

TABLE 21.1 ❧ **Suborders and great groups in the Alfisol order**

Suborder	Great Group
Aqualfs:	Natraqualfs—presence of a natric horizon
	Tropaqualfs—MAT* >8 C (47 F) and summer and winter temperatures differ less than 5 C (9 F) at 50 cm
	Fragiaqualfs—presence of a fragipan
	Glossaqualfs—albic horizon tonguing into the argillic horizon and no duripan present
	Albaqualfs—abrupt textural change from albic to argillic horizon
	Ochraqualfs—presence of an ochric epipedon
	Umbraqualf—presence of an umbric epipedon
Boralfs:	Paleboralfs—upper boundary of argillic horizon deeper than 60 cm (24 in.) and textures finer than loamy fine sand in some layer above the argillic horizon
	Fragiboralfs—presence of a fragipan
	Natriboralfs—presence of a natric horizon
	Cryoboralfs—mean summer temperature at 50 cm (20 in.) or shallower lithic or paralithic contact of less than 15 C (59 F) without an O horizon [<8 C (47 F) MST† with an O horizon]
	Eutroboralfs—base saturation ≥60% in all parts of argillic horizon and dry in some horizon part of each year
	Glossoboralfs—either never dry or have a base saturation <60% in some part of the argillic horizon
Ustalfs:	Plinthustalfs—plinthite present within 1.25 m (50 in.) of the surface
	Durustalfs—duripan present below argillic or natric horizon but within 1 m (40 in.) of the surface
	Natrustalfs—presence of a natric horizon
	Paleustalfs—presence of a petrocalcic horizon within 1.5 m (60 in.) of the surface or a thick dense argillic horizon
	Rhodustalfs—argillic horizon color redder than 5YR
	Haplustalfs—other Ustalfs
Xeralfs:	Plinthoxeralfs—presence of plinthite within 1.25 m (50 in.) of the surface
	Durixeralfs—presence of a duripan within 1 m (40 in.) of the surface
	Natrixeralfs—presence of a natric horizon
	Rhodoxeralfs—argillic horizons redder than 5YR
	Palexeralfs—solum thicker than 1.5 m (60 in.) or presence of a petrocalcic layer within 1.5 m (60 in.) of the surface
	Haploxeralfs—other Xeralfs
Udalfs:	Agrudalfs—presence of an agric horizon
	Fragiudalfs—presence of a fragipan
	Natrudalfs—presence of a natric horizon
	Tropudalfs—mean summer and mean winter temperatures at 50 cm (20 in.) or lithic or paralithic contact, if shallower, differ by less than 5 C (41 F)
	Ferrudalfs—presence of a discontinuous albic horizon and discrete iron nodules 2.5 to 5 mm in diameter in the argillic horizon
	Glossudalfs—presence of an albic horizon which tongues into the argillic horizon
	Paleudalfs—sola deeper than 1.5 m (60 in.) have a clay distribution that decreases less than 20% of its maximum within the 1.5 m (60 in.) depth
	Hapludalfs—other Udalfs

* MAT = mean annual temperature.
† MST = mean summer temperature.

LITERATURE CITED

Allan, R. J., and F. D. Hole. 1968. Clay accumulation in some Hapludalfs as related to calcareous till and incorporated loess on drumlins in Wisconsin. Soil Sci. Soc. Am. Proc. 32:403–8.

Arnaud, R. J. St., and E. P. Whiteside. 1964. Morphology and genesis of a Chernozemic to Podzolic sequence of soil profiles in Saskatchewan. Can. J. Soil Sci. 44:88–99.

Arnold, R. W., and F. F. Riecken. 1964. Grainy gray ped coatings in some Brunizem soils. Proc. Iowa Acad. Sci. 71:350–60.

Bartelli, L. J., and R. T. Odell. 1960a. Field studies of a clay-enriched horizon in the lowest part of the solum of some Brunizem and Gray-Brown Podzolic soils in Illinois. Soil Sci. Soc. Am. Proc. 24:388–90.

———. 1960b. Laboratory studies and genesis of a clay-enriched horizon in the lowest part of the solum of some Brunizem and Gray-Brown Podzolic soils in Illinois. Soil Sci. Soc. Am. Proc. 24:390–95.

Borchardt, G. A., F. D. Hole, and M. L. Jackson. 1968. Genesis of layer silicates in representative soils in a glacial landscape of southeastern Wisconsin. Soil Sci. Soc. Am. Proc. 32:399–403.

Brown, I. C., and J. Thorp. 1942. Morphology and composition of some soils of the Miami family and the Miami catena. U.S. Dept. Agr. Tech. Bull. 834.

Buol, S. W., and F. D. Hole. 1959. Some characteristics of clay skins on peds in the B horizon of a Gray-Brown Podzolic soil. Soil Sci. Soc. Am. Proc. 23: 239–41.

Cline, M. G. 1949. Profile studies of normal soils of New York. I. Soil profile sequences involving Brown Forest, Gray-Brown Podzolic, and Brown Podzolic soils. Soil Sci. 68:259–72.

Finney, H. R., N. Holowaychuk, and M. R. Heddleson. 1962. The influence of microclimate on the morphology of certain soils of the Allegheny Plateau of Ohio. Soil Sci. Soc. Am. Proc. 26:287–92.

Gardner, D. R., and E. P. Whiteside. 1952. Zonal soils in the transition region between the Podzol and Gray-Brown Podzolic regions in Michigan. Soil Sci. Soc. Am. Proc. 16:137–41.

Gooding, A. M., J. Thorp, and E. Gamble. 1959. Leached, clay-enriched zones in post-Sangamon drift in southwestern Ohio and southeastern Indiana. Geol. Soc. Am. Bull. 70:921–25.

Guerrero, R. 1963. Soils of Colombia and their relation to the new classification system of the United States. M.S. thesis. Cornell Univ., Ithaca, N.Y.

Hugie, V. K., and H. B. Passey. 1963. Cicadas and their effect upon soil genesis in certain soils in southern Idaho, northern Utah, and northeastern Nevada. Soil Sci. Soc. Am. Proc. 27:78–83.

Nettleton, W. D., K. W. Flach, and B. R. Brasher. 1969. Argillic horizons without clay skins. Soil Sci. Soc. Am. Proc. 33:121–25.

Nielsen, G. A., and F. D. Hole. 1964. Earthworms and the development of coprogenous Al horizons in forest soils of Wisconsin. Soil Sci. Soc. Am. Proc. 28:426–30.

Ranney, R. W., and M. T. Beatty. 1969. Clay translocation and albic tongue formation in two Glossoboralfs of west-central Wisconsin. Soil Sci. Soc. Am. Proc. 33:768–75.

Thorp, J., L. E. Strong, and E. Gamble. 1957. Experiments in soil genesis—the role of leaching. Soil Sci. Soc. Am. Proc. 21:99–102.

Wallace, D. L., and J. B. Fehrenbacher. 1969. Soil survey of Gallatin County, Illinois. Soil Conserv. Serv., U.S. Dept. Agr. U.S. Govt. Printing Office, Washington.

Wilde, S. A. 1950. Cryto-mull humus: Its properties and growth effects (a contribution to the classification of forest humus). Soil Sci. Soc. Am. Proc. 15: 360–62.

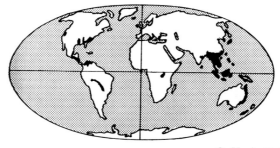

Ultisols: Low Base Status Forest Soils   ❧

In MANY PARTS of the world, between the trop-
ics and the more temperate latitudes, are extensive areas of hardwood-pine
forests. Geologically, these are equatorward from the limit of advances of
the more recent glaciations. Thus, many of the parent materials and landscapes are geologically old compared to glaciated areas. Where relatively
high amounts of precipitation have fallen on these geologically old areas,
deep and weathered soils have developed.

Active forces of soil formation over the long periods of time have
served to deepen soil profiles while leaching and weathering the minerals
present (Fig. 22.1). Krebs and Tedrow (1958) have pointed out that a
significant soil boundary exists at the end of the glacial material in New
Jersey. Older, more weathered soils (Ultisols) are present to the south of
the glacial drift, while to the north, on the glacial drift, soils have higher
base saturation and are considered in the Alfisol order.

Associated with a warm humid climate and old terrain, these soils often
take on a redder color than the soils of the cooler regions in the temperate
zone. A common designation of Ultisol areas in the United States is "red
clay hills."

These soils are not confined to temperate latitudes. Similar soils extend
into tropical areas in both forested and nonforested areas. Ahmad and
Jones (1969a, b) have reported Ultisols from savannas in northern Trinidad.

Several soil associations are present within this zone. On areas of extremely sandy parent material, Spodosols (Podzols) have formed. On
steeper slopes, areas of Inceptisols, especially Dystrochrepts, are present,
and in the recent floodplains, Udifluvents are present (Fig. 22.2). In tropical areas the Ultisols are commonly on the younger surfaces, whereas associated Oxisols occupy more stable landscape positions. In areas where the
Ultisols are associated with Alfisols, the Ultisols are on the more stable
landscapes and the Alfisols on the sideslopes (Fig. 22.3).

No satisfactory data are available concerning the global acreage of
Ultisols. This is in part due to the recent redefinition of the order and a
lack of identification of soils of this order in tropical areas. Ultisols have

been reported as Latosols, Laterites, etc., because of their red color and location in intertropical regions.

In areas of low relief, drainage catenas are present in much the same fashion as in Alfisol areas. Gley conditions form in depressions. The edges of the depressions tend to be favorable sites for fragipan and plinthite formation (Daniels et al. 1966). Intrenched drainage systems are bordered by well-drained soil (Fig. 15.2). Gley conditions dominate on the broad, flat interfluves (Daniels and Gamble 1967). In rolling piedmont areas, only a small percentage of the landscape is occupied by soils with gleyed conditions (Kaster 1960).

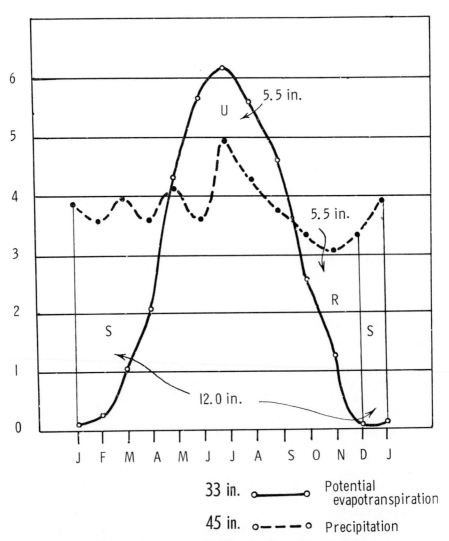

FIG. 22.1. Average water budget of soils at Winston-Salem, North Carolina.

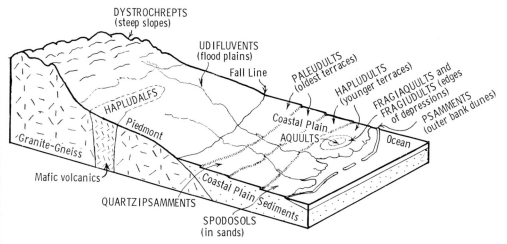

FIG. 22.2. Idealized block diagram showing distribution of Ultisols in a portion of the Carolinas.

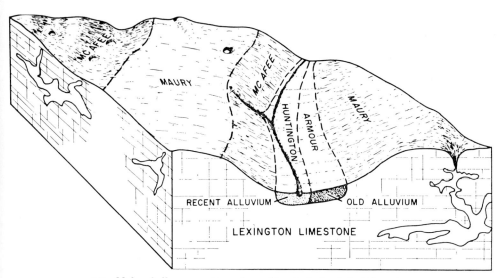

FIG. 22.3. Soilscape pattern for a Humic Paleudult (Maury) and a Humic Hapludult (Armour), with Mollic Hapludalf (McAfee) and Fluventic Hapludoll (Huntington) in Fayette County in northeast central Kentucky. (After Sims et al. 1968.)

PROCESSES OF ULTISOL FORMATION. Several individual reactions and processes (Chapter 6) are involved in the formation of Ultisols. McCaleb (1959) discussed the formation of the Red-Yellow Podzolic soils which are presently recognized as Ultisols. Extensive leaching is prerequisite and it goes on for a substantial part of each year (Fig. 22.1). Temperature of the soil [MAST > 8 C (47 F)] is a diagnostic criterion of the Ultisol order. When temperatures of this magnitude and extensive leaching are combined over a long period of time, the result is rapid and fairly complete alteration of weatherable minerals into secondary clays and oxides. A clay mineral suite results that is stable under the ambient environment. The suite is dominated by kaolinite, associated with gibbsite and chlorite-vermiculite intergradational clays (Southern Regional Project S-14, 1959). Minor amounts of mica are commonly present, but these are probably remnants of incompletely weathered primary minerals and tend to be more prominent in the coarse clay and silt fractions than in the finer clay fractions.

Lessivage, leading to the formation of albic and argillic horizons, is very pronounced. Some of the clay accumulation in the argillic horizon is in the form of argillans. The clay in the argillans appears to be poorly crystalline kaolinite derived from the A2 horizon (Khalifa and Buol 1968). McCaleb (1959) postulated that argillan development progressed upward in the profile and that the position of the lower boundary of the B horizon was limited by the supply of weatherable mineral in the overlying A horizon. Later evidence indicates that the upper part of the B horizon, as well as the A2 horizon, serves as a site for clay release in the lessivage process. This suggests that the argillans of the upper B horizon are being destroyed and removed to lower subhorizons. In this way, the A2 (albic) is actually encroaching on the B (argillic) horizon (Daniels and Gamble 1967). Less well-developed albic horizons are observed to develop in soils formed from saprolite from basic parent rocks, such as diorite gneiss and hornblende schist, than in soils formed from saprolite from granite (England and Perkins 1959).

Inability to reconstruct enough A horizon to account for the massive buildup of clay in the argillic horizon induced Simonson (1949) to discount lessivage in Ultisols and place more emphasis on clay formation in situ in the B horizon. Clay formation as a result of in situ weathering probably is significant in Ultisols.

Also coextensive with lessivage is podzolization in these profiles. The albic horizon is low in content of free iron oxide. The reverse is true of the argillic horizon. The free iron oxide:clay ratio is usually the same in A2 and B horizons. This led Simonson to discount "pure" podzolization in Ultisols. The fact remains that iron has moved from the albic to the argillic horizon. In some cases, a thick sandy albic horizon is the locus of formation of a Spodosol. The resulting profile is bisequal (see Chapter 20).

Although the extensive leaching of the Ultisols has lead to a severe removal of bases, their concentration decreases with depth. This suggests that biocycling successfully counters the leaching process.

The surface soil of the well-drained Ultisols is light colored (ochric epipedon). There is usually a slight darkening of the upper 10 cm or so of most Ultisols through melanization. This process is counteracted by rapid mineralization in well-drained Ultisols. Relatively high organic matter contents are commonly observed in the poorly drained members of the Ultisol order, namely the Humults and Umbraquults. When a dark surface develops under natural conditions, its base saturation is normally less than 50%, and it is classified as an umbric epipedon. Many of these umbric epipedons have been limed and are now classified as mollic epipedons. Mollic epipedons, created by liming, are allowed in the Ultisol order if the underlying material has a sufficiently low base status. The diagnostic depth for determining base saturation in Ultisols has been set at 125 cm (50 in.) below the top of the argillic horizon or 180 cm (72 in.) below the surface, whichever is shallower, providing there are no intervening lithic or paralithic contacts. The base saturation at this level must be less than 35% by sum of cations method. These criteria were selected to reflect the extensive leaching of the Ultisols and also placed deep enough to negate changes in classification due to normal management practices.

Two other features common to, but not definitive for, Ultisols are plinthite and fragipans. Plinthite appears to form in the subsoils of Ultisols developed in the most stable and hence oldest parts of the landscape. A precursor of the plinthite seems to be bright red mottles, commonly in a reticulate pattern in the soil. Upon repeated wetting and drying, some of these red mottles harden irreversibly. When such features, in an indurated state, are found on the surface or exposed in a cut bank, they are commonly called "laterite." But not all red mottles in the soil will harden into indurated plinthite material. Numerous observations of plinthite in Ultisols indicate that it starts at a depth in the soil that is subjected to a seasonal fluctuation of the water table. Although the incipient plinthite is observed in many Ultisols, only in those cases where the plinthite acts to impede drainage is it recognized in the taxonomic system. This has been defined to be when about 10 to 15% of the volume of a soil horizon is occupied by indurated plinthite or red, iron-cemented material capable of becoming indurated upon repeated wetting and drying.

The fragipan is a feature found in many Ultisols, especially those with some indication of poor drainage. Fragipans, like plinthite layers, act to restrict water movement in the soil. In Ultisols, fragipans have often been confused with plinthite layers where gray mottles occur in the zone of reticulate red plinthitelike mottles. Fragipans are also observed to occur in the absence of plinthite, in which case they are gray in color and create a perched water table. The occurrence of fragipans in Ultisols has been described by several authors (Daniels et al. 1966; Nettleton, McCracken, and Daniels 1968; Porter et al. 1963; Soil Survey Staff 1960; Steele et al. 1969), but the genesis of fragipans remains obscure.

USE OF ULTISOLS. Ultisols represent a vast potential for agricultural production. They develop in climates that have long frost-free seasons and an abundance of rainfall. Where rainfall is inadequate for short periods of time during the growing season, irrigation is feasible because of good supplies of water at shallow depth. The water is usually of good quality. Since small quantities are required, cost of irrigation is relatively low.

Historically, Ultisols have invited agricultural development. They usually produced good crops for the first few years, or about the time it took for the nutrient reserve in the biocycled organic matter to decompose and be taken up by the crop plant or be leached from the profile. At that point, either the farmers moved to another location or restored nutrients to the soils by the use of manure or mineral fertilizer. Intensive cultivation on the steeper parts of such soils led to erosion of the A horizons. The B horizon, which is usually quite clayey, has been exposed widely and performs badly as a medium for water infiltration and plant growth. This has led to the abandonment of many areas of Ultisols.

The low fertility and low base status of Ultisols has been, and in many areas continues to be, the major limitation to agricultural use. These limitations can be overcome by modern agricultural practices of liming and fertilization. It is necessary, however, to have adequate quantities of lime, fertilizer, and management talent available when cropping these soils.

Timber production is a valuable asset of the Ultisols. The vegetation of deciduous and conifer forests has been managed and replanted to make forestry operations profitable. Because of the dependence of these soils on nutrient recycling by deep-rooted plants for maintenance of fertility in the surface soil, cutting of native forests often leads to rapid degradation in soil fertility. It has been postulated that vast areas of savanna in the tropics have been formed by deforestation of Ultisols. Once the nutrient recycling from the trees was interrupted, the bases leached beyond the rooting depth of grasses and shrubs.

CLASSIFICATION OF ULTISOLS. By definition, Ultisols are confined to areas where the mean annual soil temperature is 8 C (47 F) or warmer; and if the mean summer and mean winter temperatures at 50 cm (20 in.) differ by more than 5 C (9 F), the mean summer soil temperature is 15 C (59 F) or warmer if with an O horizon or 8 C (47 F) or warmer if without an O horizon. Ultisols are acid, low base status soils. Their base saturation (by sum of cation method) is less than 35% at 1.25 m (50 in.) below the upper boundary of the argillic horizon or 1.8 m (72 in.) below the surface, or at a lithic or paralithic contact whichever is shallower.

The Ultisol order is subdivided into five suborders, using criteria of profile wetness and organic matter content (Fig. 22.4).

Aquults are either saturated with water at some period of the year or are artificially drained. Since it is not practical to observe a soil throughout the year in order to classify it, certain other morphological features associated with wetness are used as criteria for defining the aquults. The criteria used are the presence of mottles, iron-manganese concretions > 2 mm in diameter, or a soil color of 2 or less chroma immediately below any Ap or A1 horizon.

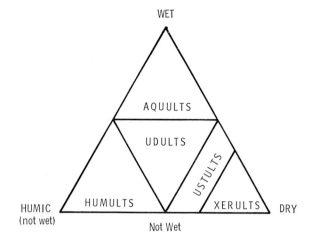

FIG. 22.4. Diagram showing some relationships between suborders of Ultisols.

Ustults are Ultisols that are dry for 90 or more cumulative days in some layer between 18 cm and 50 cm (10 and 20 in.) in most years. This dry period does not exceed 60 consecutive days unless the mean annual temperature is greater than 22 C (72 F) or the mean summer and mean winter temperatures differ by less than 5 C (9 F). These soils are low in organic matter content, having less than 1.5% in the upper 15 cm (6 in.) of the argillic horizon, after the Ap horizon has been excluded, and contain less than 20 kg of organic matter in a unit volume of 1 cubic meter from below the O horizon to a depth of 1 meter.

Xerults are the soils of the Ultisol order in areas of prolonged dry seasons and mean annual temperatures colder than 22 C (72 F). Organic matter contents are low (same as Ustults), and these soils are dry for 60 consecutive days or more in all parts of the soil between a depth of 18 and 50 cm (7 and 20 in.) or lithic or paralithic contact, whichever is shallower, in more than 7 out of 10 years.

Humults are the Ultisols that have high organic matter contents but do not have other characteristics of wetness. By definition they contain more than 1.5% organic matter in the upper 15 cm (6 in.) of the argillic horizon or have over 20 kg of organic matter in a cubic meter of the upper 1 meter of the pedon, excluding an O horizon.

Udults are Ultisols of the humid regions where dry periods are of short duration, organic contents are low but the water table remains below the solum most of the year, and the gray, mottled colors of the Aquults do not appear directly below the A horizon. The great groups of the Ultisol order are listed in Table 22.1.

Ultisols have most often been classified as Red-Yellow Podzolic or Reddish Brown Lateritic. The one chief characteristic of a base status less than 35% at a depth of 1.25 meters used to separate Ultisols from Alfisols is at best an arbitrary limit of doubtful genetic or practical significance. Addition of agricultural lime soon changes the base saturation percentage.

Nevertheless, some kind of distinction needs to be made between a soil of low base status (10 to 30%) and one of high base saturation (50 to 70%).

TABLE 22.1   &   **Suborders and great groups in the Ultisol order**

Suborder	Great Group
Aquults:	Plinthaquults—plinthite forms over one-half of a horizon within 1.25 m (50 in.) of the surface
	Fragiaquults—presence of a fragipan
	Tropaquults—mean summer and mean winter temperatures differ less than 5 C (9 F)
	Paleaquults—less than 10% weatherable minerals in the upper 1 m (40 in.) and a clay distribution such that the clay content does not decrease more than 20% within 1.5 m (60 in.) of the surface
	Ochraquults—presence of an ochric epipedon
	Umbraquults—presence of an umbric or mollic epipedon
Ustults:	Plinthustults—plinthite layer within 1.25 m (50 in.) of the surface
	Paleustults—less than 10% weatherable minerals in the 20–200 μ separate within 1 m (40 in.) and a clay distribution with less than 20% decrease within 1.5 m (60 in.) of the surface
	Rhodustults—moist color value less than 4 in all parts and argillic horizon color values less than 5
	Tropustults—mean summer and mean winter temperatures differ less than 5 C (9 F)
	Haplustults—all other Ustults
Xerults:	Paleudults—less than 10% weatherable minerals in the 20–200 μ separate of the upper 1 m (40 in.) and less than 20% clay content decrease within 1.5 m (40 in.) of the surface
	Haploxerults—other Xerults
Humults:	Palehumults—less than 10% weatherable minerals in the 20–200 μ separate of the upper 1 m (40 in.) and less than 20% decrease in clay content to a depth of 1.5 m (60 in.)
	Tropohumults—mean summer and mean winter temperatures differ by less than 5 C (9 F)
	Haplohumults—other Humults
Udults:	Fragiudults—presence of a fragipan
	Plinthudults—plinthite layer within 1.25 m (50 in.) of the surface
	Paleudults—less than 10% weatherable minerals in the 20–200 μ separate of the upper 1 m (40 in.) and less than 20% decrease in clay content to a depth of 1.5 m (60 in.)
	Rhodudults—moist color value of epipedon less than 4 and dry argillic horizon color values less than 5
	Tropudults—mean summer and winter temperatures differ by less than 5 C (9 F)
	Hapludults—all other Udults

LITERATURE CITED

Ahmad, N., and R. L. Jones. 1969a. A Plinthaquult of the Aripo Savannas, North Trinidad. I. Properties of the soil and chemical composition of the natural vegetation. Soil Sci. Soc. Am. Proc. 33:762–65.
———. 1969b. A Plinthaquult of the Aripo Savannas, North Trinidad. II. Mineralogy and genesis. Soil Sci. Soc. Am. Proc. 33:765–68.
Daniels, R. B., and E. E. Gamble. 1967. The edge effect in some Ultisols in the North Carolina coastal plain. Geoderma 1:117–24.

Daniels, R. B., W. D. Nettleton, R. J. McCracken, and E. E. Gamble. 1966. Morphology of soils with fragipans in parts of Wilson County, North Carolina. Soil Sci. Soc. Am. Proc. 30:376–80.

England, C. B., and H. F. Perkins. 1959. Characteristics of three Reddish Brown Lateritic soils of Georgia. Soil Sci. 88:294–302.

Kaster, D. L. 1960. Soil survey of Alamance County, North Carolina. Series 1956, No. 9. U.S. Dept. Agr. U.S. Govt. Printing Office, Washington.

Khalifa, E. M., and S. W. Buol. 1968. Studies of clay skins in a Cecil (Typic Hapludult) soil. I. Composition and genesis. Soil Sci. Soc. Am. Proc. 32:857–61.

Krebs, R. D., and J. C. F. Tedrow. 1958. Genesis of Red-Yellow Podzolic and related soils in New Jersey. Soil Sci. 85:28–37.

McCaleb, S. B. 1959. The genesis of the Red-Yellow Podzolic soils. Soil Sci. Soc. Am. Proc. 23:164–68.

Nettleton, W. D., R. J. McCracken, and R. B. Daniels. 1968. Two North Carolina coastal plain catenas. II. Micromorphology, composition, and fragipan genesis. Soil Sci. Soc. Am. Proc. 32:582–87.

Porter, H. C., J. F. Derting, J. H. Elder, E. F. Henry, and R. F. Pendleton. 1963. Soil survey of Fairfax County, Virginia. Series 1955, No. 11. U.S. Dept. Agr. U.S. Govt. Printing Office, Washington.

Simonson, R. W. 1949. Genesis and classification of Red-Yellow Podzolic soils. Soil Sci. Soc. Am. Proc. 14:316–19.

Sims, R. P., D. G. Preston, A. J. Richardson, J. H. Newton, D. Isgrig, and R. L. Blevins. 1968. Soil survey of Fayette County, Kentucky. Soil Conserv. Serv., U.S. Dept. Agr. U.S. Govt. Printing Office, Washington.

Soil Survey Staff. 1960. Soil classification, a comprehensive system—7th approximation. U.S. Dept. Agr. U.S. Govt. Printing Office, Washington.

Southern Regional Project S-14. 1959. Certain properties of selected southeastern United States soils and mineralogical procedures for their study. Southern Regional Bull. 61. Va. Agr. Exp. Sta., Blacksburg.

Steele, F., R. B Daniels, E. E. Gamble, and L. A. Nelson. 1969. Fragipan horizons and Be masses in the middle coastal plain of north central North Carolina. Soil Sci. Soc. Am. Proc. 33:752–55.

✥

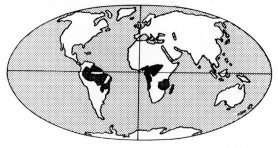

Oxisols: Sesquioxide-Rich, Highly
Weathered Soils of the Intertropical Regions ❧

T HESE are mineral soils with an oxic horizon within 2 m (80 in.) of the surface or plinthite that forms a continuous phase within 30 cm (12 in.) of the mineral surface of the soil, and with no spodic or argillic horizon overlying the oxic horizon (Soil Survey Staff 1967). They include most of the soils previously called Laterites and Groundwater Laterites, and a majority of the soils formerly designated as Latosols (Baldwin, Kellogg, and Thorp 1938; Thorp and Baldwin 1940; Thorp and Smith 1949). There are two misconceptions or possible sources of confusion:

1. Not all soils previously classed as Latosols or Lateritic (Reddish Brown Lateritic, Yellowish Brown Lateritic, etc.) can be placed in the Oxisol order. A relatively large number of the ones previously classed as Lateritic, with a qualifying or descriptive adjective, do not possess an oxic horizon and therefore must be classed with the Ultisols, Alfisols, and even Inceptisols in a few instances. The same is true, but to a much lesser extent, in the case of soils previously called Latosols. The requirements of a high degree of weathering, low cation exchange capacity, and low water-dispersible clay content in order for a soil horizon or layer to qualify as an oxic horizon eliminate from the Oxisol order many "red tropical" or subtropical soils.

2. Another important point is that not all soils between the Tropic of Cancer and the Tropic of Capricorn (the intertropical area) are Oxisols. On the contrary, Oxisols comprise only a small percentage of the soils of the intertropical areas. With the present requirements of the oxic horizon, true Oxisols are rare to absent in many tropical countries. This is especially true in Central and South America, where large areas of old landscape surfaces at low elevations and/or ferromagnesian (basic) rock initial materials are not common. These are two main prerequisites for Oxisol formation in addition to year-long warm temperatures and significant rainfall, as will be discussed later. Indeed, a rather common soil geographic situation in many intertropical regions is to find a few small islands or patches of Oxisols preserved on old fluvial terraces, pediments, or high-

lying old erosion surfaces, with the sideslopes occupied by Ultisols (Alfisols, if a strong dry period exists), Inceptisols on the nose slopes and backslopes, plus any fairly recent depositional surfaces (volcanic ash-covered, for example), with Entisols on the floodplains and the steepest, youngest hillslopes (Fig. 23.1). Vertisols, too, are important components of the smoother or depressional landscapes in these areas (where calcium, magnesium, and silica concentrations of the soil and saprolite solution are such that formation and preservation of montmorillonite is promoted and an alternating wet-dry moisture regime promotes shrink-swell). Oxisols do occupy significant parts of the landscape (and contribute to agriculture) in a few parts of the intertropical regions, but one should avoid or get rid of the common misconception that Oxisols are everywhere present and of large areal extent in the tropics. Furthermore, in some places where Oxisols do occur, they are paleosols, relict from a previous, more moist climate.

SETTING. Oxisols are confined to the intertropical regions (the few of which we are aware in the temperate zones are fossil, visible only in road cuts, cut banks, or in small outcrop spots in sideslopes). Generally, Oxisol distribution tends to be somewhat independent of present rainfall patterns (Soil Survey Staff 1967); this is especially true of great groups of Oxisols

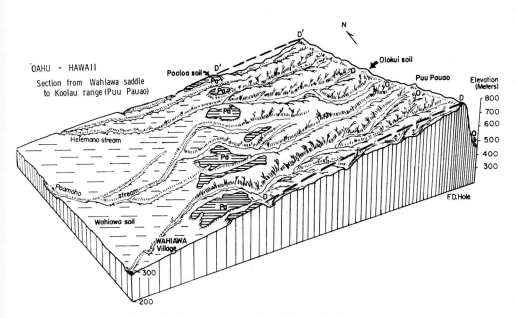

FIG. 23.1. Soilscape pattern of a Tropeptic Eutrustox (Wahiawa soil), with a Humoxic Tropohumult (Paaloa silty clay, Pa) and a Typic Placaquept (Olokui silty clay, O), on the Island of Oahu, Hawaii. (After Jackson et al. 1971.) D-D and D'-D' represent the former extent of an old geomorphic surface.

differentiated on the basis of apparent highest degree of weathering or strength of plinthite development and expression. This suggests that many may have formed under conditions of higher rainfall under a previous climate, which is entirely possible in the case of the old landscapes in which some of them occur. It has been noted that Oxisols generally are found at elevations of less than 1,500 to 2,000 m (Soil Survey Staff 1960). Thus at these low altitudes they are within regions of rather uniform air temperatures from month to month, with comparatively low diurnal fluctuation, and hence may be classed as isothermal. Mean air temperatures are of the order of 22 to 24 C in the lower elevations of Oxisol occurrence (*Encyclopaedia Britannica World Atlas* 1959, Plates 23 to 30). Solar radiation is highest near the equator because the sun's rays come nearly vertical to the soil surface. The index of radiation at the outer surface of the atmosphere based on one unit for one day of radiation at the equator when the sun is directly overhead ranges from about 350 at the equator to 330 at 20° north and south latitudes, 275 at 40° latitudes, and 145 at the poles (Sellers 1965, Fig. 5). With this high intensity of solar radiation and the year-round high temperatures, vegetational growth potential is great in the intertropical zones compared to temperate zones, but actual type and growth are controlled by moisture and nutrient supply. Rainfall is quite variable in areas of Oxisols, ranging from only a few cm to more than 1,000 cm per year. Largest areas of Oxisols occur in regions with moderate to strong dry seasons.

Oxisols support a rather wide range of vegetational-ecologic zones, but by far the greatest areas are in tropical rain forest, scrub and thorn forest, semideciduous forest, and savanna vegetational zones (Mohr and van Baren 1954; UNESCO 1961). Due to extreme weathering, very low nutrient reserve, and low exchange capacity of the soils, practically all the nutrients in the natural ecosystems are within living or dead plant tissue, with a rapid turnover nutrient cycle between the vegetation and the litter, with little involvement of the soil except for deep feeding by tree roots in lower C horizons. For this reason, shifting cultivation (Kellogg 1963; Nye and Greenland 1960) and limited grazing are the common practices in the native, indigenous cultures, especially on the most highly weathered, sterile soils of the oldest landscapes.

As previously indicated, Oxisols occur at lower elevations. Here they are either on relatively stable upland summit positions, relict from a previous regional erosion surface, or preserved remnants of an old alluvial terrace or pediment (Jongen 1960; Ollier 1959; Ruhe 1956; Soil Survey Staff 1967). In fact, several observers (including ourselves) who have studied Oxisols in the field believe they may have formed almost entirely in transported soil materials except for those formed in saprolite from very ferromagnesian (basic) rocks rich in easily weatherable iron-containing minerals.

However, Oxisols have formed in a wide variety of initial materials, from siliceous granites and sediments to basalts and gabbros. Oxisols of the younger landscapes (apparently mid to late Pleistocene) seem almost universally to have formed in the saprolite from "basic" ferromagnesian rocks or in translocated material derived from such rocks.

PEDOGENIC PROCESSES. Desilication and concentration of free iron, and in some cases gibbsite, are major processes in Oxisols, along with essentially complete decomposition of weatherable primary minerals and 2:1 clay minerals associated with nearly complete leaching of exchangeable bases. A good deal of the desilication apparently is geochemical, with loss of silica from the system in the first flush of decomposition in the weathering crust near or at the rock surface in the initial material (Cady 1951). The balance of the desilication appears to result from long-term decomposition of the alumino-silicates with continual loss of silica from the system continuing the reaction despite the low solubility of silica. Ferritization appears to be a process of relative accumulation of iron in the better-drained Oxisols; that is, concentration of the relatively insoluble and stable free iron oxides due to losses from the soil system of the more easily weatherable and soluble compounds (D'Hoore 1954). In Oxisols which now have or have had a fluctuating water table relatively near the soil surface, there is segregation of iron, with some iron translocation (absolute accumulation of D'Hoore 1954), to form the red-and-gray mottled soft material called plinthite (Chapter 2, Chapter 22) which has in the past been designated as laterite, first by Buchanan (1807), with later studies continuing the use of this term (Alexander and Cady 1962; Du Preez 1949; Maignien 1959; Prescott and Pendleton 1952; Sivarajasingham et al. 1962; Soil Survey Staff 1967). If large amounts of plinthite have accumulated, it tends to form a continuous phase in the soil. If subjected to repeated wetting and drying, as in exposure by clearing of forests and erosion of overlying material, it becomes indurated to ironstone or beds of irregular modular aggregates.

Lessivage (physical downward translocation of clay) apparently has taken place in a few (but not all) Oxisols, based on presence of argillans (clay films), clay increase with depth, and narrow coarse clay:fine clay ratios in some profiles. But presence of an argillic horizon is not a requirement for Oxisols, and most of them lack illuviation cutans. In general, the high stability of the clay (due to presence of the free iron), associated low content of water-dispersible clay, absence of weatherable primary minerals to produce additional clay, and possibly the great age of many Oxisols combine to preclude presence of argillans. Argillans may be present in the younger, less weathered Oxisols intergrading to other orders (D'Hoore 1968).

Melanization and humification take place in all Oxisols to some degree. They are especially prominent and significant processes in those Oxisols with high rainfall and that are usually moist. The high biomass resulting from the higher solar radiation and year-round warm temperatures also humifies and mineralizes rapidly, but a high equilibrium level of organic matter is maintained. This especially pertains to the *Humox* suborder. The organic matter content of Oxisols is indirectly proportional to soil temperature (D'Hoore 1968). In general, Oxisols are not as dark in color at similar organic matter contents as soils of the other orders. Indeed, it is difficult if not impossible to judge the organic matter content of Oxisols by their color.

Gleization is a major process in those Oxisols that are or have been saturated with water at some period or periods during the year. It is possible for the gleization process to operate without formation of plinthite, and there are significant areas of Oxisols with characteristics associated with wetness but lacking plinthite.

Pedoturbation is a major process in those Oxisols without continuous plinthite very near the surface or without currently active strong gleization with its associated high water table. The main kind is faunal pedoturbation, most prevalent in the better drained and more clayey Oxisols associated with the older erosion surfaces, to the extent the entire upper solum may be reworked by insects and animals (Nye 1955; Watson 1962). The most active fauna in this process are termites, especially in central and western Africa, where they build numerous mounds as high as 3 m or more and as much as 10 m in diameter, composed of particles carried up from the soil (Chapter 11). A graphic description of a cross section of a termite mound in the Ivory Coast and of their activities is given by Carroll (1969). Figure 23.2 is a reproduction of his diagram of a "termitarium," illustrating the extensive reworking of the upper part of the soil by fauna in the Oxisols.

USES OF OXISOLS. Main uses of Oxisols are for shifting cultivation, subsistence farming, low-intensity grazing, and intensive plantation agriculture—mainly of sugarcane, pineapples, bananas, and coffee. A few areas of the wet Oxisols are used for paddy rice.

These soils have special and unique use and management requirements, limitations, and possibilities. This is due to their very low nutrient reserve and extremely low native fertility associated with their high degree of weathering, very low active acidity, and low exchangeable aluminum, high permeability, and low erodibility (Soil Survey Staff 1967). On this unique collection of properties and qualities is superimposed the variable rainfall distribution and soil moisture regimes over their wide area of occurrence. Therefore, it is difficult to be specific about their potential uses.

The most extremely weathered Oxisols in the drier areas are not used by indigenous native cultivators if soils of other orders are available (Soil Survey Staff 1967). Oxisols and their associated Ultisols and Inceptisols constitute the largest reserve of uncultivated soils available for development to meet world food needs (Kellogg and Orvedal 1968). Shifting cultivation (bush fallow) is practiced on many areas of Oxisols, a low-intensity cultivation practice which may be roughly described as clearing a patch of forest, growing crops a few years until the nutrients obtained by breaking into the nutrient biocycle are exhausted, after which the forest is allowed to revegetate (Kellogg 1963; Nye and Greenland 1960).

Bradfield (1968), working in the Philippines, has noted that whereas enough rice for a year's supply for 235 persons can be produced by modern methods on one hectare of paddy soil, about 3,000 hectares of tropical forest soils under shifting cultivation would be required to support the same number of people. He suggests that the shifting cultivation system might better be replaced by a stable system in which 90% of the land is left in valuable tree crops and 10% in intensively farmed vegetable and forage crop patches.

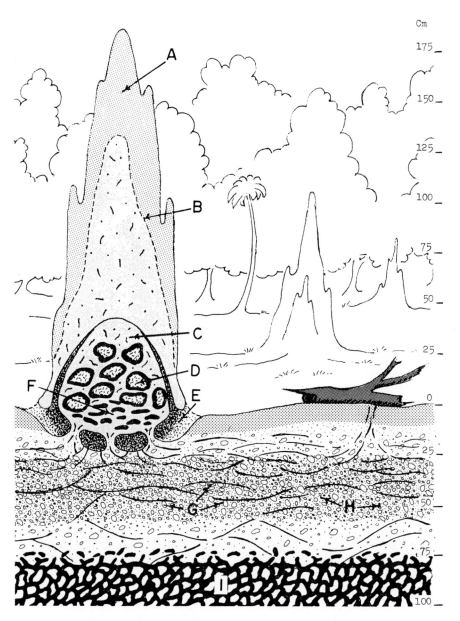

FIG. 23.2.   Schematic drawing of a representative termitarium
or nest of the earth-dwelling *Macrotermes natalensis* and
the soil profile below. Shown in the drawing are the: *(A)*
outer wall of the earthen nest, *(B)* interior of mound exposed
through sectioning along dashed line, *(C)* comb, *(D)* fungus
gardens, *(E)* columned basement, *(F)* chambers or cells
grouped around the queen cell and serving in part as
nurseries, *(G)* termite runways, *(H)* ironstone gravel, and
(I) plinthite.

At the present time, less is known about the genesis and appropriate classification of Oxisols than any other order of soils. Yet they offer a great opportunity and challenge for development to increase food production.

CLASSIFICATION OF OXISOLS. Five suborders of Oxisols are presently recognized, using as a primary basis the soil moisture regime throughout the year (Soil Survey Staff 1967). Their relationships are illustrated in Figure 23.3. A large number of accessory characteristics are associated with variations in content of soil moisture in Oxisols. For example, the base saturation increases with increasing dry periods and decreasing mean annual rainfall, whereas organic matter in the soil has an inverse relationship.

It appears that the Ferrisols and Ferrallitic soils of the Soil Map of Africa and the INEAC Tropical Soil Classification (D'Hoore 1968) correlate with the Oxisols, as would the better developed and more highly weathered portions of the Sols Ferrugineux Tropicaux.

The following is a summary of the main features of the five suborders of Oxisols, based on tentative criteria proposed by the Soil Survey Staff (1967).

1. Aquox: Either with continuous phase plinthite within 30 cm of the mineral surface; or are either water saturated during some period of the year or artificially drained and have an oxic horizon with characteristics associated with wetness.

2. Humox: Always moist or have no period when the soil is dry in any horizon below the surface 18 cm for 60 consecutive days or more in most years; have 20 or more kg of organic carbon per m^3 within the upper 1 m (exclusive of surface litter); mean annual soil temperature less than 22 C; base saturation (by NH_4OAc) less than 35% in the oxic horizon.

3. Orthox: Oxisols other than Aquox with mean annual soil temperature of 22 C or more or have less than 20 kg organic carbon per m^3 within the upper 1 m, and have no period when the soil is dry in any horizon below the surface 18 cm.

FIG. 23.3. Diagram showing some relationships of suborders of Oxisols.

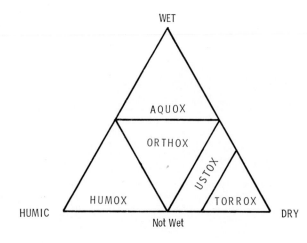

4. Torrox: Usually dry in most years in all parts of the soil between the Ap horizon or 18 cm and 2 m (or a lithic or paralithic contact, whichever is shallower) and have an ochric epipedon that has moist color values of 4 or more in all subhorizons.

5. Ustox: Have some subhorizon below the surface 18 cm that is dry for 90 cumulative days or more in most years; have mean annual soil temperature of 15 C or more; either have moist values of less than 4 in some part of the epipedon or are usually moist.

A summary of the great groups of Oxisols is presented in Table 23.1.

TABLE 23.1 ❧ **Suborders and great groups in the Oxisol order**

Suborder	Great Group
Aquox:	Gibbsiaquox—no continuous phase plinthite but have sheets containing 30% or more gibbsite within 1 m (40 in.) of the surface
	Plinthaquox—continuous phase plinthite within 1.25 m (50 in.) of the surface
	Ochraquox—have an ochric epipedon
	Umbraquox—have an umbric epipedon
Torrox:	No great groups at present
Ustox:	Acrustox—have less than 1 meq/100 g clay base retention from NH_4Cl in some part of the oxic horizon
	Eutrustox—mollic or umbric epipedon and a base saturation in the oxic horizon of >50% if clayey or >35% if loamy
	Haplustox—other Ustox
Humox:	Sombrihumox—have an oxic horizon darker in color and containing more o.m. than overlying horizon
	Gibbsihumox—some layer within 1 m (40 in.) of surface containing >30% gibbsite
	Haplohumox—all parts of oxic horizon retain more than 1 meq bases/100 g of clay
	Acrohumox—other Humox
Orthox:	Gibbsiorthox—have some layer within 1.25 m (50 in.) of the surface containing >30% gibbsite
	Acrorthox—some part of the oxic horizon retains < 1.5 meq bases/100 g of clay from NH_4Cl extraction
	Eutrorthox—no anthropic epipedon and have a base saturation >35% in all horizons to a depth of 1.25 m (50 in.)
	Umbriorthox—have either an umbric or ochric epipedon that contains more than 1% organic matter in all subhorizons to a depth of 75 cm (30 in.)
	Haplorthox—other Orthox

LITERATURE CITED

Alexander, L. T., and J. G. Cady. 1962. Genesis and hardening of laterite in soils. U.S. Dept. Agr. Tech. Bull. 1282.

Baldwin, M., C. E. Kellogg, and J. Thorp. 1938. Soil classification, pp. 979–1001. In Yearbook of agriculture. U.S. Dept. Agr. U.S. Govt. Printing Office, Washington.

Bradfield, R. 1968. The role of organic matter in soil management and the maintenance of soil fertility. In Study week on organic matter and soil fertility. John Wiley & Sons, New York, pp. 107–21.

Buchanan, F. 1807. Journey from Madras through Mysore, Canara and Malabar. Geol. Mag., vol. 2, p. 436.

Cady, J. G. 1951. Rock weathering and soil formation in the North Carolina Piedmont region. Soil Sci. Soc. Am. Proc. 15:337–42.

Carroll, P. H. 1969. Soil-dwelling termites in the southwest region of the Ivory Coast. Soil Survey Horizons 10:3–16.

D'Hoore, J. L. 1954. Proposed classification of the accumulation zones of free sesquioxides in a genetic basis. African Soils 3:66–81.

——. 1968. The classification of tropical soils, pp. 7–23. In R. P. Moss (ed.), The soil resources of tropical Africa. Cambridge Univ. Press, London.

Du Preez, J. W. 1949. Laterite: A general discussion with a description of Nigerian occurrences. Bull. Agr. Congo Belge 40:53–60.

Encyclopaedia Britannica World Atlas. 1959. The world scene, plates 23–30. Chicago, Ill.

Jackson, M. L., T. W. M. Levelt, J. K. Syers, R. W. Rex, R. N. Clayton, G. D. Sherman, and G. Uehara. 1971. Geomorphological relationships of tropospherically derived quartz in the soils of the Hawaiian Islands. Soil Sci. Soc. Am. Proc. 35:515–25.

Jongen, P. 1960. Relations entre faits geomorphologiques et pedogenese des sols tropicaux (Congo Belge). (English summary.) Trans. 7th Intern. Congr. Soil Sci. (Madison, Wis.) 4:335–46.

Kellogg, C. E. 1963. Shifting cultivation. Soil Sci. 95:221–30.

Kellogg, C. E., and A. C. Orvedal. 1968. World potentials for arable soils. War on Hunger 2(9):14–17. A report from the Agency for Intern. Development, Washington.

Maignien, R. 1959. Soil cuirasses in tropical West Africa. African Soils 4:5–42.

Mohr, E. C. J., and F. H. van Baren. 1954. Tropical soils: A critical study of soil genesis as related to climate, rock, and vegetation. Interscience, New York.

Nye, P. H. 1955. Some soil-forming processes in the humid tropics. IV. Action of the soil fauna. J. Soil Sci. 6:73–83.

Nye, P. H., and D. J. Greenland. 1960. The soil under shifting cultivation. Commonwealth Bur. Soils Tech. Comm. 51.

Ollier, C. D. 1959. A two-cycle theory of tropical pedology. J. Soil Sci. 10:137–48.

Prescott, J. A., and R. D. Pendleton. 1952. Laterite and Lateritic soils. Commonwealth Bur. Soils Tech. Comm. 47.

Ruhe, R. V. 1956. Landscape evolution in the High Ituri Belgian Congo. INEAC Series Sci. 66. (Scientific report from former Belgian Congo agronomic research institute.)

Sellers, W. D. 1965. Physical climatology. Univ. Chicago Press, Chicago.

Sivarajasingham, S., L. T. Alexander, J. G. Cady, and M. G. Cline. 1962. Laterite. Adv. Agron. 14:1–60.

Soil Survey Staff. 1960. Soil classification, a comprehensive system—7th approximation. U.S. Dept. Agr. U.S. Govt. Printing Office, Washington.

——. 1967. Supplement to Soil classification, a comprehensive system—7th approximation. U.S. Dept. Agr. U.S. Govt. Printing Office, Washington.

Thorp, J., and M. Baldwin. 1940. Laterite in relation to soils of the tropics. Ann. Assoc. Am. Geog. 30:163–94.

Thorp, J., and G. D. Smith. 1949. Higher categories of soil classification: Order, suborder and great soil groups. Soil Sci. 67:117–26.

UNESCO. 1961. Tropical soils and vegetation. Proc. Abidjan Symp. 1959. United Nations Educational, Scientific and Cultural Organization, Rome, Italy.

Watson, J. P. 1962. The soil below a termite mound. J. Soil Sci. 13:46–51.

&

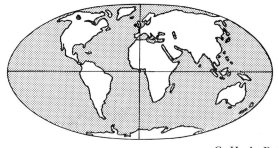

Histosols: Organic Soils

A̲LTHOUGH comprising only a small portion of the world's land area, organic soils are widely distributed. They form wherever production of organic matter exceeds its mineralization, usually under conditions of almost continuous saturation with water which curtails circulation of oxygen through the soil. The resulting slowing of the decomposition of organic matter permits its accumulation (Davis and Lucas 1959; Nikonov and Sluka 1964).

Although maritime climates and slowly permeable underlying materials favor their formation, Histosols may develop quite independently of climate and substratum. Depressions that descend below the water table provide sites favorable to accumulation of organic materials. Histosols also form at seepage sites on concave upland slopes where cool and anaerobic conditions are persistent. Some Histosols form in convex upland positions in cool moist climates such as parts of Ireland, Scotland, etc. These are often known as blanket peats, blanket or hanging bogs, or climatic bogs.

Aclimatic Histosols are found in association with a wide range of soils. Most often, especially in the areas where they form in depressions, these associates are hydromorphic soils of an *Aqu* suborder, that is, Aquods, Aquolls, Aquepts, etc., that are typically high in content of organic matter, as in Humaquepts, Umbraquults, Humaquods, etc. Small bodies of Histosols may even develop in concave positions on steep slopes in association with Aridisols. Figure 24.1 is a block diagram illustrating the range of positions that Histosols may occupy in landscapes. Figure 24.2 shows the content of organic matter necessary for organic soil classification.

Most Histosols have bulk densities of less than 1 g/cc. Farnham and Finney (1965) have reported values as low as 0.06 g/cc on dry weight basis. Bulk density tends to increase with decomposition (Boelter 1965). The amount of mineral material and type of vegetation contributing to the Histosol will cause major variations in the bulk density. Histosols are commonly saturated and have an extremely high water-holding capacity both on a weight and volume basis. Much of the water is either in the larger pores (gravitational water) or in such fine pores that it is unavailable for

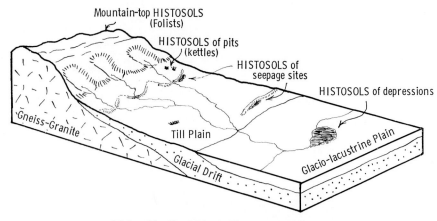

FIG. 24.1. Idealized block diagram showing some relationships of Histosols to topographic position.

plant growth (Boelter and Blake 1964). Because Histosols shrink considerably, their moisture characteristics are best expressed on a wet bulk volume basis or measured in situ (Boelter 1964).

The cation exchange capacity of Histosols is derived from carboxyl, phenolic, and probably other functional groups (Broadbent and Bradford 1952). The number of functional groups increases as decomposition progresses and cation exchange capacities of 200 me/100 g and higher have been reported for some organic matter (Broadbent 1953). These are pH-dependent charges and the cation exchange capacity of organic soils has been known to change from 10 to 20 meq/g at soil pH (about 3.7 in 1:1 water) to over 100 me/100 g at pH 7 (Dolman and Buol 1967).

PROCESSES OF HISTOSOL FORMATION. The initial process by which Histosols form is termed paludization (Chapter 6). The "growth" of a Histosol is from the bottom up. This was demonstrated with [14]C. In some Histosols of the Everglades of Florida (McDowell, Stephens, and Stewart 1969), the organic material a few centimeters above a limestone contact is about 4,300 years old and the material 1.26 meters above the limestone is about 1,250 years old. Since genesis of Histosols depends on organic matter deposition, the process is often considered to be geogenic rather than pedogenic. In this sense, one can consider the initial deposit of organic materials the "parent material" in which Histosols can form by alteration from recognizable organic forms of leaves, stems, etc., to unrecognizable organic material.

The decomposition of organic matter is controlled by a number of interrelated factors of which moisture content, temperature, composition of the deposit, acidity, microbial activity, and time are the most important (Broadbent 1962). The alterations and reactions taking place during decomposition are numerous, complicated, and only partially understood.

In the Netherlands, Heuvelen, Jongerius, and Pons (1960), Jongerius and Pons (1962), and Pons (1960) have considered pedogenesis of

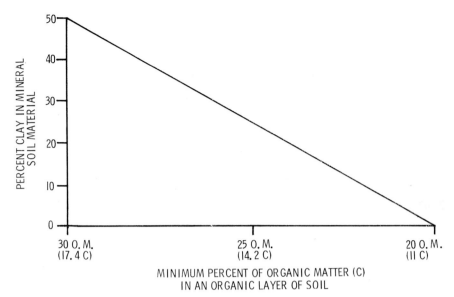

FIG. 24.2. Organic matter (carbon) contents required for soil horizons of different clay contents to all qualify as organic horizons.

Histosols to begin as soon as microbial activity is promoted by entry of air into the organic deposit. They have referred to the processes involved as physical, chemical, and biological ripening.

Physical ripening[1] primarily involves a decrease in volume. The amount of physical ripening depends upon the nature of the plant remains, the content of mineral matter, and height of the water table.

Chemical ripening represents the chemical decomposition of the organic components. This involves both the complete decomposition of some components and the partial conversion of others to compounds of intermediate complexity that may reunite to form new organic substances of higher resistance called humus (Kononova 1961).

Biological ripening involves a reduction in particle size and the mixing of the organic material by living organisms.

The state of decomposition present in the organic material that composes the Histosols is an important characteristic that needs to be quantitatively evaluated for classification purposes. In Sweden, Post (1924) developed a ten-interval scale based on the proportion of organic material that remains in the hand after squeezing a sample of wet organic soil. If colorless liquid is produced upon squeezing, the soil is said to be little decomposed; if all the organic material escapes between the fingers the soil is classified at the highest stage of decomposition. With practice, such a technique works well in identifying fibric, hemic, and sapric horizons (Chapter 2). A fibric horizon emits only slightly turbid water; a hemic

1. This ripening process also goes on in dominantly mineral sediments that undergo dewatering, as in polders.

horizon emits turbid water, and up to 2/3 of the original handful escapes between the fingers upon squeezing; that is, > 1/3 of the handful is retained. Upon squeezing wet sapric material, more than 2/3 of the sample extrudes between the fingers.

In Finland, Kaila (1956) determined the degree of decomposition by the color of the liquid produced from mixing the organic material in saturated sodium pyrophosphate solution at ambient temperatures. This is a convenient field test. A color of the liquid, when absorbed on white filter paper, lower in value or higher in chroma than 10YR 7/3 (Munsell notation) is diagnostic for a sapric horizon. Lighter colors and values and chromas greater than 7 and 3, respectively, are produced by the less decomposed materials of hemic and fibric horizons.

DESCRIBING HISTOSOLS. Conventions useful in describing mineral soils (Chapter 2) are not wholly applicable to Histosols. Each layer of a Histosol is described in the following terms: color, fiber content, structure, consistence, roots, additional features, reaction, and boundary (Soil Survey Staff 1968). Wet and dry colors are determined, when possible, and in addition the color of the material is determined after it has been pressed or rubbed in the hands.

The estimate of fiber content excludes live roots. Two determinations are made, one by observation of a freshly broken surface and a second by examining the material after it has been rubbed between the fingers to break down extremely rotten fibers.

Structure, consistence, content of live roots, reaction, and boundaries are described as in mineral soils.

Additional notes include results of the sodium pyrophosphate test, botanical origin of the fibers, notes on included thin strata, presence of logs and stumps, mineral content, and any other features observed by the soil scientist. All depths are recorded from the soil-air interface rather than from the mineral-organic boundary used in the description of mineral soils.

Horizon nomenclature has been developed[1] that relates directly to diagnostic layers in the Comprehensive Soil Classification System (Soil Survey Staff 1968) (Chapter 2). O is used for organic horizons, and subhorizons are designated as i (Oi) for fibric, e (Oe) for hemic, and a (Oa) for sapric. Limnic layers (Chapter 2) are designated L. Subhorizon designations are Lca for marl, Lco for coprogenous earth, and Ldi for diatomaceous earth. Other subhorizon symbols are f for frozen layers, p for cultivated layers, and cn for bog iron. Prefixed Roman numerals are used for lithologic discontinuities and consecutive arabic numbers are suffixed for horizon subdivisions as in mineral soils (Chapter 2).

USE OF HISTOSOLS. Basically the management of Histosols for growing crops involves counter practices in water management (Stephens 1955). It is necessary to drain the soil for plant growth and to facilitate the operation of equipment (Roe 1936). However, when the soils are drained, they oxidize and subside, making further drainage difficult.

---

1. TSC Advisory Soils, LI-4 Feb. 26; USDA-SCS, Lincoln, Nebraska, 1969.

Subsidence in Florida was found by Stephens (1956) to be about 3 cm (1.25 in.) per year under cultivation. He reported that the rate of subsidence was directly related to the depth of artificial drainage and could be predicted by the following formula: $x = (y - 2.45)/14.77$ where x is the annual subsidence in inches and y is the average depth of drainage in inches. He further found that this was about twice as fast as subsidence of drained Histosols in Indiana where the formula $x = (y - 9.6)/23$ was applicable. Undoubtedly, the difference is related to higher temperatures and lack of freezing in Florida.

Fire hazard increases after the Histosols are drained. Peat and muck fires are difficult to control and may burn for several months, polluting the air as well as destroying the organic material. In many cases when the organic matter is destroyed, farming is forced to proceed on the remaining mineral substratum that is usually poorly drained (Davis and Engberg 1955; Jongedyk, Hickok, and Mayer 1954; Jongedyk et al. 1950; Mirza and Irwin 1964; Neller 1944). In some areas, resistant, hard logs and stumps concentrate on the surface as the other organic matter oxidizes. These create difficulties for the operation of equipment (Dolman and Buol 1967).

Often when Histosols of advanced decomposition are "over-drained" or otherwise dried, they harden irreversibly so as to restrict water movement and root penetration (Chapek 1934). Breaking of the hardened layer will facilitate water movement through the layer, but since the resulting clods do not readily absorb water, they do not contribute to the water-holding capacity of the soil. A droughty soil results. Structureless "pastelike" horizons also restrict drainage, causing ponding during wet seasons, with corresponding damage to crops. The most desirable water management practices on Histosols involve the strict control of surface and subsurface drainage. Fields can be shaped to provide "short-run" removal of surface water during wet seasons. Drainage canals can be provided with gates so that groundwater levels can be held as high as possible during dry seasons to reduce oxidation and also to subsurface irrigate the crop at the same time. Windbreaks are usually needed to control wind erosion in large open cultivated areas of Histosols.

The type of crop grown on Histosols is largely regulated by the climate, and since these soils are largely aclimatic, almost no generalizations can be made. The low bulk density of these soils, however, does appear to favor the production of vegetables of ideal shape. The muck of Figure 24.3 is used for production of onions, potatoes, carrots, and salad crops.

Bodies of Histosols have been, and in many areas still are, sources of fuel. Construction of structures and roads on Histosols is usually difficult. The organic material has very low bearing capacity, and structures must be placed on foundations that extend to underlying mineral material. If an area is drained just prior to the time of construction, subsidence of the soils becomes apparent only gradually. Ultimately, the structures are left standing far above the ground level. In the case of house construction on piles driven into the mineral soil, the building becomes a "house on posts" perched a foot or two above the lawn. The garage stands above the driveway and is inaccessible to the car.

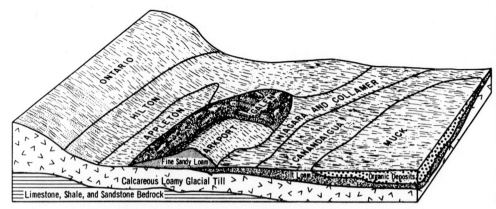

FIG. 24.3. Soilscape pattern of Histosols (muck) with a Glossoboric Hapludalf (Ontario), Glossoboric Hapludalfs (Hilton, Collamer), Psammentic Hapludalfs (Galen, Arkport), Aeric Ochraqualfs (Appleton, Niagara), and Mollic Haplaquept (Canadaigua) in Genesee County in western New York. (After Wulforst, Wertz, and Leonard 1969.)

CLASSIFICATION OF HISTOSOLS. The classification of the Histosols has lagged behind that of most of the mineral soils in the United States soil taxonomic system. Three important elements are of major concern in defining the order: (1) a standard minimum content of soil organic matter; (2) a required depth or thickness of this organic soil; and (3) avoidance of criteria that would necessitate reclassification of the soil as a result of common agricultural practices such as drainage.

The following definitions have been formulated on the basis of these elements. Organic soil material, if saturated with water for prolonged periods or if artificially drained, is defined as having more than 30% organic matter if the mineral fraction has 50% or more clay; more than 20% organic matter if no clay is present, or proportionately more organic matter than a line connecting these points for intermediate clay contents (see Fig. 24.2). If the soil is almost never saturated with water, it must have more than 35% organic matter. The depth of organic material, as measured from the soil-air interface, must be one of the following:

1. Greater than 60 cm (24 in.) if 75% or more of the volume is fibric or if the bulk density is less than 0.1 g/cc.

2. Greater than 40 cm (16 in.) of organic material having at least 75% by volume of moss and a bulk density greater than 0.1 g/cc if saturated more than 6 months per year or if artificially drained.

3. Greater than 1 m (36 in.) if predominantly organic material with less than 1/3 fiber and not saturated with water for long periods or artificially drained.

4. Greater than 10 cm (4 in.) over a lithic or paralithic contact, provided the organic materials are twice as thick as mineral materials above that contact.

5. Any depth if underlain by fragmental (greater than 2 mm) material in which the interstices are filled with organic material.

6. Have no mineral layer thicker than 40 cm (16 in.) present at the surface or with an upper boundary of the mineral layer less than 40 cm (16 in.) below the surface.

7. The sum of the mineral layers in the upper 80 cm (32 in.) cannot be greater than 40 cm (16 in.).

An arbitrary control section of 130 cm (51 in.) or 160 cm (63 in.) is defined for use in classifying Histosols, providing no lithic or paralithic layer occurs in that depth. The thicker limit is used only where the surface 60 cm (24 in.) is more than 75% fibric moss. This control section is then subdivided into three layers referred to as surface, subsurface, and bottom tiers. The surface tier is 30 cm (12 in.) thick, except where 75% or more of it is of fibric composition, in which case it is considered 60 cm (24 in.) thick. The subsurface tier is 60 cm (24 in.) thick and may include mineral material, provided it is not lithic or paralithic. The bottom tier is 40 cm (16 in.) thick or to the top of a lithic or paralithic layer, whichever is shallower.

The Histosol order is divided into four suborders (Fig. 24.4). Those Histosols that are not saturated with water more than a few days per year are in the suborder Folists, provided a lithic or paralithic contact or fragmental material with interstices filled with organic material is present within 1 m (40 in.) of the surface. They must also have less than 75% of the thickness occupied by fibric sphagnum moss.

The other three suborders separate on the degree of decomposition present in the organic material in the subsurface tier. Fibrists (Fig. 20.2) have fibric material dominant in the subsurface tier; Hemists have a subsurface tier dominated by hemic organic material; and Saprists have a subsurface tier dominated by sapric material.

A listing of the great groups recognized in the Histosol order is given in Table 24.1. Mean annual soil temperature is the primary criterion for great group separation. The presence of a 2-cm or thicker humilluvic layer is used as a criterion for Histosols with mean annual soil temperatures above 8 C (47 F), provided they have a mean summer temperature more

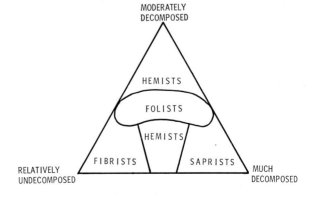

FIG. 24.4. Diagram showing some relationships between suborders of Histosols.

TABLE 24.1 ❧ **Suborders and great groups in the Histosol order**

Suborder	Great Group
Folists:	No great groups presently recognized
Fibrists:	Cyrofibrists—MAT less than 8 C (47 F) and frozen in top 5 cm (2 in.) or part of control section 2 months after solstice Sphagnofibrists—75% or more fibric sphagnum moss in top 90 cm (36 in.) Borofibrists—MAT less than 8 C (47 F) Tropofibrists—mean summer and mean winter temperatures differ less than 5 C (9 F) Medifibrists—other Fibrists with less than 2 cm (0.8 in.) of humilluvic material Luvifibrists—other Fibrists
Hemists:	Cryohemists—frozen in control section or top 5 cm (2 in.) 2 months after summer solstice Borohemists—MAT less than 8 C (47 F) Tropohemists—mean summer and mean winter temperatures differ less than 5 C (9 F) Medihemists—other Hemists with less than 2 cm (0.8 in.) of humilluvic materials Luvihemists—other Hemists Sulfihemists—sulfidic material present within 1 m (40 in.) of surface; potentially acid sulfate soils (cat clays) Sulfohemists—sulfuric acid horizon within 50 cm (20 in.) of surface (organic cat clays)
Saprists:	Cryosaprists—frozen in control section or top 5 cm (2 in.) 2 months after summer solstice Borosaprists—MAT less than 8 C (47 F) Troposaprists—mean summer and mean winter temperatures differ by less than 5 C (9 F) Medisaprists—other Saprists with no humilluvic horizon thicker than 2 cm (0.8 in.) Luvisaprists—other Saprists

than 5 C (9 F) warmer than their mean winter temperature. A humilluvic layer is an accumulation of illuviated humus, usually near a contact with sandy material, that has a high solubility in sodium pyrophosphate and is difficult to re-wet after drying.

The nature of the bottom tier is considered in the subgroup classification.

Although the underlying mineral material is presently quite deep, it is important to realize that once the ecosystem of the Histosols is disturbed and a drainage and cropping system imposed, mineralization is usually greatly accelerated. Much or even all the organic material will eventually be oxidized. One study in Florida indicated that in 50 years of drainage, 1.8 meters of Histosol were lost, or according to the researchers' calculations, about as much as had been created in 1,200 years (McDowell, Stephens, and Stewart 1969). Therefore, a description of Histosols is incomplete without reference to their susceptibility to change.

LITERATURE CITED

Boelter, D. H. 1964. Water storage characteristics of several peats in situ. Soil Sci. Soc. Am. Proc. 28:433–35.
———. 1965. Hydraulic conductivity of peats. Soil Sci. 100:227–31.

Boelter, D. H., and G. R. Blake. 1964. Importance of volumetric expression of water contents of organic soils. Soil Sci. Soc. Am. Proc. 28:176–78.

Broadbent, F. E. 1953. The soil organic fraction. Adv. Agron. 5:153–83.

———. 1962. Biological and chemical aspects of mineralization. Intern. Soil Conf., New Zealand, pp. 220–22. Soil Bureau, Wellington, New Zealand.

Broadbent, F. E., and G. R. Bradford. 1952. Cation-exchange groupings in the soil organic fraction. Soil Sci. 74:447–57.

Chapek, M. V. 1934. The physico-chemical phenomenon of the wetting resistance of peat to water. Kolloid Ztschr. 66:91–95.

Davis, J. F., and C. A. Engberg. 1955. A preliminary report of investigations of subsidence of organic soils in Michigan. Mich. Agr. Exp. Sta. Quart. Bull. 37, pp. 498–505.

Davis, J. F., and R. E. Lucas. 1959. Organic soils, their formation, distribution, utilization and management. Mich. Agr. Exp. Sta. Spec. Bull. 425.

Dolman, J. D., and S. W. Buol. 1967. A study of organic soils (Histosols) in the Tidewater region of North Carolina. N.C. Agr. Exp. Sta. Tech. Bull. 181.

Farnham, R. S., and H. R. Finney. 1965. Classification and properties of organic soils. Adv. Agron. 17:115–62.

Heuvelen, B. van, A. Jongerius, and L. J. Pons. 1960. Soil formation in organic soils. Trans. 7th Intern. Congr. Soil Sci. (Madison, Wis.) 4:195–204.

Jongedyk, H. A., R. B. Hickok, I. D. Mayer. 1954. Changes in drainage properties of a muck soil as a result of drainage practices. Soil Sci. Soc. Am. Proc. 18:72–76.

Jongedyk, H. A., R. B. Hickok, I. D. Mayer, and N. K. Ellis. 1950. Subsidence of muck soil in northern Indiana. Purdue Univ. Agr. Exp. Sta. Circ. 366.

Jongerius, A., and L. J. Pons 1962. Soil genesis in organic soils. Auger and Spade 12:156–68.

Kaila, A. 1956. Determination of the degree of humification in peat samples. J. Sci. Agr. Soc., Finland 23:18–35.

Kononova, M. M. 1961. Soil organic matter. (Transl. from Russian by T. Z. Nowakowski and G. N. Greenwood.) Pergamon Press, London.

McDowell, L. L., J. C. Stephens, and E. H. Stewart. 1969. Radiocarbon chronology of the Florida Everglades peat. Soil Sci. Soc. Am. Proc. 33:743–45.

Mirza, C., and R. W. Irwin. 1964. Determination of subsidence of an organic soil in southern Ontario. Can. J. Soil Sci. 44:248–53.

Neller, J. R. 1944. Oxidation loss of lowmoor peat in fields with different water tables. Soil Sci. 58:195–204.

Nikonov, M. N., and V. P. Sluka. 1964. Distribution of peat bogs. Soviet Soil Sci., 1964, No. 10, pp. 1042–47. (Transl. 1965 from Pochvovedeniye, 1964, 10:44–50.)

Pons, L. J. 1960. Soil genesis and classification of reclaimed peat soils in connection with initial soil formation. Trans. 7th Intern. Congr. Soil Sci. (Madison, Wis.) 4:205–11.

Post, L. von. 1924. Das genetische system der organogenen Bildungen Schwedens. Comm. Inter. Pedologie, 4, Comm. 22:287–304.

Roe, H. B. 1936. A study of influence of depth of ground-water level on yields of crops grown on peat lands. Minn. Agr. Exp. Sta. Bull. 330.32P.

Soil Survey Staff. 1968. Supplement to Soil classification, a comprehensive system—7th approximation, Histosols. U.S. Dept. Agr. U.S. Govt. Printing Office, Washington.

Stephens, J. C. 1955. Drainage of peat and muck lands, pp. 539–57. In Yearbook of agriculture. U.S. Dept. Agr. U.S. Govt. Printing Office, Washington.

———. 1956. Subsidence of organic soils in the Florida Everglades. Soil Sci. Soc. Am. Proc. 20:77–80.

Wulforst, J. P., W. A. Wertz, and R. P. Leonard. 1969. Soil survey of Genesee County, New York. Soil Conserv. Serv., U.S. Dept. Agr. U.S. Govt. Printing Office, Washington.

Natural Soil Bodies and Soilscapes

$\mathbf{A}$ SOIL BODY is a representative specimen of a taxon in place on the natural terrain. It is a unit of the mosaic of soil bodies that constitute the "soil cover" (term of Fridland 1965; Gerasimov and Glazorskaya 1960). A soilscape[1] is the assemblage of soil bodies on a land surface in a particular landscape. In previous chapters consideration has been given to the soil unit column (pedon) as related to the soil body (polypedon of Johnson 1963). In this chapter the relation of the soil body to the soilscape (polypedon cluster) will be explored in a preliminary way. This is in line with the proposal of Neustruyev (1915) that the concept of zonal soils be replaced by the concept of zonal soil "combinations," which are associations and patterns of soils, that is, soilscape fabrics.

Prasolov (1922) gives this hierarchy of soil geographic units: soil zone, soil province, soil region, elementary soil landscape (called soilscape in following paragraphs), soil complex, and soil variety ("soil areal" of Fridland 1965; called soil body in this discussion). Figure 25.1 depicts a portion of a soilscape and three component soil bodies.

SOIL BODIES RELATED TO SOILSCAPES. A typical soil body depicted as a mapping unit on a detailed soil map (Powell and Springer 1965; Wilding, Jones, and Schafer 1965) may actually include other soils to the extent of about 35% by area rather than the prescribed 5 to 15% of accepted definitions (Fridland 1965; Soil Survey Staff 1951). However, most of the included soils are usually similar to the one used to designate the mapping unit. Soil bodies range in size from 1.5 m² (0.00015 ha) for a Solonetz hummock to hundreds of thousands of hectares for some Chernozems (Fridland 1965).

Soil bodies can be characterized not only by profile properties but also as to natural drainage index, soil body pattern index (degree of departure from the shape of a circle), and soilscape position (Hole 1953). Fridland (1965) used a "coefficient of dissection" in reporting degree of sinuosity

1. The term "soilscape" is a contraction of the term "soil landscape" (Walker, Hall, and Protz 1968a) and analogous to "townscape" (Esser 1968) of architects.

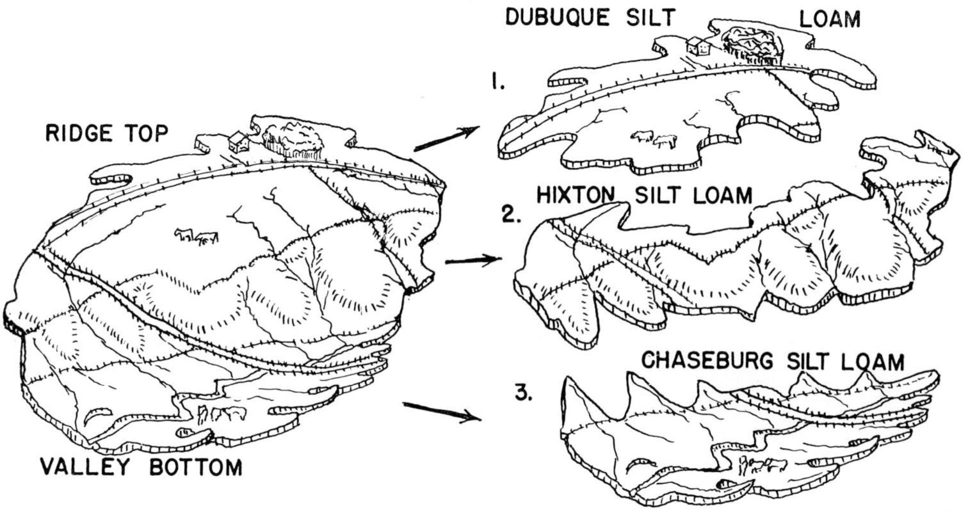

FIG. 25.1. Sketch *(left)* of a sequence of soils from a ridge crest to the bottom of a small valley. This portion of a soilscape consists of three soil bodies. The Dubuque soil is a Typic Hapludalf formed in shallow leached loess over cherty residual clay over dolomite. The Hixton soil is a Typic Hapludalf formed in soft Cambrian sandstone and siltstone strata. The Chaseburg soil is a Typic Udifluvent formed in silty colluvium and local alluvium. (After Hole and Lee 1955.)

(extent to which they are winding and irregular) of soil body boundaries and discussed soil body forms and degree of distinctness of boundaries of soil bodies. He suggested criteria for establishing degree of contrast between component bodies in a given soilscape, defined relative proportions of components of a soilscape by areal extent, and characterized relative coarseness or fineness of soilscape fabric in terms of sizes of soil bodies.

Soil bodies of a given soil series or other taxa have a fixed natural drainage index, but the soil body pattern index and landscape position may vary from region to region. One can say that a particular kind of soil body occupies a somewhat different ecological or soilscape niche in each region in which it occurs (see Fig. 25.2). To the present classification of soil bodies based on soil profile characteristics, a soilscape expression may be added.

It is interesting to note that in the *7th Approximation* (Soil Survey Staff 1960), Humic-Gley (Haplaquoll, Argiaquoll) soils are scattered throughout the classification, whereas some pedologists, including Duchaufour and Aubert of the French school (Duchaufour 1963) would prefer to group these poorly drained soils together (as they are in Fig. 25.3) on the basis of landscape position and degree of gleization. Here again, the significant differences in soilscapes in which such depressional soils occur should enter into a geographic classification of soils with which this chapter is concerned.

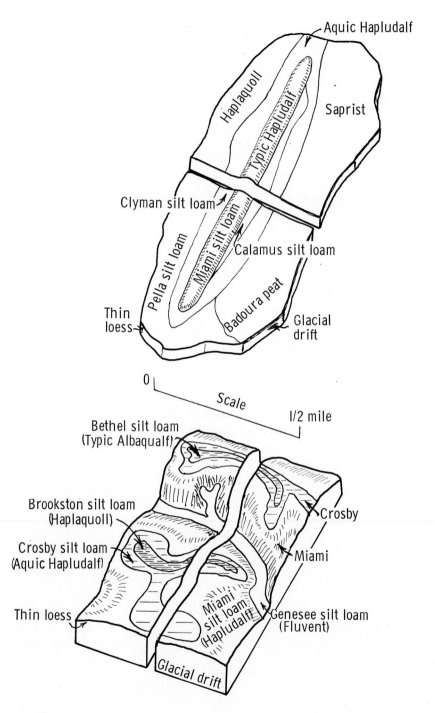

Aquic Hapludalf

Haplaquoll

Typic Hapludalf

Saprist

Clyman silt loam

Pella silt loam

Miami silt loam

Calamus silt loam

Badoura peat

Thin loess

Glacial drift

0

Scale

1/2 mile

Bethel silt loam
(Typic Albaqualf)

Brookston silt loam
(Haplaquoll)

Crosby silt loam
(Aquic Hapludalf)

Crosby

Miami

Thin loess

Miami silt loam,
(Hapludalf)

Genesee silt loam
(Fluvent)

Glacial drift

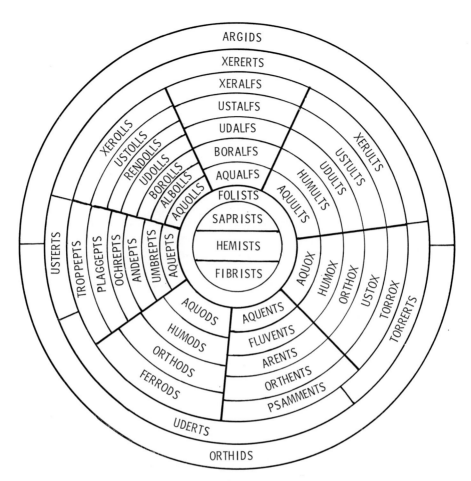

FIG. 25.3. Index to the suborders of the new
United States soil taxonomy.

FIG. 25.2. Contrasting landscape positions of Miami silt
loam, a Typic Hapludalf, in southern Wisconsin (Hole, Lee,
and Brickbauer 1953) and east central Indiana (Bushnell
et al. 1930). The sketch above is of a drumlin about ¾ mile
long on which the Miami soil occurs, with another Typic
Hapludalf, the moderately well-drained Calamus, on a
footslope, and surrounded by wetter soils. The diagram
below shows the Miami soil on marked slopes (exaggerated
in the sketch), with wetter soils on the upland flats.

A consideration of the soil body pattern index leads us to inquire into the origin of the shapes of soil bodies. To what extent are they residual, in the geologic sense, and to what extent are they depositional or erosional? Can we determine whether the shape of a soil body was established before, during, or after the development of the major horizons? Through the action of repeated fires that maintain a large opening in a forest on a hillslope, a body of prairie soil (Hapludoll, Argiudoll) could conceivably acquire its shape at the same time that profile characteristics evolved. Initial material may dictate the shape of the future soil body as in the case of one on the crest of a drumlin, where an Entisol polypedon may gradually differentiate into an Alfisol. In a soilscape of Hapludalfs, several bodies of Albaqualfs may become distinct from the others because of clay pan formation. Bodies of Aquolls may change in size and shape as the wetness and dryness of the climate change. In a sense the boundary of a body of Aquoll is a "high-water mark" of amphibious conditions that have occurred at the site. Ruhe (1969) describes the relation of soil patterns in Iowa to superimposed glacial and loess deposits and their paleosols on landscapes of variable relief and degree of dissection. A thorough understanding of soilscapes involves an investigation of the landforms and their soils (Gerasimov and Glazovskaya 1960).

SOILSCAPES. A soilscape is the pedologic portion of a landscape. It is what a soil surveyor discovers progressively as he works his way across a terrain in mapping soil bodies on an aerial photograph. In this process of mapping, the field soil scientist combines observation of surface soil and subsoil materials at sites of auger holes with observations in perspective of topography and vegetative cover. The map depicting the soilscape is free from the distortions of perspective which the mapper experienced in the course of his traverse. In this sense a soilscape is the true assemblage of soil bodies on the land, as distinct both from the piece-meal discovery of them and from the soil scientist's cartographic depiction of them.

The term soilscape denotes the pedologic quality of a terrain, as distinct from the more inclusive word, landscape, and from the units of soil map legends such as soil association and soil phase. Very different soilscapes may occur on widely separated landscapes that appear strikingly similar, especially where native vegetation has been removed to make way for agricultural activity. For example, nearly level uplands are occupied by Argiustolls (Chestnut soils) in western Nebraska, by Albaqualfs (Planosols) in southern Illinois, by Paleudults (deep, ancient Red-Yellow Podzolics) in eastern North Carolina, and Acrustox (Latosols) in portions of Brazil. Differences in bioclimatic factors, time, and initial material account for soilscape variability that in some cases is virtually independent of land form. Such situations make obvious the importance of characterizing the soilscape.

NATURAL DRAINAGE INDEX
OF SOIL REGIONS. Hole (1953) has suggested how a soil profile can be described in terms of a natural soil drainage index. An entire soilscape

can be given a soil-drainage index expressing the average wetness or dryness of the terrain. The index alone is not adequate, but rather a histogram is needed to show the full sweep of variation present in the soilscape. Soilscapes can thus be numerically characterized as xeric, mesic, hydric, or as combinations of them or intergrades between them.

NATURAL SOILSCAPE POSITION
INDEX OF SOILS. We may assign to a soil body an index number designating its position relative to the crest of a hill. The soil body on the crest is in "soilscape position number one." It may be possible to distinguish several crest positions in the case of a long undulating ridge: 1a, knob; 1b, ridge-line; 1c, saddle. Below this first position are soil bodies to which numbers are assigned that indicate their relative distance downhill (measured in terms of numbers of soil bodies) from the first or crest position. The minimum count or, better, a statistically meaningful count is made of soil bodies encountered along transects down the slope. Data of this kind are strictly comparable from one soil region to another only if the soil maps are made at the same scale and by soil scientists using the same concepts and techniques. Soil regions in which relief is great may have a larger number of soil landscape positions than soil regions with low relief. A coupling of data of this sort with information about actual topographic relief is valuable.

The greater the relief in a soilscape the greater the differences in conditions on different aspects (exposures relative to insolation and precipitation) and the more the distribution of soil bodies is related to this feature (Fridland 1965).

It is to be noted that a particular kind of soil (taxon) may occur in more than one soilscape position. A soil occurs wherever the factors of soil formation provide the requisite state and this may be in more than one niche (Arnold 1965). For example, a Typic Hapludalf, the Miami silt loam, occurs in at least three soilscape positions: the first on 2-km-long drumlin hills, and the second on irregular hills of ground moraine, both in Dodge County, Wisconsin (Hole, Lee, and Brickbauer 1953); and the third on valley sides in Wayne County, Indiana (Bushnell et al. 1930) (Fig. 25.2).

NATURAL SOIL BODY PATTERN INDEX.
Juday and Birge (Juday 1914) noted that with time the shape of a lake approaches that of a circle. They assigned shape indices to lakes. Silver Dollar Lake in Florence County, Wisconsin (Hole et al. 1962) is a fine example of a lake with a nearly perfect circular outline, produced by the encroachment of peat from the original shores.

The soil body pattern index expresses the degree of irregularity of a soil body. A simple index formula is $I = P/A$ where I is the index, P is the perimeter, and A the circumference of a circle of the same area as that of the soil body. Where soil body patterns are largely determined by dendritic drainage systems, the patterns can be expected to be as alike as the surface drainage patterns are. For example, soil body patterns of Barron County, Wisconsin, in a glaciated region and those of Richland and Grant counties, Wisconsin, in a much older, mature landscape, virtually devoid of glacial till, are surprisingly similar. However, bodies in the older landscape tend to be much larger than those in the younger one. The larger a soil body is, the more intricacies of the boundaries are observed and recorded.

A particular kind of soil may have a different soil body pattern in one soilscape from that in another soilscape. Pattern indices of bodies of Miami and Dodge soils in the Town of Beaver Dam, Dodge County, Wisconsin, are lower when developed on ground moraine than when developed on drumlin hills (Hole 1953; Shields 1955).

CATENAL ELUVIATION AND ILLUVIATION. The movement of material from the oromorphic to the hydromorphic portions of a catena via groundwater, seepage water, and surface runoff has a bearing on the evolution of soil bodies. For example, hydromorphic soils may become finer in texture, higher in content of montmorillonite (Ciolkosz 1967), more cemented, either more alkaline or more acid, and more or less toxic than soils on higher ground, depending on whether $Ca^{++} + Mg^{++}$ and carbonates or hydronium and aluminum (toxic substances) are available in the substrata of the soilscape to be translocated through and over soils to depressions.

CATENAL VARIATIONS
IN MICROTOPOGRAPHY. The microtopography of soil bodies may vary from highland to depression. Microrelief within a square meter may be 5 cm on a hilltop soil, 10 cm on a footslope soil, and 20 cm on a wetland soil in an adjacent lowland. "Cat-step" microtopography characterizes some soil bodies on steep slopes slightly affected by mass wasting.

GENERAL FUNDAMENTAL PROCESSES AND
CONDITIONS OF GENESIS OF SOILSCAPES. General processes and conditions may provide a framework for study of a relatively undeveloped phase of pedology, i.e., soilscape analysis. Two overlapping trends in genesis of soilscapes may be termed complication and simplification, corresponding to horizonation and haploidization of soil pedon development (Table 6.1). The soilscape patterns of parts of southwestern Australia are intricate mosaics of very different soils (Thorp 1957), including Oxisols, Mollisols, Alfisols, Inceptisols, and Entisols. Coexistence of bodies of contrasting initial materials, including some ancient soils, on a variety of landforms makes this multidifferentiated soilscape possible. On the other hand, the extensive loess blanket in the eastern part of the prairie triangle of the north central region of the United States from Kansas to Iowa provides the condition for a regionally simple soil pattern. Mollisols clearly dominate this vast terrain.

Simplicity at the order level does not, however, mean that intricacy does not exist at lower levels of classification and at larger scales of mapping.

Pedologists of Kansas and Iowa, over which Mollisols predominate, are keenly aware of the complexity of soil patterns in those states (Walker, Hall, and Protz 1968a, b). Pedoturbations, which in soil pedon development are considered processes of haploidization (Table 6.1), may effect complication of soilscapes at the scale of the pedon and polypedon by operating in different intensities and to different depths from point to point. Three scales may be noted in this introductory discussion: (1) regional, (2) polypedonic, and (3) pedonic. The first may range from 2 to as much as 1,500 km across. The second varies from a few meters to 2 km in the longest dimension. The third has been defined by the *7th Approximation* (Soil Survey Staff 1960) to be less than 1 to 3.5 m in diameter. The first two categories include the macro-, meso-, and micro- "combinations" proposed by Fridland (1965) on the basis of relief and relative coarseness of soilscape fabric.

The soilscape of the eastern portion of the prairie triangle centered in Iowa is viewed here as a result of simplification on a regional scale and, at the same time, of complication on polypedonic and pedonic scales, wherever geomorphology,[2] vegetational complexity, and variations in pedoturbation have made this possible. In Illinois the eastern tip of the prairie triangle is complicated regionally because of the interpenetration of forest and grassland communities over thousands of years.

A soilscape may possibly fit into more than one of these categories. Numerical limits to these categories have not yet been set, so that a complete description of a soilscape in these terms is not yet possible.

An ambitious field soil scientist can discover enough detail in almost any soilscape to cover an aerial photograph with as many lines and symbols as the scale will allow. But the soilscapes depicted on two field sheets of apparently similar detail may fall into different categories in the classification of soilscapes suggested here.

PRACTICAL APPLICATIONS OF
SOILSCAPE CHARACTERIZATION.
Soilscape characterization is useful in scientific description and explanation of landscape habits of specific soil units. In some instances management of soils for agricultural purposes may be rationalized in terms of soil body and soilscape properties. For example, soil bodies with soil body pattern indices greater than two are more susceptible to contour and strip-crop farming than are soil bodies with lower indices (Shields 1955).

Productivity of soilscapes varies with the character and areal distribution of component soil bodies. For example, two Spodosol-Histosol soilscapes will differ markedly in productivity of lumber, potatoes, or peat moss if the proportionate areal extents of the two major kinds of soil are very different. Riecken (1963) notes that a soilscape consisting entirely of level phases of Tama and Muscatine silt loams (Typic Hapludolls) can produce twice as many bushels of corn (maize) per hectare as a soilscape con-

---

2. Ruhe (1969) calls attention to the complicated "staircase" landscape of the loess country of Iowa.

sisting of 25% by area of level Tama soil and 75% of rolling Tama. Osch-wald (1966) reports similar influences of soilscape characteristics on crop yields per unit area.

LITERATURE CITED

Arnold, R. W. 1965. Multiple working hypotheses in soil genesis. Soil Sci. Soc. Am. Proc. 29:717–24.

Bushnell, T. M., F. E. Barnes, E. D. Fowler, and J. Thorp. 1930. Soil survey of Wayne County, Indiana, part I. Series 1925, No. 21. Bur. Chem. Soils. U.S. Dept. Agr.

Ciolkosz, E. J. 1967. I. The mineralogy and genesis of the Dodge catena of south-eastern Wisconsin. Ph.D. thesis. Univ. Wis.

Duchaufour, P. 1963. Soil classification: A comparison of the American and the French systems. J. Soil Sci. 14:149–55.

Esser, A. H. 1968. Use of space by animals and men. Science 162:701.

Fridland, V. M. 1965. Make-up of the soil cover. Soviet Soil Sci., 1965, No. 4. pp. 343–54. (Transl. 1965 from Pochvovedeniye, 1965, 4:15–27.)

Gerasimov, I. P., and M. A. Glazovskaya. 1960. Fundamentals of soil science and soil geography. (Transl. from Russian by A. Gourevich.) Israel Prog. for Sci. Trans., Jerusalem, 1965. Available U.S. Dept. Commerce, Springfield, Va.

Hole, F. D. 1953. Suggested terminology for describing soil as three-dimensional bodies. Soil Sci. Am. Proc. 17:131–35.

Hole, F. D., and G. B. Lee. 1955. Introduction to soils of Wisconsin. Univ. Wis. Geol. Nat. History Survey Bull. 79.

Hole, F. D., G. B. Lee, and E. A. Brickbauer. 1953. Soils of Town of Beaver Dam, Dodge County, Wisconsin. Univ. Wis. Geol. Nat. History Survey. (Folder map.)

Hole, F. D., G. W. Olson, K. O. Schmude, and C. J. Milfred. 1962. Soil survey of Florence County, Wisconsin. Univ. Wis. Geol. Nat. History Survey Bull. 84.

Johnson, W. M. 1963. The pedon and the polypedon. Soil Sci. Soc. Am. Proc. 27:212–15.

Juday, C. 1914. The inland lakes of Wisconsin. Univ. Wis. Geol. Nat. History Survey Bull. 27.

Neustruyev, S. S. 1915. Soil combinations in plains and mountainous countries. Pochvovedeniye, 1915, No. 1.

Oschwald, W. R. 1966. Quantitative aspects of soil survey interpretation in ap-praisal of soil productivity, pp. 152–59. In L. J. Bartelli, A. A. Klingebiel, J. V. Baird, and M. R. Heddleson (eds.), Soil surveys and land use planning. Soil Sci. Soc. Am. and Am. Soc. Agron., Madison, Wis.

Powell, J. C., and M. E. Springer. 1965. Composition and precision of classifica-tion of several mapping units of the Appling, Cecil, and Lloyd series in Wal-ton County, Georgia. Soil Sci. Soc. Am. Proc. 29:454–58.

Prasolov, L. I. 1922. Pochvennye oblasti Europeiskoi Rossii (Soil regions of Euro-pean Russia) Otdelenie zemledeliya i pochvovedeniya Selskokhozyaistven-nogouchenogo Komiteta Norodnogo Komissariata Zemledeliza, Comm. 38.

Riecken, F. F. 1963. Some aspects of soil classification in farming. Soil Sci. 96: 49–61.

Ruhe, R. V. 1969. Quaternary landscapes in Iowa. Iowa State Univ. Press, Ames.

Shields, L. G. 1955. A cartographic analysis of landscape, Dodge County, Wis-consin. M.S. thesis. Univ. Wis.

Soil Survey Staff. 1951. Soil survey manual. U.S. Dept. Agr. Handbook No. 18. U.S. Govt. Printing Office, Washington.

————. 1960. Soil classification, a comprehensive system—7th approximation. U.S. Dept. Agr. U.S. Govt. Printing Office, Washington.

Thorp, J. 1957. Report on a field study of soils of Australia. Sci. Bull. I. Earlham College, Richmond, Ind.

Walker, P. H., G. F. Hall, and R. Protz. 1968a. Soil trends and variability across selected landscapes in Iowa. Soil Sci. Soc. Am. Proc. 32:97–101.

————. 1968b. Relation between landform parameters and soil properties. Soil Sci. Soc. Am. Proc. 32:101–4.

Wilding, L. P., R. B. Jones, and G. M. Schafer. 1965. Variation of soil morphological properties within Miama, Celina and Crosby mapping units in west-central Ohio. Soil Sci. Soc. Am. Proc. 29:711–17.

❧

Interpretations of Soil
Classifications and Maps

$\mathscr{A}$

INTERPRETATION of the soil classifications and accompanying maps is the "proof of the pudding," for it is through this process that the theories, hypotheses, and assumed "knowledge" about soils as expressed in the taxonomic system and graphically displayed in the soil maps are put to the test through applied, practical uses. The best test of our principles and theories of classification and genesis is to have them applied in the making of soil maps and interpreted for a number of different applied, technical purposes.

For our discussion, we will use the definition of "interpretation" of Aandahl 1958: "Soil survey interpretation comprises the organization and presentation of knowledge about characteristics, qualities, and behavior of soils as they are classified and outlined on maps."

As aptly pointed out by Riecken (1963), the United States Soil Classification System "has long had the two functions of (1) organizing, developing, and stimulating the search for knowledge about the origin and genesis of soils; and (2) serving as a base for the application of technology in farming." The main purpose of soil classification activity in the U.S. Department of Agriculture is for use in soil surveys, according to Kellogg (1960), and the principal uses of the soil survey in the USDA have been for interpretations for technical assistance in soil conservation programs, for planning agricultural programs, and as a basis for credit (Smith 1968). And there are many other uses for soil survey interpretation, as we shall enumerate later. It would not be possible to justify the many detailed soil maps which have been made in the United States solely on the basis of their scientific value, but we have been able to make many advances in the field of soil genesis and classification through use and study of these maps as a by-product activity to the main purpose of preparing county soil maps.

Soil map and taxonomic information is put to a number of other agricultural interpretive uses. These include development of productivity ratings of soils, soil management advisory activities, forest production advisory work, crop suitability information, determining feasibility of projected irrigation programs, and many others.

Increasing use is being made of interpretations of soil maps with their associated soil classification for nonfarm activities. We note some of these activities as examples without attempting to develop a comprehensive list.

1. Planning and zoning—local, state, and regional.

2. Suitability of areas for septic tank filter fields where the area is not served by a central sewage system.

3. Highway route location, plus reduction of construction site erosion with associated off-site sedimentation, and stabilization and maintenance of the right-of-ways.

4. Building and real estate development site location.

5. Location of underground pipelines to minimize corrosion losses and excavation costs.

6. Valuation of land for tax assessment purposes.

7. Planning location and layout of outdoor recreation facilities, especially parks.

A comprehensive set of discussions and examples is presented by Bartelli et al. (1966). For all such purposes, we feel it is much more desirable and economical to prepare a general soil map using the classification we have described in the previous chapters as a base from which the different interpretations can readily be made. If a single-purpose technical classification were to be made of the soils for one specific objective, it is highly probable that such a classification would not be suitable for other types of planning and site evaluation, such as the kinds we have described. Then the mapping and classification would need to be redone at considerable unnecessary expense. A well-prepared soil map, based on a sound general classification system that has quantitatively defined taxa, is useful as a base for a number of interpretation activities for technical and specific objectives over a long period of time.

INTERPRETATION WITH CATEGORICAL
AND CARTOGRAPHIC GENERALIZATION.
However, for the soil survey and its associated classification to be most useful, they must be appropriately interpreted and generalized. Interpretation and technical application of the soil survey and classification have, in the first step, these two interrelated but differing phases.

One of these is the interpretation and generalization (as needed) of the taxonomic units used as the basis of the most detailed maps which are being prepared. In the United States, this map is the standard county soil survey map, and the basic taxonomic units are the soil series. Once interpretations for the desired technical objective(s) are made for the basic taxa (usually the soil series, as we have said), these may be generalized (grouped) for application to the broader and more general units of the higher categories of the taxonomic system, if it is necessary and/or desirable to reduce taxonomic detail. For example, the soil families are particularly well suited for use and management generalizations, as their differentiating characteristics of textural, mineralogy, and soil temperature classes have been selected because of their importance to plant root growth. However, if regional interpretations of a broad scale are being made, it may be necessary to generalize to a higher categorical level, such as the great soil group, for example.

The second part of the initial interpretation phase is to provide for interpretation of the map units. This interpretation must provide for phases of the taxonomic units, as well as take into account taxonomic inclusions within the map units. Often, if the map units in a survey area represent only a portion of the range of characteristics defined in the entire taxonomic unit, the interpretations can be more specific than those prepared for the taxa in general. But frequently the interpretations must be broader than those for the pure, abstract taxa in order to provide for significant mapping inclusions. These points must be considered in any generalization (omission) of the boundaries between soil series and their phases (shown as soil bodies on the map) as needed to meet the needs of the specific technical objective(s). That is, to simplify the maps for greater ease of interpretation or to prepare maps of smaller scale (larger areas on one map), map lines are omitted from the basic soil map but one must take into account the extent and nature of the map unit inclusions.

Thus, we recognize two separate but similar units: *taxonomic* (categorical) and *mapping* (cartographic) units. The interpretation, and at times generalization, of the taxonomic units are essential to provide extrapolation of information from area to area, whereas the mapping unit interpretation provides information about areas on a specific map, at specific locations. It may or may not be necessary to generalize the taxonomic and the cartographic (map) units. Sometimes one needs to be generalized but not the other.

Following the terminology proposed by Orvedal and Edwards (1941), we designate *cartographic generalization* as the practice of omitting lines between soil bodies as shown on detailed soil maps to reduce the detail on the map—or between different taxonomic units if taxa of levels higher than the soil series are used to define and describe the basic map units. That is, it results in a lowering of the number of soil delineations. We take as the best example of a cartographically detailed soil map those county soil maps produced in the United States National Cooperative Soil Survey at a scale of 4 in. (10 cm) per mile (1:15,840) and those with a scale of 1:20,000 on which are shown soil series, together with textural type, slope, erosion, and other phases. Omitting certain lines from this type of map, with consequent reduction in detail, is generalizing cartographically.

We designate as *categorical generalization* the identification and description of soil map units in terms of taxa at levels of abstraction higher than the soil series. That is, it involves use of taxonomic units from the family through the order levels to designate the mapping unit name and thus reduce the classification or taxonomic detail. In true cartographic and categorical generalization, it is necessary to specify the proportions or percentages if more than one different taxonomic unit is contained in any of the generalized areas.

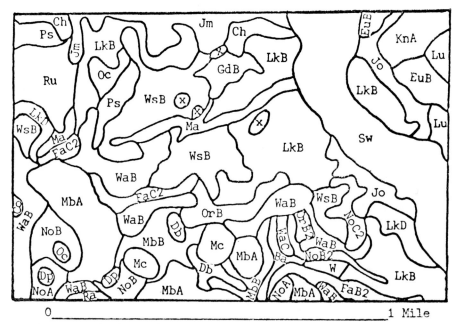

FIG. 26.1. Example of a standard county soil survey map, cartographically detailed, categorically detailed. (After Horton, 1967.)

LEGEND

Ba = Bibb soils, local alluvium. Typic Fluvaquent, coarse loamy, siliceous, acid, thermic.
Ch = Chipley loamy sand. Aquic Quartzipsamment, thermic, coated.
Co = Coxville loam. Typic Paleaquult, clayey, kaolinitic, thermic.
Db = Dunbar fine sandy loam. Aeric Paleaquult, clayey, kaolinitic, thermic.
Dp = Duplin sandy loam. Aquic Paleudult, clayey, kaolinitic, thermic.
EuB = Eustis sand, 0–6% slopes. Psammentic Paleudult, sandy, siliceous, thermic.
FaB2 = Faceville loamy sand, 2–6% slopes, eroded. Typic Paleudult, clayey, kaolinitic, thermic.
FaC2 = Faceville loamy sand, 6–10% slopes, eroded. Typic Paleudult, clayey, kaolinitic, thermic.
GdB = Gilead loamy sand, 2–6% slopes. Typic Fragiudult, kaolinitic, thermic.
Jm = Johnston mucky loam. Cumulic Humaquept, coarse loamy, siliceous, acid, thermic.
Jo = Johns loamy sand. Aquic Hapludult, fine loamy, siliceous, thermic.
KnA = Kenansville loamy sand, 0–2% slopes. Arenic Hapludult, loamy, siliceous, thermic
LkB = Lakeland sand, 0–10% slopes. Typic Quartzipsamment, thermic, coated.
LkD = Lakeland sand, 10–20% slopes. Typic Quartzipsamment, thermic, coated.
Lu = Lumbee loamy sand. Typic Ochraquult, fine loamy, siliceous, thermic.
Ma = Mantachie soils, local alluvium. Aeric Fluvaquent, fine loamy, siliceous, acid, thermic.
MbA = Marlboro loamy sand, 0–2% slopes. Typic Paleudult, clayey, kaolinitic, thermic.
MbB = Marlboro loamy sand, 2–6% slopes. Typic Paleudult, clayey, kaolinitic, thermic.
Mc = McColl loam. Typic Fragiaquult, clayey, kaolinitic, thermic.
NoA = Norfolk loamy sand, 0–2% slopes. Typic Paleudult, fine loamy, siliceous, thermic.
NoB = Norfolk loamy sand, 2–6% slopes. Typic Paleudult, fine loamy, siliceous, thermic.
NoC2 = Norfolk loamy sand, 6–10% slopes, eroded. Typic Paleudult, fine loamy, siliceous, thermic.
Oc = Ocilla loamy sand. Aquic Arenic Paleudult, loamy, siliceous, thermic.
OrB = Orangeburg loamy sand, 2–6% slopes. Typic Paleudult, fine loamy, siliceous, thermic.
OrB2 = Orangeburg loamy sand, 2–6% slopes, eroded. Typic Paleudult, fine loamy, siliceous, thermic.
Ps = Plummer sand, Grossarenic Paleaquult, loamy, siliceous, thermic.
Ra = Rains fine sandy loam. Typic Paleaquult, fine loamy, siliceous, thermic.
Ru = Rutlege loamy sand. Typic Humaquept, sandy, siliceous, thermic.
Sw = Swamp.
WaB = Wagram loamy sand, 2–6% slopes. Arenic Paleudult, loamy, siliceous, thermic.
WaC = Wagram loamy sand, 6–10% slopes. Arenic Paleudult, loamy, siliceous, thermic.
WsB = Wagram sand, thick surface, 0–6% slopes. Arenic Paleudult, loamy, siliceous, thermic.
x = Gravel pit.
w = Water.

It is possible to have different combinations of cartographic and categorical generalization and detail, as can be seen in the following combinations.

1. Cartographically detailed, categorically detailed.
2. Cartographically detailed, categorically generalized.
3. Cartographically generalized, categorically detailed.
4. Cartographically generalized, categorically generalized.

Combination (1) is that used for the standard county soil survey maps.
Combination (3) is that used for so-called "soil association maps" when the
units are defined and described by soil series. Statewide and regional maps
are ordinarily of the type indicated by combination (4). In some regions
and states where soil patterns are relatively uniform and there is no intricate
intermingling, it is possible to prepare maps of the type indicated by combination (2). Examples of these four types of maps are shown in Figures 26.1
to 26.4.

The first operation in interpreting soil maps and their associated classification is to decide on the combination of generalization and/or detail
needed for the technical objective or interpretation purpose, followed by the
generalization operation if such is needed, using the cartographically and
categorically detailed soil survey as a base. This can be done by tracing a
new map, omitting the unnecessary lines, and coloring the new map, or by
use of more sophisticated cartographic techniques. Part of the same opera-

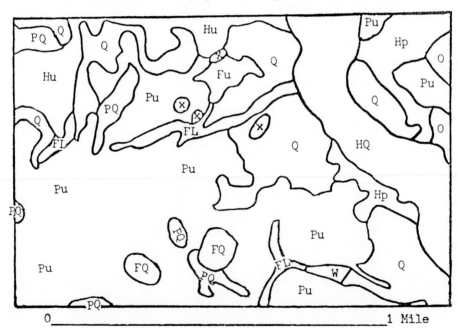

FIG. 26.2. Same area as Figure 26.1 but cartographically detailed, with categorical generalization.

LEGEND

FL = Fluvaquents		O = Ochraquults
Fu = Fragiudults		Pu = Paleudults
FQ = Fragiaquults		PQ = Paleaquults
Hp = Hapludults		Q = Quartzipsamments
HQ = Haplaquepts		x = Gravel pit
Hu = Humaquepts		w = Water

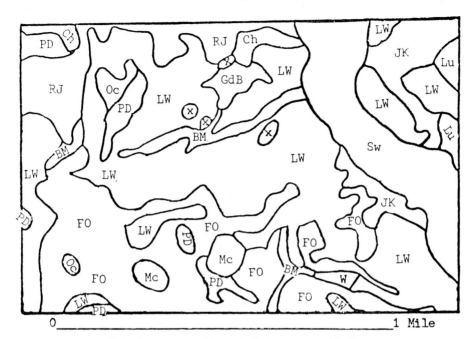

FIG. 26.3. Same area as Figure 26.1 but cartographically generalized; categorically detailed.

LEGEND

BM = Bibb-Mantachie association
Ch = Chipley loam
FO = Faceville-Marlboro-Norfolk-Orangeburg association
GdB = Gilead loamy sand
JK = Johns-Kenansville association
Lu = Lubee loamy sand
LW = Lakeland-Wagram-Eustis association
Mc = McColl loam
Oc = Ocilla loamy sand
PD = Plummer-Rains-Coxville-Dunbar association
RJ = Rutlege-Johnston association
Sw = Swamp
x = Gravel pit
w = Water

tion is to place the soil series in the appropriate taxa if such categorical generalization is needed (ordinarily the county soil survey reports carry tables and descriptive sections indicating the classification of the soils of the area in the higher categorical levels).

The next step is to prepare tables indicating the qualities, expected or predicted performance, use limitations, and hazards of each of the taxonomic units (including any phases) being used with respect to the objective. This can be done in terms of comparative classes ("good," "fair," "poor," for example) or quantitative limits, if such information is available. We obtain information and knowledge for preparing such tables through actual research, experience, and observations on soils of the taxa under consideration, or by application of the principles of soil science to predict how a soil with properties (clay mineral species, clay content, acidity, etc.) similar to the ones which have been studied and characterized will perform (Kellogg 1966). The preparation of the tables is best done by a soil scientist working in close cooperation with a person(s) with experience and expertise with the technical objective for which the interpretations are being made. Most of the recent county soil survey reports carry some types of interpre-

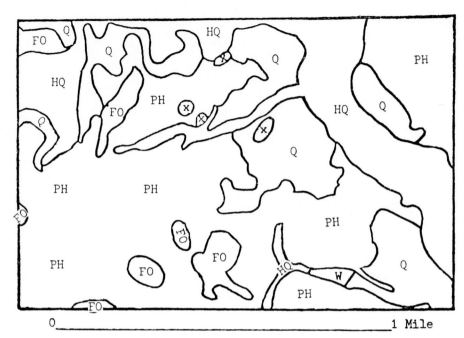

FIG. 26.4. Same area as Figure 26.1 but both cartographically and categorically generalized.

LEGEND
FO = Fragiaquult-Paleaquult-Ochraquult association
HQ = Humaquept-Fluvaquent-Swamp association
PH = Paleudult-Hapludult-Fragiudult association
Q = Quartzipsamment
x = Gravel pit
w = Water

tive tables. We present in Tables 26.1 to 26.3 some examples of interpretive tables for different objectives. In the use of the interpreted maps and tables in operational programs, it is essential to keep in mind the points discussed in Chapter 25. The map shows location of soil bodies, to be sure, but these bodies have inclusions of different soils which occupy as much as 30 to 35% of the area, as indicated in the previous chapter. Fortunately, most of such inclusions are of soils not greatly different from the dominant soil in the soil body. Nevertheless, for specialized, detailed projects, more

TABLE 26.1 ❧ Estimated average yields per acre under high-level management

Soil	Corn	Cotton (lint)	Tobacco	Soybeans	Oats	Wheat
	(bu)	*(lb)*	*(lb)*	*(bu)*	*(bu)*	*(bu)*
Gilead loamy sand 0–2% slopes	65	600	2,100	30	60	30
Johns loamy sand	80	650	2,200	40	70	35
Lakeland sand 0–10% slopes	40	325	1,200	...	...	...
Marlboro loamy sand 0–2% slopes	90	950	2,400	40	85	40
McColl loam	75	625	...	40	75	35
Wagram loamy sand 2–6% slopes	60	550	1,900	24	55	28

SOURCE: Horton 1967.

TABLE 26.2 ❧ Woodland suitability

Soil Type	Potential Production*			Plant Completion	Hazards and Limitations			
	Loblolly pine	Longleaf pine	Slash pine		Seedling mortality	Equipment limitations	Water erosion	Windthrow hazard
Gilead loamy sand	81±7	68±6	84	severe	slight	slight to moderate	slight to severe	slight
Johns loamy sand	90	70	90	severe	slight or moderate	moderate	slight	slight
Lakeland sand	75±8	63±12	78±10	severe	moderate or severe	moderate or severe	slight or moderate	slight
Marlboro loamy sand	86±4	70	85	moderate	slight	slight	slight or moderate	slight
McColl loam	87±6	69	95	severe	slight† to severe	moderate‡ to severe	slight	slight
Wagram loamy sand	85	70	85	moderate or severe	slight or moderate	slight	slight	slight

Source: Horton 1967.
* Site index or height of dominant trees at 50 years of age.
† Slight on drained areas; severe on ponded areas.
‡ Moderate on drained areas; severe on ponded areas.

TABLE 26.3 ❧ Engineering interpretations

Soil Series	Suitability as a Source of:		Limitations for Sewage Disposal			Features Affecting		
	Topsoil	Road fill	Filter fields	Lagoons	Highway Location	Farm ponds	Agricultural drainage	Sprinkler Irrigation
Gilead	fair	fair	moderate to severe	moderate: 2–6% slope; severe: >2% slope	seepage	moderately slow permeability	generally not needed	slow infiltration
Johns	fair	good	severe: high water table	moderate	seasonal high water table	moderate permeability	seasonal water table	medium water-holding capacity
Lakeland	poor	good	moderate: limited filtering action	severe high permeability	excessively drained moderately unstable	excessive seepage	not needed	very low water-holding capacity
Marlboro	fair	fair	moderately slow permeability	slight to 2%; moderate on >2%	few problems	moderately slow permeability	not needed	medium water-holding capacity
McColl	poor	poor	severe	moderate	high water table	moderate to slow permeability	slow permeability	slow infiltration
Wagram	poor	poor	moderate	severe	moderately unstable	excessive seepage	not needed	very low water-holding capacity

SOURCE: Horton 1967.

detailed soil information, both as to location and properties, must be gotten than can be shown on the interpretive maps and in the charts.

Some examples of specific uses of soil survey and classification in which the soils information is put to work are described in the following paragraphs.

Experience in east central Florida (Doyle 1966) showed the following uses for soil survey interpretations in regional land use and resource conservation:

1. Predicting urban growth areas through analyses of favorable and unfavorable soil conditions.

2. Reserving of open areas through use of soil information to delineate greenways, open space corridors, and buffer areas.

3. Identifying water recharge areas by interpreting soils in terms of their hydrologic characteristics.

4. Recognition of potential use conflict areas as high quality, high-producing soils in the path of suburban and urban development.

5. Refining broad land use categories through analysis of soil suitabilities.

Uses of soil survey interpretations in subdivision design have been described by Quay (1966). He indicates the most useful soil interpretations were found to be (1) percolation rate, (2) flood potential, (3) water table, (4) bearing strength, (5) corrosion potential, (6) shrink-swell, (7) AASHO ratings (in connection with street and highway planning and construction), (8) erosion potential, (9) frost action, and (10) agricultural capabilities. In addition, interpretations were made for (1) tree and shrub growth, (2) grass growth, and (3) wildlife conservation.

Thornburn (1966) pointed out the ways in which soil survey interpretation is of use in highway planning and construction.

The application of soil surveys to problems of health, sanitation (as in suitability and design of septic tank filter fields), and engineering has been well described by Olson (1964), based on his work in New York State.

As urban and suburban areas expand, the number of highways enlarges, and demand for greater efficiency in agricultural production increases, need for soil survey interpretations will greatly increase. These will be useful and sound in proportion as the interpretation is carefully done and as the basic classification and mapping have been sound.

Agricultural interpretations of soil surveys have in the past included these types of information or predictions:

1. Yield predictions or potentials for major crops on soil types—at defined levels of management inputs.

2. Resource input requirements for different crops of agriculture and forestry. That is, the amount of input required on a given soil type for a desired level of production or type of use (we formerly called this soil suitability—such as well or poorly suited—for crops, but now prefer to make the analyses and estimates so the operator or manager can make a decision as to the input level to be provided a certain soil).

3. Probable behavior of different soils under irrigation.

4. Artificial drainage potential.

5. Estimates of probable response to fertilizers and lime as controlled by the more permanent properties of soils, considering the fertility or lime level as revealed by soil test.

In the future, agricultural interpretations probably will also include:

1. Suitability for and probable response to land forming (reshaping the soil to produce a more desirable slope gradient, to remove depressional areas, and to allow use of large equipment with precision placing of agricultural chemicals and fertilizers).

2. Probable behavior and productivity of the soil in a complete "systems approach" to high-intensity farming, using heavy machinery, minimum tillage, chemical weed control, narrow rows, and continuous row crop culture.

Crop yield predictions (especially corn) for major Illinois soil series have been developed by Odell and his associates (Odell 1958; Rust and Odell 1957). The Illinois approach has been to obtain extensive farm yield and management records over a period of at least 10 years for fields of the soil series (or associations of soil series) of interest. This approach assumes the properties and definition of the soil series as fixed and given; the mean yields are calculated for classes of management levels (such as low, medium, moderately high, and high). Obtaining these yield predictions allows calculations to be made on the value of the soils, to indicate to farmers and advisors the "productive capacity" of their soils, and to advise them of the management inputs needed to obtain different yield levels on the specific soil series or associations of soil series (cartographically generalized) of their farms.

Emphasis in soil survey interpretation in Iowa in recent years has been on obtaining not only estimates of corn yields for the more important soil series of that state but also an analysis-synthesis evaluation of the fabric and interrelationship of associated soils on the landscape (Oschwald 1966; Riecken 1963), a topic introduced in the previous chapter on soil bodies. In this type of interpretation process, an analysis is made of the soil bodies on a portion of the landscape making up a field or a farm—of size and shape suitable for an operational unit in modern farming technology, a technological soilscape unit. Yield predictions and feasible management practices are then synthesized or specified for these soilscape units on the basis of the kind and extent of different soil bodies composing them.

An approach used in North Carolina to this problem of needed information for agricultural interpretation has been analytical and quantitative (Sopher 1969). In this approach, extensive on-farm corn plot yield record data have been taken, together with complete soil analysis, management, and rainfall data for each of the plots. By multiple regression techniques and the use of a high-speed computer, it has been possible to deduce which soil and environmental properties actually contribute most to variation in corn yields. These analyses have been related to specific soil types in the North Carolina coastal plain and to comparisons of Aquults and

Udults. The results have been used not only to establish corn yield levels and productivity potentials but also to evaluate the taxonomic units themselves and to aid in readjusting their defined limits. That is, the soil series has not been accepted as "given" and fixed, as in the earlier Illinois studies.

It is predicted that there will be more quantitative and comprehensive approaches to soil interpretation for agricultural production in the future, as modern agricultural and agribusiness technology will increasingly demand more information of this type.

LITERATURE CITED

Aandahl, A. R. 1958. Soil survey interpretation—theory and purpose. Soil Sci. Soc. Am. Proc. 22:152–54.

Bartelli, L. J., A. A. Klingebiel, J. V. Baird, and M. R. Heddleson (eds.). 1966. Soil surveys and land use planning. Soil Sci. Soc. Am. and Am. Soc. Agron., Madison, Wis.

Doyle, R. H. 1966. Soil surveys and the regional land use plan, pp. 8–14. In L. J. Bartelli, A. A. Klingebiel, J. V. Baird, and M. R. Heddleson (eds.), Soil surveys and land use planning. Soil Sci. Soc. Am. and Am. Soc. Agron., Madison, Wis.

Horton, R. E. 1967. Soil survey of Scotland County, North Carolina. Soil Conserv. Serv., U.S. Dept. Agr. U. S. Govt. Printing Office, Washington.

Kellogg, C. E. 1960. In Soil classification, a comprehensive system—7th approximation. U.S. Dept. Agr. U.S. Govt. Printing Office, Washington.

———. 1966. Soil surveys for community planning, pp. 1–7. In L. J. Bartelli, A. A. Klingebiel, J. V. Baird, and M. R. Heddleson (eds.), Soil surveys and land use planning. Soil Sci. Soc. Am. and Am. Soc. Agron., Madison, Wis.

Odell, R. T. 1958. Soil survey interpretation-yield prediction. Soil Sci. Soc. Am. Proc. 22:157–60.

Olson, G. W. 1964. Application of soil survey to problems of health, sanitation, and engineering. Cornell Univ. Memoir 387.

Orvedal, A. C., and M. J. Edwards. 1941. General principles of technical grouping of soils. Soil Sci. Soc. Am. Proc. 6:386–91.

Oschwald, W. R. 1966. Quantitative aspects of soil survey interpretation in appraisal of soil productivity, pp. 152–59. In L. J. Bartelli, A. A. Klingebiel, J. V. Baird, and M. R. Heddleson (eds.), Soil surveys and land use planning. Soil Sci. Soc. Am. and Am. Soc. Agron., Madison, Wis.

Quay, J. R. 1966. Use of soil surveys in subdivision design, pp. 76–86. In L. J. Bartelli, A. A. Klingebiel, J. V. Baird, and M. R. Heddleson (eds.), Soil surveys and land use planning. Soil Sci. Soc. Am. and Am. Soc. Agron., Madison, Wis.

Riecken, F. F. 1963. Some aspects of soil classification in farming. Soil Sci. 96: 49–61.

Rust, R. H., and R. T. Odell. 1957. Methods used in evaluating the productivity of some Illinois soils. Soil Sci. Soc. Am. Proc. 21:171–75.

Smith, G. D. 1968. Soil classification in the United States, pp. 6–24. In World soil resources report 32. FAO, Rome, Italy.

Sopher, C. D. 1969. Factors affecting corn production on selected North Carolina coastal plain soils. Ph.D. thesis, N.C. State Univ.

Thornburn, T. H. 1966. The use of agricultural soil surveys in the planning and construction of highways, pp. 87–103. In L. J. Bartelli, A. A. Klingebiel, J. V. Baird, and M. R. Heddleson (eds.), Soil surveys and land use planning. Soil Sci. Soc. Am. and Am. Soc. Agron., Madison, Wis.

❧

A P P E N D I X : *Soil Profiles*

Each of the following ten profiles illustrates one kind of soil in each order. As such, they do not represent the entire range of the soils in each order.

Profile No. 1. TYPIC USTIPSAMMENT: MIXED, THERMIC

SOIL TYPE: Tivoli fine sand.

LOCATION: Reno County, Kansas. 800 feet south and 225 feet east of the northwest corner of section 9, T 22S, R1OW.

DATE OF SAMPLING: January 19, 1959.

COLLECTORS: Ratcliff and Bouse.

PHYSIOGRAPHIC POSITION: Undulating upland on recent sandy aeolian deposits.

CLIMATE: Average annual precipitation approximately 27 inches.

TOPOGRAPHY: Hummocky topography with hummocks ranging from 4 to 30 feet in height.

DRAINAGE: Runoff very slow; permeability very rapid.

VEGETATION: The native vegetation on this soil is big bluestem, little bluestem, Indiangrass, and switchgrass.

USE: Grassland.

DESCRIPTION BY: J. J. Rockers and I. W. Ratcliff, Jr.

SOIL NO: S59Kans-78-1.

Horizon	Depth	Description
	(cm)	
A1	0–23	Yellowish brown (10YR 5/3, moist; 5/4, dry); fine sand; loose; single grain to very weak fine granular; many fine roots; noncalcareous; grades within 5 inches to
AC	23–38	Yellowish brown (10YR 5/4, moist; 5/4 dry); fine sand; single grain; many fine roots; noncalcareous; wavy boundary to
C1	38–71	Light yellowish brown (10YR 6/4, moist; 6/4, dry); fine sand; single grain; fewer roots than above; noncalcareous; grades within 6 inches to
C2	71–122	Light yellowish brown (10YR 6/4, moist; 6/4, dry); fine sand; very porous and massive breaking to single grain with the least amount of pressure; irregular horizontal bands 1/4 to 1/2 inch wide and approximately 6 inches apart which are of slightly higher clay content. These bands are of about two chips less value in color than the matrix; noncalcareous; grades within 6 inches to
C3	122–172	Light yellowish brown (10YR 6/4, dry); fine sand; single grain; few roots and few horizontal bands; noncalcareous; grades to
C4	173–245	Light yellowish brown (10YR 6/4, dry); same as horizon above except for fewer roots and fewer horizontal bands. Augered.

SOURCE: Data and description from Soil Survey Laboratory Data and Descriptions for Some Soils of Kansas, pp. 70–71. Soil Survey Investigations Report No. 4, 1966. Soil Conservation Service, United States Department of Agriculture.

SOIL SURVEY LABORATORY *Lincoln, Nebraska*

DATE *May, 1959*

SOIL TYPE *Tivoli fine sand*

LOCATION *Reno County, Kansas*

SOIL NO. *S59Kans-78-1*

Depth (cm)	Horizon	Particle Size Distribution (mm) 1B1a*						(%)	3A1		Textural Class
		Very Coarse Sand	Coarse Sand	Medium Sand	Fine Sand	Very Fine Sand	Silt	Clay			
		(2–1)	(1–0.5)	(0.5–0.25)	(0.25–0.10)	(0.10–0.05)	(0.05–0.002)	(<0.002)	(0.2–0.02)	(0.02–0.002)	
0–23	A1	0.1	5.7	22.5	57.1	9.3	3.2	2.1	35.7	1.4	fs
23–38	AC	0.4	6.7	23.3	57.4	8.4	1.8	2.0	34.0	1.2	fs
38–71	C1	0.1	4.7	21.8	60.0	9.6	1.1	2.7	37.8	0.8	fs
71–122	C2	<0.1	3.7	20.1	61.5	10.7	0.9	3.1	40.4	0.6	fs
122–173	C3	<0.1	0.4	3.4	71.3	19.8	1.3	3.8	69.8	0.8	fs
173–245	C4	0.1	3.6	17.0	57.2	14.7	2.0	5.4	44.0	0.7	fs

pH	Organic Matter			Moisture Tensions			CEC	Extractable Cations 5B1a					Base Sat. 5C1 NH₄Ac Exch.	Base Sat. 5C3 Sum Cat.	Sum Bases 5B1a	Sum Cations 5A3a
8C1a	OC 6A1a	N 6B1a	C/N	4B1a 1/10 Atmos.	4B1a 1/3 Atmos.	4B2 15 Atmos.	5A1a NH₄Ac	Ca 6N2b	Mg 6O2b	H 6H1a	Na 6P2a	K 6Q2a				
(1:1)	(%)	(%)		(%)	(%)	(%)		(Milliequivalents per 100 g soil)					(%)	(%)	(me/100 g)	
6.5	0.27	0.026	10	4.4	2.7	1.1	2.3	1.7	0.4	0.8	<0.1	0.2	100	74	2.3	3.1
6.2	0.14	0.017	...	3.2	2.3	1.2	2.0	1.2	0.6	2.0	<0.1	0.2	100	50	2.0	4.0
5.8	0.09	0.008	...	3.5	2.9	1.3	2.1	1.3	0.4	1.2	<0.1	0.1	86	60	1.8	3.0
6.1	0.05	...	...	3.6	2.7	1.2	2.1	1.4	0.6	1.2	<0.1	0.1	100	64	2.1	3.3
6.3	0.03	...	...	5.5	3.6	1.7	2.6	1.7	0.7	1.2	<0.1	0.1	96	68	2.5	3.7
6.3	0.04	...	...	6.7	4.8	1.8	3.3	2.1	1.1	1.2	<0.1	0.1	100	73	3.3	4.5

* Code numbers in data column headings can be found in Soil Survey Laboratory Methods and Procedures for Collecting Soil Samples. Soil Survey Investigations Report No. 1, 1967. Soil Conservation Service, United States Department of Agriculture. United States Government Printing Office, Washington.

Profile No. 2. TYPIC CHROMUDERT: VERY FINE, MONTMORILLONITIC, THERMIC

SOIL TYPE: Houston clay.
LOCATION: 7.3 miles southwest of Arkadelphia on Hwy. 67; Clark County, Arkansas.
COVER: Cotton; cultivated over 25 years.
PHYSIOGRAPHY: Region: Coastal Plain.
Position: Upland.
RELIEF: Gently sloping.
EROSION: Slight.
SLOPE: 2–3%.
DRAINAGE: Moderate.
GREAT SOIL GROUP: Grumusol.
PARENT MATERIAL: Weathered chalk.
CORRELATOR: Marvin Larson.
DATE SAMPLED: August, 1951.

SOURCE: Data and description from Certain Properties of Selected Southeastern United States Soils and Mineralogical Procedures for Their Study, p. 61. Southern Regional Bulletin 61 for Cooperative Regional Research Project S-14, 1959. Virginia Agricultural Experiment Station, Blacksburg, Virginia.

Profile Description: by C. L. Garey

No.	Horizon	Depth (cm)	Color	Texture	Structure	Consistence
1	A_{1p}	0–10	2.5Y 3/2	c	2msbk	mfi-wvp
2	A_{12}	10–30	5Y 3/2	c	3msbk	mfi-wvp
3	A_{13}	30–50	2.5Y 4/4	c	3msbk	mfi-wvp
4	AC	50–66	5Y 5/4	c	2m-fabk	mfi-wvp
5	C	66–102	5Y 5/6	c	msbk	mfi–wvp

Chemical Data: C. L. Garey, analyst

No.	pH	CEC (me/100 g)	Exchangeable Cations (me/100 g)				Organic Matter (%)	P (ppm)
			Ca*	Mg	K	Na		
1	6.9	32.9	35.0	1.85	0.59	0.18	2.3	13
2	7.0	32.1	38.0	1.90	0.67	0.21	1.9	15
3	7.2	29.0	34.0	1.95	0.60	0.18	1.2	6
4	7.3	28.2	33.0	2.05	0.50	0.20	0.6	3
5	7.4	29.5	36.0	2.20	0.50	0.15	0.5	2

* Exchangeable and sparingly soluble Ca.

Physical Data: C. L. Garey, analyst

No.	Water Retained at: 1/3 Atm. (%)	15 Atm. (%)	Particle Size Distribution (%) Sand	Silt	Clay	Bulk Density (g/cc)	Total Pore Space (%)
1	39.7	24.3	12	22	66	1.81	31.6
2	40.7	24.2	13	21	66	1.91	27.9
3	39.2	23.8	10	20	70	1.91	27.9
4	38.8	22.2	11	19	70	1.82	31.6
5	41.1	23.1	8	20	72	1.91	27.9

Mineralogical Data: C. L. Garey, analyst

No.	$2-0.2\mu$ clay	$<0.2\mu$ clay	Chemical Data for Clay and Silt Fractions CEC (me/100 g) $2-0.2\mu$ clay	$<0.2\mu$ clay
1	$K_1M_2I_3$	$M_1K_2I_3$	25.2	89.2
2	$K_1M_2I_3$	$M_1K_2I_3V_3$	27.2	87.4
3	$K_1M_2I_2$	$M_1K_2I_3$	28.2	84.6
4	$K_1M_2I_2$	$M_1K_2I_3$	30.3	86.8
5	$K_1M_2I_2$	$M_1K_2I_3$	31.3	88.2

Profile No. 3. TYPIC XEROCHREPT: COARSE LOAMY, MIXED, MESIC

SOIL TYPE: Vista coarse sandy loam.

LOCATION: San Diego County, California. A soil profile pit located 2¼ miles south and ¾ mile east of Fallbrook in the NE¼ of the SW¼ of section 31, T9S, R3W.

DATE OF SAMPLING: February 17, 1964.

DESCRIPTION BY: David C. Estrada and Gerald Kester.

COLLECTORS: Klaus Flach, Gerald Kester, George Borst, Clifford Henry, Roy Bowman, and Gerald Anderson.

VEGETATION: Annual weeds and grasses.

CLIMATE: Annual precipitation 15 inches with a mean annual temperature of 60 F, a mean January temperature of 54 F, and a mean July temperature of 70 F.

PARENT MATERIAL: Tonalite grus.

TOPOGRAPHY: Southwest-facing, smooth 8% slopes in hilltop and upper backslope positions on rolling upland.

ELEVATION: 590 feet above sea level.

DRAINAGE: Well drained; medium runoff; moderately rapid permeability.

SOIL MOISTURE: Profile moist to about 15 inches.

REMARKS: There is much rodent activity by ground squirrels. The irregular boundary between the B22 and C1 horizons may be the result of this activity. In a nearby pit, burrow holes occur in the C1 horizon. Isolated bodies of C horizon occur within, or surrounded by, material of the B horizon.

SOIL NO.: S64Calif-37-1.

Horizon	Depth	Description
	(cm)	
A11	0–8	Dark grayish brown (10YR 4/2) coarse sandy loam, dark brown (10YR 3/3) moist; moderate, fine to medium crumb structure; soft, very friable, nonsticky, and nonplastic; plentiful very fine and fine roots; many very fine and fine random pores; neutral (pH 6.7); abrupt smooth boundary.
A12	8–23	Dark brown (10YR 4/3) coarse sandy loam, dark brown (10YR 3/3) moist; weak fine to medium granular structure; slightly hard, very friable, nonsticky, and nonplastic; plentiful very fine and fine roots; common very fine and fine tubular and interstitial pores; neutral (pH 6.7); diffuse to clear irregular boundary. Krotovinas filled with this material, 3 to 4 inches wide, extend into the underlying horizons, in places extending into the upper part of the C horizon.
A13	23–48	Dark brown (10YR 4/3) coarse sandy loam, dark brown (10YR 3/3) moist; weak fine to medium granular structure; slightly hard, very friable, nonsticky, and nonplastic; few very fine and fine roots; common very fine and fine tubular and interstitial pores; slightly acid (pH 6.5); clear wavy boundary.
B21	48–71	Dark brown (10YR 4/3) coarse sandy loam, dark brown (10YR 3/3) moist; massive; hard, friable, nonsticky, and nonplastic; few very fine and fine roots; common very fine and fine tubular and interstitial pores; slightly acid (pH 6.3); clear smooth boundary.
B22	71–89	Yellowish brown (10YR 5/4) coarse sandy loam, dark yellowish brown (10YR 3/4) moist; massive; hard, friable, nonsticky, and nonplastic; very few very fine and fine roots; common very fine and fine interstitial, few fine tubular pores; slightly acid (pH 6.3); abrupt irregular to broken boundary.
C1	89–112	Yellowish brown (10YR 5/4) and very pale brown (10YR 7/4) tonalite grus composed mostly of quartz, feldspar, hornblende, and accessory dark-colored minerals; clear irregular boundary.
C2	112–162+	Brown (10YR 5/3) and very pale brown (10YR 7/3) tonalite grus composed mostly of quartz, feldspar, hornblende, and other dark-colored minerals, mostly unweathered.

SOURCE: Data and description from A Toposequence of Soils in Tonalite Grus in the Southern California Peninsular Range, pp. 38–39, by Wiley D. Nettleton, Klaus W. Flach, and George Borst. Soil Survey Investigations Report No. 21, 1968. Soil Conservation Service, United States Department of Agriculture.

SOIL SURVEY LABORATORY Riverside, California

SOIL TYPE *Vista coarse sandy loam*

SOIL NO. *S64Calif-37-1*

LOCATION *San Diego County, California*

Depth (cm)	Horizon	1B1b* Total Sand (2–0.05)	1B1b* Total Silt (0.05–0.002)	1B1b* Total Clay (<0.002)	Very Coarse (2–1)	Coarse (1–0.5)	Medium (0.5–0.25)	Fine (0.25–0.1)	Very Fine (0.1–0.05)	Silt (0.05–0.02)	Int. III (0.02–0.002)
0–8	A11	72.7	17.1	10.2	19.5	21.1	9.0	14.8	8.2	8.4	8.7
8–23	A12	70.9	18.5	10.6	17.8	18.0	9.0	17.2	8.9	9.0	9.5
23–48	A13	70.3	18.2	11.5	18.2	18.4	8.9	16.0	8.8	8.4	9.8
48–71	B21	68.7	19.4	11.9	16.2	17.0	8.6	16.3	10.6	9.5	9.9
71–89	B22	70.6	18.5	10.9	19.5	18.0	8.5	16.1	8.5	8.7	9.8
89–112	C1	82.3	12.5	5.2	35.3	23.2	6.9	11.8	5.1	6.0	6.5
112–162+	C2	82.9	12.5	4.6	32.5	23.9	7.8	12.9	5.8	6.0	6.5

Column header: Size Class and Particle Diameter (mm) (% of <2mm) 3A1 — Sand and Silt classes.

* Code numbers in data column headings can be found in Soil Survey Laboratory Methods and Procedures for Collecting Soil Samples. Soil Survey Investigations Report No. 1, 1967. Soil Conservation Service, United States Department of Agriculture. United States Government Printing Office, Washington.

Size Class and Particle Diameter (mm) (% of <2mm) 3A1

Int. II (0.2–0.02)	(2–0.1)	Clay‡ (<0.002)	Clay§ (<0.002)
24.2	64.5	11.5	9.3
26.8	62.0	10.9	8.5
25.8	61.5	12.2	9.3
28.2	58.1	12.5	9.5
26.0	62.1	12.4	8.9
17.0	77.2	4.1	2.4
18.1	77.1	4.5	2.7

Coarse Fragments† 1A2a

>2 (%)	2–19 (% of <76 mm)	19–76 (% of <76 mm)
19	19	0
22	22	0
23	23	0
24	24	0
28	28	0
41	41	0
40	40	0

6A1a‡ Organic Carbon (%)	6B1a‡ Nitrogen (%)	C/N	6C2a‡ Ext. Iron as Fe (%)	Bulk Density 4A1f 1/3 bar (g/cc)	Bulk Density 4A1h Oven dry (g/cc)	3B2 C' (g/cc)
0.96	0.080	12	1.0	1.55	1.89	0.88
0.56	0.055	10	0.9	1.51	1.86	0.86
0.34	...	...	1.0	1.46	1.51	0.86
0.28	...	...	0.9	1.43	1.52	0.85
0.10	...	...	1.1	1.46	1.52	0.82
0.02	...	...	0.7	1.92	1.97	0.67
0.03	...	...	0.8	...	...	...

† From characterization sample; determined by shaking overnight in a sodium hexametaphosphate solution.
‡ Analysis of ground whole soil; results expressed on less than 2-mm basis.
§ Analysis of whole soil; results expressed on whole-soil basis.

Water Content / pH / Extensibility / Extractable Bases / CEC

	Water Content		pH				Extractable Bases 5B1a					CEC	
	4B1c	4B2‡	8C1b	8C1a	4D1	4D1	6N2a‡	6O2a‡	6P2a‡	6Q2a‡	6H2a‡	5A2a‡	5A3a‡
	1/3 bar	15 bar	Saturated Paste	H_2O	Extensibility COLEF	Extensibility COLE	Ca	Mg	Na	K	Ext. Acidity	NaOAc	Sum
	(%)	(%)		(1:1)	(in./in.)	(in./in.)	(meq/100 g)						
	8.7	5.4	6.7	6.9	0.066	0.059	10.5	3.4	1.4	0.2	1.7	16.9	17.2
	9.8	4.0	6.5	6.6	0.072	0.061	10.5	2.8	0.4	0.1	2.3	18.6	16.1
	12.0	3.1	6.6	6.8	0.010	0.010	13.0	4.0	0.4	0.1	2.2	19.8	19.7
	10.1	3.7	6.6	6.7	0.021	0.017	11.1	4.0	0.4	0.1	1.8	19.1	17.4
	10.9	6.4	6.7	6.9	0.014	0.011	13.5	5.1	0.4	0.1	2.1	22.8	21.2
	6.7	4.6	7.2	7.1	0.008	0.006	9.1	3.7	0.5	tr.	1.2	14.9	14.5
	…	4.2	7.3	7.4	…	…	9.0	3.7	0.7	0.2	0.8	16.2	14.4

‡ Analysis of ground whole soil; results expressed on less than 2-mm basis.

Clay Mineralogy (<0.002 mm) 7A
Relative abundance based on height of diagnostic X-ray peak 7A2

8A‡ Water at Sat. (%)	5D2 Exch. Na (%)	Base Sat. Cat. (%)	Base Sat. NaOAc CEC (%)	Mont.	Verm.	Hydro-biotite	Illite	Kaolin	7A3 Kaolin DTA (%)	Kaolin in Ground Whole Soil (DTA) (%)
						(X-ray)				
…	8	90	92	—	—	t	xxxx	xx	7	—
…	2	86	74	—	xxx	x	—	xx	…	…
…	2	89	89	—	xxx	x	—	xx	…	…
…	2	89	81	—	xxx	x	t	xxx	12	2
44.4	3	90	84	—	xx	x	t	xxx	6	—
…	4	92	90	—	xxxx	t	—	xx	16	t
…		92	84	—	xxxx	t	—	xx		

‡ Analysis of ground whole soil; results expressed on less than 2-mm basis.

— = looked for but not found
t = trace
x = small
xx = moderate
xxx = abundant
xxxx = dominant

Profile No. 4. TYPIC HAPLARGID: FINE LOAMY, MIXED, THERMIC

Soil type: Mohave sandy loam.
Area: Maricopa County, Arizona.
Vegetation: Sparse cover of galleta and other grasses, creosotebush, paloverde, cane cholla, and annual weeds.
Parent material: Alluvium washed from materials that weathered from granite, schist, and rhyolite.
Topography: Less than 1% slope, convex; near the low end of a large fan; elevation about 1,690 feet.

Horizon	Depth	Description
	(cm)	
A1	0–10	Brown (7.5YR 4/4) coarse sandy loam, reddish yellow (7.5YR 6/6) when dry; on surface is a thin veneer of coarse sand about ⅛ inch thick; weak, medium, and thick, platy breaking to weak, fine, granular structure; slightly hard, very friable, sticky, plastic; common, fine and medium roots; many fine pores; abrupt, wavy boundary.
B1	10–25	Dark brown (7.5YR 4/4) coarse sandy loam, strong brown (7.5YR 5/6) when dry; massive, breaking to weak, medium, subangular blocky structure; slightly hard, very friable, sticky, plastic; few thin, patchy, clay skins on peds; common, fine roots follow ped faces; common, fine and medium, continuous, tubular pores in peds; clear, wavy boundary.
B21t	25–48	Reddish brown (5YR 4/4) sandy clay loam, brown (7.5YR 5/4) when dry; moderate, medium, and coarse, prismatic, breaking to moderate, medium and coarse, angular blocky structure; very hard, friable, very sticky, very plastic; thin, continuous, clay skins; common, fine roots follow ped faces; few fine pores; clear, wavy boundary.
B22t	48–69	Reddish brown (5YR 4/4) clay loam or sandy clay loam, brown (7.5YR 5/4) when dry; moderate, medium, and coarse, subangular blocky structure; very hard, friable, very sticky, very plastic; thin, continuous, clay skins; common, fine roots follow ped faces; numerous, fine mycelia of lime; calcareous; clear, wavy boundary.
B3tca	69–94	Brown (7.5YR 4/4) loam, strong brown (7.5YR 5/6) when dry; common, medium, and distinct, pinkish white (7.5YR 8/2) mottles; massive, breaking to weak, fine, and medium, subangular blocky structure; slightly hard, very friable, sticky, plastic; thin, patchy, clay skins on ped faces; common, fine and medium, continuous pores in peds; pink (7.5YR 7/4) segregations of lime; calcareous; clear, wavy boundary.
IIC1ca	94–137	Pinkish gray (7.5YR 7/2) gravelly coarse sandy loam, pinkish white (7.5YR 8/2) when dry; mottled with pinkish gray (7.5YR 6/2) and reddish yellow (7.5YR 8/6) when dry; massive; extremely hard, firm, nonsticky, nonplastic; few fine, continuous pores; calcareous.
IIC2ca	137–193	Same description as for IIC1ca horizon except color is pink (7.5YR 7/4) when moist; gradual, wavy boundary.
IIC3	193–250+	Light brown (7.5YR 6/4) gravelly coarse sand, pinkish gray (7.5YR 7/2) when dry; massive; soft, loose, slightly sticky, nonplastic; few carbonate concretions; calcareous.

Source: Data and description from Soil Classification, a Comprehensive System—7th Approximation, p. 165. 1960. Soil Survey Staff, Soil Conservation Service, United States Department of Agriculture. United States Government Printing Office, Washington.

Climatic data (Wittman, Arizona)

	J	F	M	A	M	J	J	A	S	O	N	D	Ann
Mean temperature, 1931–47 (°F)	49	53	58	66	75	84	91	89	83	71	58	52	69
Mean precipitation, 1931–47 (inches)	0.9	1.3	0.9	0.6	0.2	0	1.0	1.8	1.0	0.4	0.7	1.3	10.0

Annual precipitation more than 2.9 and less than 17.1 inches during 9 years out of 10.

Particle Size Distribution (mm) (%) Pipette and Hexametaphosphate

Depth (cm)	Horizon	Very Coarse Sand (2–1)	Coarse Sand (1–0.5)	Medium Sand (0.5–0.25)	Fine Sand (0.25–0.10)	Very Fine Sand (0.10–0.05)	Silt (0.05–0.002)	Clay (<0.002)	(0.2–0.02)	(0.02–0.002)	(>2)
0–10	A1	17.4	19.9	5.8	6.8	8.9	30.6	10.6	32.5	10.7	5
10–25	B1	12.8	18.0	6.4	7.8	10.1	31.1	13.8	35.1	10.3	8
25–48	B21t	11.7	17.9	6.6	7.4	8.3	26.5	21.6	29.8	8.9	9
48–69	B22t	9.6	15.7	5.7	5.7	7.7	27.7	27.9	28.8	9.7	10
69–94	B3tca	12.3	15.1	5.4	5.8	6.7	33.2	21.5	29.9	13.1	9
94–137	IIC1ca	19.1	18.1	6.5	5.9	4.1	29.7	16.6	20.8	15.9	26
137–193	IIC2ca	15.8	20.8	9.5	8.3	4.3	26.4	14.9	21.9	12.7	21
193–250+	IIC3	24.6	33.7	15.1	8.9	2.0	10.0	5.7	11.4	4.0	20

Organic Matter C (%)	C/N	CaCO₃ Equiv. (%)	CEC	Extractable Cations (meq/100 g) Ca	Mg	Na	K	pH Sat. Paste	pH 1:10	EC (mmhos per cm 25 C)	HCO₃	Exch. Na (%)	Saturation Extract Soluble (meq/l) Na	Ca	Mg	Cl	H₂O at Sat. (%)
0.16	6	...	8.5	7.7	8.0	0.1	0.9	7.2	7.8	0.3	1.0	1.2	0.4	1.8	0.8	0.5	28.6
0.12	6	...	14.7	9.6	8.7	0.3	0.6	7.5	7.4	0.4	1.6	2.0	0.6	2.1	0.9	0.5	23.3
0.14	6	...	20.7	13.9	10.8	0.5	0.4	7.6	8.2	0.6	0.6	2.4	1.8	2.7	0.9	2.2	32.6
0.16	7	2	23.3	29.1	13.5	0.6	0.4	8.1	8.9	0.9	0.6	2.6	3.1	3.7	1.3	4.5	41.2
0.16	8	10	17.1	27.9	9.6	0.7	0.3	8.1	8.9	0.9	0.4	4.1	3.7	3.0	1.1	5.0	44.2
0.05	...	22	6.3	21.6	8.0	0.8	0.2	8.6	9.2	0.9	2.4	12.7	13.5	1.0	0.6	5.7	40.3
0.02	...	15	6.1	24.5	11.4	1.2	0.2	8.8	9.6	1.2	1.8	19.7	18.9	0.5	0.6	5.1	37.7
0.01	...	6	4.0	19.5	9.1	1.2	0.2	9.0	9.7	1.2	2.6	30.0	20.0	0.4	0.5	3.4	30.8

Profile No. 5.   TYPIC HAPLUDOLL: FINE SILTY, MIXED, MESIC

SOIL TYPE: Marshall silty clay loam.

SOIL NO.: S57Nebr-89-9-(1-7).

LOCATION: 180 feet south and 135 feet west of northeast corner of southeast quarter of northeast quarter of section 25, T17N, R10E, Washington County, Nebraska.

PHYSIOGRAPHY: Slightly rounded ridge top is dissected loess plain.

RELIEF: 60 to 80 feet.

SLOPE: 1%, smooth, slightly convex.

PARENT MATERIAL: Peorian loess.

DRAINAGE: Well drained.

PERMEABILITY: Moderately slow.

SALT or ALKALI: No carbonates to 120 inches.

LAND USE: 1957, oats and seeding of alfalfa.

DESCRIBED BY: J. A. Elder and H. L. Kollmorgen.

Horizon	Depth	Description
	(cm)	
A1p	0–18	Very dark brown (10YR 2/2, moist) to dark grayish brown (10YR 3.5/2, dry) silty clay loam; moderate subangular clods breaking to fine and very fine granules; friable, moist; slightly hard, dry; no effervescence; abrupt smooth boundary.
AB	18–28	Very dark brown (10YR 2/2, moist) (10YR 3/3, moist crushed) to dark grayish brown (10YR 4.5/2, dry) silty clay loam; moderate fine granular; friable, moist; slightly hard, dry; abundant roots, few worm casts; no effervescence; clear smooth boundary.
B21	28–43	Dark brown (10YR 3/3, moist) (10YR 4/3, dry) silty clay loam; weak medium prismatic breaking to fine and very fine subangular blocky; slightly firm, moist; hard, dry; abundant roots, few worm casts and openings, numerous fine and very fine pores; no effervescence; clear smooth boundary.
B22	43–70	Dark brown (10YR 3/3.5, moist) (10YR 4/3, moist crushed) to brown (10YR 5.5/3, dry) silty clay loam; moderate medium prismatic breaking to moderate fine and very fine subangular blocky; slightly firm, moist; slightly hard, dry; abundant roots and very fine pores within aggregates; thin continuous coating on large and small aggregates; fine gray and yellowish brown mottles in lower 2 inches of the horizon that are few, faint, and fine; no effervescence; clear smooth boundary.
B23	70–102	Dark brown (10YR 3.5/3, moist) to brown (10YR 5.5/3, dry) silty clay loam; weak coarse prismatic breaking to weak medium subangular blocky; slightly firm, moist; slightly hard, dry; numerous fine and very fine pores; abundant roots on faces of large aggregates; common faint fine yellowish brown mottles; common distinct medium strong brown mottles; no effervescence; clear smooth boundary.
B3	102–122	Dark brown (10YR 3.5/3, moist) to brown (10YR 5.5/3, dry) silty clay loam; weak coarse prismatic breaking to weak medium and coarse subangular blocky; friable, moist; slightly hard, dry; numerous fine and very fine pores, few medium pores; common faint fine yellowish brown mottles; common distinct medium strong brown mottles; a few dark brown soft fine manganese concretions; few roots; no effervescence; gradual smooth boundary.
C	122–152	Brown (10YR 4/3, moist) to pale brown (10YR 6/3, dry) silty clay loam; weak coarse prismatic, no smaller aggregates; thin coating on prisms; mottling as in horizon above with gray mottles common, faint, and coarse; no effervescence.

SOURCE: Data and description from Soil Survey Laboratory Data and Descriptions for Some Soils of Nebraska, pp. 122–23. Soil Survey Investigations Report No. 5, 1966. Soil Conservation Service, United States Department of Agriculture.

SOIL SURVEY LABORATORY *Lincoln, Nebraska*   DATE *April 9, 1958*

SOIL TYPE *Marshall silty clay loam*   LOCATION *Washington County, Nebraska*

SOIL NO. *S57Nebr-89-9-(1-7)*

Depth (cm)	Horizon	1Bla* Very Coarse Sand (2-1)	Coarse Sand (1-0.5)	Medium Sand (0.5-0.25)	Fine Sand (0.25-0.10)	Very Fine Sand (0.10-0.05)	Silt (0.05-0.002)	Clay (<0.002)	(0.2-0.02)	(0.02-0.002)	Textural Class 4A3a
0-18	Alp	0.2	0.1	0.1	0.2	1.6	69.0	28.8	43.1	27.6	sicl
18-28	AB	..	..	..	0.2	1.3	65.3	33.2	38.0	28.7	sicl
28-43	B21	..	..	0.1	0.2	1.5	64.7	33.5	37.5	28.8	sicl
43-70	B22	..	0.1	0.1	0.2	2.4	64.6	32.6	41.0	26.1	sicl
70-102	B23	..	..	0.1	0.3	3.2	64.5	31.9	43.2	24.7	sicl
102-122	B3	..	..	..	0.2	3.0	66.6	30.2	40.8	28.9	sicl
122-152	C	..	..	..	0.3	3.1	69.0	27.6	42.6	29.7	sicl

Heading: Particle Size Distribution (mm) (%) 3A1

* Code numbers in data column headings can be found in Soil Survey Laboratory Methods and Procedures for Collecting Soil Samples. Soil Survey Investigations Report No. 1, 1967. Soil Conservation Service, United States Department of Agriculture. United States Government Printing Office, Washington.

pH 8Cla (1:1)	Organic Matter 6A1a OC (%)	6B1a N (%)	C/N	Moisture Tensions 4B2 15 Atm. (%)	CEC 5A1a NH4Ac	Extractable Cations 51Ba 6N2b Ca	6O2b Mg	6H1a H	6P2a Na	6Q2a K	Base Sat. 5C1 NH4Ac Exch. (%)	Base Sat. 5C3 Sum Cat. (%)	Sum Bases 5B1a (me/100 g)	Sum Cations 5A3a	Vol. Wt. 4A3a (g/cc)
5.8	1.78	0.160	11.1	11.5	20.8	11.0	3.6	11.0	..	0.8	74	58	15.4	26.4	...
5.9	1.64	0.149	11.0	13.9	23.5	13.5	4.8	9.9	..	0.5	80	66	18.8	28.7	...
6.1	1.11	0.105	10.6	13.8	24.0	14.2	5.9	9.1	0.1	0.5	86	69	20.7	29.8	...
6.3	0.54	0.058	9.3	13.8	24.4	15.4	6.4	7.2	0.1	0.4	91	76	22.3	29.5	1.32
6.5	0.22	..	..	14.4	24.9	16.3	6.9	5.6	0.1	0.5	96	81	23.8	29.4	...
6.8	0.17	..	..	14.0	24.8	16.6	6.9	5.6	0.2	0.4	97	81	24.1	29.7	...
6.8	0.10	..	..	14.1	24.6	17.0	7.0	4.4	0.3	0.4	100	85	24.7	29.1	1.36

Extractable Cations expressed in (Milliequivalents per 100 g soil)

Profile No. 6. AERIC HAPLAQUOD: SANDY, SILICEOUS, THERMIC

SOIL TYPE: Leon fine sand.

LOCATION: Long County, Georgia, wooded area 5 miles south of Allenhurst on paved road.

COVER: Long-leaf pine with understory of sawtooth palmetto, huckleberry, runner oak, and wiregrass.

RELIEF: Level.

DRAINAGE: Poor.

PARENT MATERIAL: Marine.

| | | | Profile Description | | | |
Horizon	Depth (cm)	Dominant Color	Texture	Structure	Consistence	Boundary
0	1–0	Thin cover of twigs, leaves, and decaying plant matter.				
A1	0–10	N4	fs	0	ml	as
A2	10–36	N8	fs	0	ml	as
B2h	36–46	5YR 3/3 to 10YR 2/2	fs	m	mfi	cw
B3	46–69	2.5Y 7/4 to 10YR 4/4	fs	lfsb	mvfr	gw
C1	69–99	10YR 6/2	fs	0	ml	cw
C2	99–114+7.5YR 5/2		fs	0	ml	...

SOURCE: Data and description from Morphological, Chemical, and Physical Characteristics of Eighteen Representative Soils of the Atlantic Coast Flatwoods, p. 26, by F. L. Long, H. F. Perkins, J. R. Carreker, and J. M. Daniels. University of Georgia, College of Agriculture Experiment Stations Research Bulletin 59, 1969.

	Particle Size Distribution (mm)					Sand	Silt	Clay
Horizon	Sand					2–0.05	0.05–0.002	<0.002
	2–1	1–0.5	0.5–0.25	0.25–0.10	0.10–0.05	(%)	(%)	(%)
		(Percent of Total Sand)						
A1	0	5.45	32.92	59.57	2.06	93	4	3
A2	0	5.44	27.66	64.59	2.31	94	3	3
B2h	0.13	6.25	28.20	62.22	3.22	93	4	3
B3	0.26	5.01	29.02	63.51	2.20	93	3	4
C1	0.10	6.08	30.62	61.02	2.18	96	1	3
C2	0.31	6.86	31.70	59.21	1.91	92	3	5

	Bulk Density and Moisture Characteristics						Available Water		
		Water at Bars Tension of:							
Horizon	Bulk Density (gm/cc)	1/10	1/3	1	4	15	Per cm (cm)	Cumulative (cm)	Hyd. Cond. (ft/day)
		(Percent by Weight)							
A1	1.43	4.48	3.27	3.08	2.76	2.54	.010	0.10	3.60
A2	1.51	1.51	1.22	0.92	0.73	0.56	.010	0.25	6.96
B2h	1.29	11.61	10.67	9.68	8.46	6.54	.052	1.31	0.96
B3	1.53	2.97	2.85	2.62	2.07	1.54	.020	1.77	...
C1	1.58	1.63	1.39	1.28	0.95	0.63	.012	2.14	...
C2	1.74	3.56	3.10	3.01	2.33	1.77	.023	2.49	...

Chemical Characteristics

Horizon	OM (%)	pH Wet	pH Dry	pH Salt	Base Sat. (%)	CEC	Ca	Mg (meq/100 g)	K	Na	Zn	Mn (ppm)	Extr. P
A1	1.55	3.8	4.0	2.8	27.7	2.2	0.38	0.08	0.01	0.14	2.6	<1.0	11.2
A2	0.19	4.3	4.5	3.4	12.3	3.1	0.25	0.01	0.02	0.10	3.1	<1.0	18.0
B2h	4.02	4.5	4.5	3.5	2.3	20.2	0.34	0.02	0.01	0.10	2.8	<1.0	220.0
B3	0.37	4.9	5.9	4.2	1.4	18.0	0.13	0.01	0.02	0.10	3.4	<1.0	1.5
C1	0.11	5.0	5.0	4.2	15.8	1.2	0.09	0.01	0.01	0.08	4.0	<1.0	<0.5
C2	0.23	4.9	4.8	4.2	12.3	2.2	0.16	0.01	0.02	0.08	2.8	<1.0	2.0

Profile No. 7.   TYPIC HAPLUDALF: FINE SILTY, MIXED, MESIC

Soil type: Fayette silt loam.

Location: NE¼ of SW¼ of section 35, T95N, R3W, Mendon Township, Clayton County, Iowa; approximately 75 feet inside the entrance to Pike's Peak State Park and 25 feet to the right.

Vegetation: Oak-hickory forest.

Parent material: Gray-Brown Podzolic soils developed on post Iowan-post Tazewell loess; Fayette soils are the forested members of a biosequence which includes the Tama series, a Brunizem, and the Downs series, a transition.

Slope: 2% convex.

Drainage: Well drained.

Described by and date: Roger B. Parsons, July 22, 1959.

Soil no.: S59Iowa-22-1.

Horizon	Depth	Description
	(cm)	
A0	1–2.5	Partly decomposed forest litter from white and red oak and shagbark hickory.
A1	0–5	Very dark gray (10YR 3/1, moist) silt loam; strong very fine granular structure; very friable when moist; abundant fine roots; abrupt and wavy horizon boundary.
A2	5–23	Pale brown (10YR 6/3, moist) silt loam; moderate to strong very thin platy structure; friable when moist; clear and wavy horizon boundary.
B1	23–43	Brown (10YR 4/3, moist) fine silt loam; moderate very fine subangular blocky structure; slightly firm when moist; vesicular; prominent very pale brown (10YR 7/3, moist) grainy coatings on ped faces; common thin discontinuous clay skins on horizontal and vertical ped faces; clear and wavy horizon boundary.
B21	43–61	Brown (10YR 4/3, moist) silty clay loam; strong fine subangular blocky structure; slightly firm to firm when moist; few grainy gray coatings on peds; abundant discontinuous clay skins; clear and wavy horizon boundary.
B22	61–89	Brown (10YR 4/3, moist) silty clay loam; strong fine subangular blocky structure; slightly firm to firm when moist; prominent continuous dark yellowish brown (10YR 4/4, moist) clay skins on horizontal and vertical ped faces; some black (2.5Y 2/0) manganese concretions; diffuse and wavy horizon boundary.
B23	89–107	Brown (10YR 4/3, moist) silty clay loam; strong fine subangular blocky structure; slightly firm to firm when moist; prominent nearly continuous clay skins on ped faces; clear and wavy horizon boundary.
B3	107–122	Yellowish brown (10YR 5/4, moist) silty clay loam; moderate fine subangular blocky structure to massive; slightly firm when moist; discontinuous dark yellowish brown (10YR 4/4, moist) clay skins mostly on vertical ped faces; clear and wavy horizon boundary.
C1	122+	Yellowish brown (10YR 5/6, moist) fine silt loam; massive; slightly firm when moist; some clay skins prominent along vertical fracture planes.

Source: Data and description from Soil Survey Laboratory Data and Descriptions for Some Soils of Iowa, pp. 40–41. Soil Survey Investigations Report No. 3, 1966. Soil Conservation Service, United States Department of Agriculture.

SOIL SURVEY LABORATORY *Lincoln, Nebraska*

SOIL TYPE *Fayette silt loam*

SOIL NO. *S59Iowa-22-1*

LOCATION *Clayton County, Iowa*

Depth (cm)	Horizon	Very Coarse Sand (2-1)	Coarse Sand (1-0.5)	Medium Sand (0.5-0.25)	Fine Sand (0.25-0.10)	Very Fine Sand (0.10-0.05)	Silt (0.05-0.002)	Clay (<0.002)	(0.2-0.02)	(0.02-0.002)	Textural Class
		1B1a*					3A1				
0-5	A1	0.1	0.3	0.2	0.3	1.5	84.6	13.0	41.1	45.2	sil
5-23	A2	<0.1	0.2	0.1	0.3	1.5	84.4	13.5	40.0	46.1	sil
23-43	B1	<0.1	0.1	0.2	0.3	1.3	81.2	16.9	37.7	45.0	sil
43-61	B21	<0.1	0.1	0.1	0.2	3.0	72.8	23.8	38.7	37.2	sil
61-89	B22	0.1	<0.1	<0.1	0.3	3.2	68.7	27.7	44.0	28.1	sicl
89-107	B23	<0.1	<0.1	<0.1	0.3	3.2	65.8	30.7	43.3	25.9	sicl
107-122	B3	<0.1	<0.1	<0.1	0.3	2.8	67.3	29.6	43.5	26.8	sicl
122+	C1	<0.1	<0.1	<0.1	0.3	3.0	70.4	26.3	46.1	27.5	sil

Particle Size Distribution (mm) (%)

* Code numbers in data column headings can be found in Soil Survey Laboratory Methods and Procedures for Collecting Soil Samples. Soil Survey Investigations Report No. 1, 1967. Soil Conservation Service, United States Department of Agriculture. United States Government Printing Office, Washington.

pH 8C1a (1:1)	OC 6A1a (%)	N 6B1a (%)	C/N	Free Iron 6C1a Fe₂O₃ (%)	Electrical Conductivity 8A1a EC × 10³ Millimhos per CM	Moisture Tensions 4B2 15 Atm. (%)	CEC 5A1a NH₄Ac	Ca 6N2b	Mg 6O2b	H 6H1a	Na 6P2a	K 6Q2a
								Extractable Cations 5B1a (Milliequivalents per 100 g soil)				
5.6	5.63	0.343	16.4	0.9	0.6	12.0	21.1	12.3	4.3	13.5	<0.1	0.6
5.0	0.66	0.079	8.0	1.0	0.2	5.3	7.3	1.2	0.7	8.4	<0.1	0.2
5.0	0.30	0.047	6.0	1.2	0.2	6.4	8.4	2.0	1.7	7.6	<0.1	0.2
5.0	0.20	0.040	...	1.6	0.1	9.5	13.0	4.9	2.9	9.2	<0.1	0.3
4.9	0.16	0.031	...	1.7	0.1	11.9	17.1	6.9	3.6	10.3	0.1	0.3
5.0	0.17	0.030	...	1.8	0.1	13.9	19.9	10.2	5.0	9.0	0.1	0.4
5.2	0.14	0.028	...	1.8	0.1	13.7	19.1	10.8	5.1	8.8	0.1	0.3
5.1	0.12	0.026	...	1.8	0.2	12.2	17.5	10.0	4.5	7.8	0.1	0.3

Base Sat. 5C1 NH₄Ac Exch. (%)	Base Sat. 5C3 Sum Cat. (%)	Sum Bases 5B1a	Sum Cations 5A3a	8D3 Ca/Mg	Moisture at Sat. 8A (%)
		(me/100 g)			
82	56	17.2	30.7	2.9	85.3
29	20	2.1	10.5	1.7	43.8
46	34	3.9	11.5	1.2	40.4
62	47	8.1	17.3	1.7	50.9
64	51	10.9	21.2	1.9	55.3
79	64	15.7	24.7	2.0	60.9
85	65	16.3	25.1	2.1	61.4
85	66	14.9	22.7	2.2	62.2

Profile No. 8. TYPIC HAPLUDULT: CLAYEY, KAOLINITIC, THERMIC

SOIL TYPE: Cecil fine sandy loam.
LOCATION: Iredell County, North Carolina.
DATE SAMPLED: June 1960.
SAMPLES COLLECTED BY: Forrest Steele, R. J. McCracken, J. W. Cawthorn, C. L. Hunt.
DESCRIPTION BY: R. J. McCracken.
SOIL CORRELATED BY: L. E. Aull.
SLOPE: Gently sloping, 3%.
DRAINAGE: Well drained.
PARENT MATERIAL: Mica gneiss.
PERMEABILITY: Moderate.
PHYSIOGRAPHIC REGION: Piedmont.
LANDFORM: Hillslope-broad secondary ridge.
CLIMATIC DATA: Statesville, North Carolina.

	Jan.	Feb.	Mar.	Apr.	May	June	
Mean Temperature, °C:	5.7	6.7	9.9	15.4	20.6	24.4	
Mean Precipitation, cm:	11.9	11.3	13.8	11.0	10.3	10.7	
	July	Aug.	Sept.	Oct.	Nov.	Dec.	Ann.
Mean Temperature, °C:	25.6	24.6	22.8	16.5	10.3	5.7	15.6
Mean Precipitation, cm:	14.9	13.6	11.0	9.7	9.0	9.6	119.6

Horizon	Depth	Description
	(cm)	
O1	5–2.5	Loose oak and hickory leaves.
O2	2.5–0	Partly decayed leaves and litter.
A1	0–1	Dark grayish brown (10YR 4/2) fine sandy loam; weak, fine granular structure; very friable, clear wavy boundary.
A2	1–10	Strong brown (7.5YR 5/6) fine sandy loam; weak, fine subangular blocky structure breaking to weak, coarse granular structure; very friable; clear smooth boundary.
B1t	10–18	Yellowish red (5YR 4/8) with coatings of strong brown (7.5YR 5/6) sandy clay loam; friable; few thin discontinuous clay films; few fine mica flakes; clear wavy boundary.
B21t	18–28	Red (2.5YR 4/8) clay, moderate, fine, and medium subangular blocky structure; friable; common nearly continuous clay films; few fine mica flakes; gradual wavy boundary.
B22t	28–60	Red (2.5YR 4/8) with coatings of red (10R 4/8) clay; moderate, medium, and fine, angular and subangular blocky structure; firm; thick nearly continuous clay films; a few fine mica flakes; gradual smooth boundary.
B23t	60–90	Red (2.5YR 4/8) with coatings of red (10R 4/8) clay; moderate, medium, and fine, granular and subangular blocky structure; firm; thick nearly continuous clay films; a few fine mica flakes; gradual smooth boundary.
B31t	90–115	Red (10R 4/8) clay; moderate medium subangular blocky structure; friable; thick but discontinuous clay films; few coarse mica flakes; few fragments of weathered mica gneiss; diffuse irregular boundary.
B32t	115–168	Red (2.5YR 4/8) with prominent spots of reddish yellow (7.5YR 6/8) clay; weak, medium subangular blocky structure breaking readily to fine subangular blocky structure; friable; thick and discontinuous clay films mostly on vertical structural faces; common fine mica flakes; common fragments of soft weathered rock; diffuse irregular boundary.
C1	168–198	Red (2.5YR 4/8) and reddish yellow (7.5YR 7/6) clay loam; massive; very friable; a few clay films on vertical structural faces; common fine mica flakes; numerous soft weathered rock fragments (saprolite).

SOURCE: Data and description from Soils of the Hayesville, Cecil, and Pacolet Series on the Southern Appalachian and Piedmont Regions of the United States, by R. J. McCracken, E. J. Pederson, L. E. Aull, C. I. Rich, and T. C. Peele. Southern Cooperative Series Bull. No. 157, 1971. North Carolina State University at Raleigh.

Horizon	Depth (cm)	Total Sand 2–0.050 (%)	Silt 0.050–0.002 (%)	Clay 0.002 (%)	Textural Class	pH Water (1:1)	Organic Matter (%)	Exch. Al (meq/100 g)	Ca	Mg	K	Exch. Acidity	CEC Sum of Cat.	PBS Sum Cat.	Mineralogy of Clay Fraction—Estimated Clay Mineral Content (<2μ)
											(meq/100 g)				
A1	0–1	...	...	...	...	...	...	3.8	0.80	0.20	0.05	14.2	15.3	...	...
A2	1–10	60.5	29.1	10.4	fsl	4.8	2.3	1.9	0.21	0.03	0.08	7.5	7.8	4	K1V/C2Q3
B1t	10–18	45.4	27.7	26.9	scl	4.7	1.8	2.4	0.21	0.01	0.08	7.2	7.5	4	...
B21t	18–28	14.6	20.2	65.2	clay	4.8	1.0	3.9	0.21	0.21	0.10	11.7	12.2	4	...
B22t	28–60	15.1	14.6	70.3	clay	5.2	0.6	2.7	0.21	0.38	0.09	10.3	11.0	5	K1V/C3Gi4
B23t	60–90	8.6	20.8	70.6	clay	5.3	0.4	4.5	0.21	0.15	0.08	11.9	12.3	3	...
B31t	90–115	18.2	28.6	53.2	clay	5.3	0.2	3.3	0.10	0.00	0.08	8.1	8.3	2	...
B32t	115–168	24.2	32.3	43.5	clay	5.3	0.1	3.2	0.03	0.03	0.07	7.2	7.3	1	K1V/C4Gi4
C1	168–198	32.6	31.6	35.8	clay loam	...	...	3.9	...	...	...	8.0	...	...	...

Profile No. 9.   TYPIC HAPLORTHOX: CLAYEY, OXIDIC,
ISOHYPERTHERMIC

Soil type: Delicias clay.

Soil no.: S61PR-14-10.

Location: Suroeste Soil Conservation District, Puerto Rico, 0.5 kilometers south (on narrow road) from kilometer marker 5.6 on Puerto Rico Highway 311, 50 feet east of fence. Photo No. GS-LR-9-160.

Vegetation and use: Guinea and molasses grasses. Other native pastures and weeds.

Slope and land form: 20%. North aspect. Located on a sideslope that leads to a drainageway.

Drainage and permeability: Well drained with medium runoff and medium internal drainage. Permeability is moderately rapid.

Parent material: Colluvium derived largely from ultrabasic rocks.

Collected by: R. B. Grossman, W. E. McKinzie, O. R. Carter, L. H. Rivera, D. A. Compton, J. E. Trigo, December 7, 1961.

Described by: W. E. McKinzie, December 7, 1961.

Horizon	Depth	Description
	(cm)	
Ap	0–15	Dark reddish brown (2.5YR 3/4) clay; moderate medium granular structure; friable when moist, slightly sticky and plastic when wet; many fine roots; many medium and fine iron concretions; clear smooth boundary.
B21	15–30	Dark red (10R 3/6) clay; moderate medium granular structure; friable when moist, sticky and plastic when wet; common fine roots; many fine and medium iron concretions; clear smooth boundary.
B22	30–56	Dark red (10R 3/6) clay; weak medium subangular blocky structure which breaks into weak fine granular; friable when moist, nonsticky, and slightly plastic when wet; common fine roots; many fine pores; common fine and medium iron concretions; gradual wavy boundary.
B23	56–86	Dark red (10R 3/6) clay; weak fine angular blocky structure; firm when moist, nonsticky and nonplastic when wet; common fine roots; many fine and medium iron concretions; gradual wavy boundary.
B24	86–114	Dark red (2.5YR 3/6) clay; weak fine angular blocky structure; friable when moist, nonsticky and nonplastic when wet; few fine roots; common fine iron concretions; gradual wavy boundary.
B25	114–152	Dark red (2.5YR 3/6) clay; weak fine angular blocky structure; friable when moist, nonsticky and nonplastic when wet; very few fine roots; common fine iron concretions.

Remarks: The Ap, B22, and B24 horizons were sampled for the Bureau of Public Roads. Colors given are for moist soil. Soil was at field capacity when sampled.

Mineralogy, Micromorphology (Methods 7B1, 4E1). The sand of the upper four horizons consists primarily of ferruginous nodules in the coarse and very coarse fractions and of angular and subangular clean, clear quartz in the medium and finer sand fractions. The sand of the B25 horizon consists largely of reddish brown, earthy-looking grains with subordinate quartz as described above; the B24 is intermediate in the proportion of earthy grains. The very fine sand consists largely of quartz with some opaques and a few zircons in the upper four horizons; feldspar was not observed. Thin section observations indicate that quartz occurs within the iron concretions. The fabric of the upper B2 horizon is highly porous with little long-range preferred clay orientation and with no clay films.

Source: Data and description from Soil Survey Laboratory Data and Descriptions for Some Soils of Puerto Rico and the Virgin Islands, pp. 162–63. Soil Survey Investigations Report No. 12, 1967. Soil Conservation Service, United States Department of Agriculture.

SOIL SURVEY LABORATORY *Lincoln, Nebraska*
SOIL TYPE *Delicias clay*
SOIL NO. *S61PR-14-10*
General Methods: 1A, 1B1a, 2A1, 2B

LOCATION *Suroeste SCD, Puerto Rico*

Size Class and Particle Diameter (mm) (% of <2mm) 3A1*

Depth (cm)	Horizon	Total			Sand					Silt	
		Sand	Silt	Clay	Very Coarse	Coarse	Medium	Fine	Very Fine		Int. III
		(2-0.05)	(0.05-0.002)	(<0.002)	(2-1)	(1-0.5)	(0.5-0.25)	(0.25-0.1)	(0.1-0.05)	(0.05-0.02)	(0.02-0.002)
0-15	Ap	37.4a	11.9	50.7	6.9	3.5	2.8	14.9	9.3	3.6	8.3
15-30	B21	32.3a	10.2	57.5	9.3	3.3	2.2	10.8	6.7	3.1	7.1
30-56	B22	31.8a	17.6	50.6	9.1	3.3	1.9	10.4	7.1	4.5	13.1
56-86	B23	27.8a	32.4	39.8	5.1	2.0	1.6	9.4	9.7	12.2	20.2
86-114	B24	29.3b	29.6	41.1	5.4	2.1	1.6	8.7	11.5	13.1	16.5
114-152	B25	30.4b	33.6	36.0	4.3	2.8	2.2	9.4	11.7	13.0	20.6

* Code numbers in data column headings can be found in Soil Survey Laboratory Methods and Procedures for Collecting Soil Samples. Soil Survey Investigations Report No. 1, 1967. Soil Conservation Service, United States Department of Agriculture. United States Government Printing Office, Washington.

NOTES: Italics indicate that the nonclay probably has strong claylike properties.

a = Fe-Mn nodules: greater than 50% (2-0.5 mm).
b = Earthy nodules: greater than 50% (2-0.05 mm).
c = 11 kg/m² to 60 inches (Method 6A).
d = Range in replicates is 0.12 g/cc.
e = Estimated.
f = Field state.
g = Percent clay is estimated (15-bar water percentage × 2.5).

Size Class and Particle Diameter (mm) (% of <2mm) 3A1

Int. II			Coarse Fragments					Bulk Density			
		3A1b Water Disp. Clay	2A2 2-19 mm	6A1a Organic Carbon (c)	6B1a Nitrogen	C/N	6C2a Ext. Iron as Fe	4A1a Field-State	4A1c 30-cm	4A1b Air-dry	4D1 COLE
(0.2-0.02)	(2-0.1)	(<0.002)		(%)	(%)		(%)	(g/cc)	(g/cc)	(g/cc)	(f)
		(%)	(%)								
24.1	28.1	27.6	tr	2.68	0.212	13	9.8	1.38	...	1.51	0.032
17.8	25.6	...	18	1.01	0.110	9	10.9	1.24	1.22	1.28	0.010
19.4	24.7	...	22	0.43	0.069	6	11.8	1.16d	1.16d	1.20d	0.010
29.1	18.1	...	21	0.25	0.038	7	13.6	1.43	1.40	1.44	0.003
31.2	17.8	...	15	0.22	...	...	17.5	1.40e	...	...	...
31.3	18.7	...	7	0.18	...	...	25.3	1.36	...	1.38	0.003

Water Content			pH		Extractable Bases 5B1a					6H1a	CEC			
4B4 Field-State	4B3 30-cm	4B2 15-Bar	8C1c N KCl	8C1a H₂O	6N2b Ca	6O2b Mg	6P2a Na	6Q2a K	Sum	Ext. Acidity	5A3a Sum Cations	5A1a NH₄OAc	6G1d KCl-Ext. Al	5A3b Bases Plus Al me/100g Clay
(%)	(%)	(%)	(1:1)	(1:1)					(meq/100 g)					
29.0	...	20.4	5.0	5.8	4.5	1.9	0.1	0.6	7.1	10.8	17.9	11.4	...	14.0
32.6	30.3	22.3	4.3	5.1	1.2	0.9	0.1	0.2	2.4	11.6	14.0	7.5	0.4	4.9
34.3	37.2	22.0	4.3	5.0	0.4	0.8	0.1	tr	1.3	10.2	11.5	5.1	0.3	3.2
29.4	31.5	23.4	5.0	5.2	0.6	0.8	0.1	tr	1.5	9.6	11.1	4.4	0.1	3.0g
...	...	26.8	5.1	5.6	0.3	0.8	0.1	tr	1.2	10.6	11.8	4.1	...	2.0g
34.8	...	28.1	5.6	5.5	0.1	0.8	0.2	tr	1.1	12.6	13.7	3.2	...	2.0g

Base Saturation		Ratios to Clay 8D1		
5C3 Sum Cations	5C1 NH₄OAc	NH₄OAc	Ext. Iron	15-Bar Water
(%)	(%)			
40	62	0.22	0.19	0.40
17	32	0.13	0.19	0.39
11	25	0.10	0.23	0.43
14	34	0.11	0.34	0.59
10	29	0.10	0.42	0.65
8	34	0.09	0.70	0.78

Profile No. 10.   TERRIC MEDISAPRIST: LOAMY, DYSIC, THERMIC

SOIL TYPE: Ponzer muck.

LOCATION: 60 feet north of Canal 1, 0.5 of a mile east of Canal A; Washington County, North Carolina.

ELEVATION: 16 feet.

NATURAL VEGETATION: Mixed hardwood.

DATE OF SAMPLING: September 7, 1965.

DESCRIPTION BY: J. D. Dolman and D. W. Goss.

Depth	Description
(cm)	
0–20	Black (10YR 2/1, moist) well-decomposed organic material; fibers constitute 20 to 30% of total material; rubbed color (wet) black (10YR 2/1); sodium pyrophosphate extract on white filter paper dark brown (10YR 3/3, moist); weak medium subangular blocky structure; friable; some charcoal; recent roots; abrupt, smooth boundary to
20–35	Black (10YR 2/1, moist) well-decomposed organic material; fibers constitute 20 to 30% of total material; rubbed color (wet) black (10YR 2/1); sodium pyrophosphate extract on white filter paper dark yellowish brown (10YR 4/4, moist); massive; friable; slightly brittle; recent roots; abrupt, wavy boundary to
35–47	Very dark brown (10YR 2/2, moist) well-decomposed organic material; fibers constitute 20 to 30% of total material; rubbed color (wet) black (10YR 2/1); sodium pyrophosphate extract on white filter paper dark yellowish brown (10YR 4/4, moist); weak medium subangular blocky structure; friable; recent roots; abrupt, wavy boundary to
47–65	Black (7.5YR 2/0, moist) mineral and organic material; fibers constitute 10 to 20% of total material; rubbed color (wet) black (7.5YR 2/0); sodium pyrophosphate extract on white filter paper very dark grayish brown (10YR 3/2, moist); massive; friable; recent roots; clear, wavy boundary to
65–80	Brown (10YR 4/3, moist) loam; massive; firm; slightly plastic, slightly sticky; recent roots; clear, wavy boundary to
80–112	Light yellowish brown (2.5Y 6/4, moist) loam; massive, firm; slightly plastic, slightly sticky; recent roots.

SOURCE: Data and description from A Study of Organic Soils (Histosols) in the Tidewater Region of North Carolina, pp. 18–20, by J. D. Dolman and S. W. Buol. North Carolina Agricultural Experiment Station Technical Bulletin No. 181, 1967.

Depth (cm)	Particle Size Distribution (%)								Class
	2–1 (mm)	1– 0.5 (mm)	0.5– 0.25 (mm)	0.25– 0.100 (mm)	0.100– 0.050 (mm)	Total Sand	Silt	Clay	
65–80	...	0.06	2.36	12.13	6.11	20.67	56.93	22.40	loam
80–112	0.08	0.24	2.91	20.17	7.92	31.31	48.18	20.50	loam

Depth (cm)	Moisture Content (% of Wt.)	Reaction (1:2)		Organic Matter				C/N
		H_2O	KCl	Loss on Ignition (% Wt.)	Mebius (% of OM)	Nitrogen (% of soil)	Nitrogen (% of OM)	
0–20	6.55	5.3	4.5	50.09	47.43	1.120	2.23	26.30
20–35	8.20	4.0	2.9	82.36	89.03	1.370	1.66	35.36
35–47	7.69	4.0	2.9	86.69	90.78	1.200	1.38	42.49
47–65	5.07	3.8	3.2	25.23	30.77	0.700	2.77	21.20
65–80	1.89	4.2	3.6	4.58	5.44	0.154	3.36	17.48
80–112	1.18	4.5	3.7	2.50	1.02	0.067	2.68	21.94

Depth (cm)	Extractable Cations (mg/100 g)			(ppm)	Exchangeable Cations (mg/100 g)					Exchange Acidity Soil pH (me/100 g)
	Ca	Mg	K	P	Ca	Mg	K	Al	H	
0–20	9.12	5.49	0.12	3.0	16.00	18.59	0.40	0.20	0.15	0.35
20–35	2.48	2.20	0.16	4.0	4.90	6.00	0.52	1.80	1.40	3.20
35–47	1.20	1.12	0.13	1.5	2.10	2.52	0.42	15.40	2.50	17.90
47–65	0.40	0.27	0.10	6.0	0.30	0.80	0.28	8.35	1.35	9.70
65–80	0.16	0.15	0.08	6.0	0.00	0.50	0.22	6.15	0.55	6.70
80–112	0.16	0.21	0.07	1.5	0.20	0.57	0.22	8.55	0.40	8.95

Depth (cm)	Cation Exchange Capacity (me/100 g)			% Base Sat. at Soil pH	% Base Sat. at pH of 7	Extr. Iron, % Fe₂O₃	Clay Mineral Suite
	CEC Perm.	CEC at Soil pH	CEC at pH of 7				
0–20	35.19	35.34	59.90	99.01	58.41	0.41	V/C,* K,† I‡
20–35	12.22	13.62	46.55	83.85	24.53	0.18	V/C, K, I
35–47	20.44	22.94	116.65	21.97	4.32	0.13	V/C, K, I
47–65	9.73	11.08	65.05	12.45	2.12	0.08	V/C, K, I
65–80	6.87	7.42	21.00	9.70	3.43	0.03	V/C, K, I
80–112	9.54	9.94	11.75	9.96	8.43	0.05	V/C, K, I

* Vermiculite-chlorite intergrade.
† Kaolinite.
‡ Illite.

AUTHOR INDEX

349

SUBJECT INDEX